# Enterprise Architecture and Integration:
## Methods, Implementation, and Technologies

Wing Lam
*U21 Global, Singapore*

Venky Shankararaman
*Singapore Management University, Singapore*

**INFORMATION SCIENCE REFERENCE**

Hershey · New York

| | |
|---|---|
| Acquisitions Editor: | Kristin Klinger |
| Development Editor: | Kristin Roth |
| Senior Managing Editor: | Jennifer Neidig |
| Managing Editor: | Sara Reed |
| Assistant Managing Editor: | Sharon Berger |
| Copy Editor: | Shanelle Ramelb |
| Typesetter: | Jennifer Neidig |
| Cover Design: | Lisa Tosheff |
| Printed at: | Yurchak Printing Inc. |

Published in the United States of America by
Information Science Reference (an imprint of IGI Global)
701 E. Chocolate Avenue, Suite 200
Hershey PA 17033
Tel: 717-533-8845
Fax: 717-533-8661
E-mail: cust@igi-pub.com
Web site: http://www.info-sci-ref.com

and in the United Kingdom by
Information Science Reference (an imprint of IGI Global)
3 Henrietta Street
Covent Garden
London WC2E 8LU
Tel: 44 20 7240 0856
Fax: 44 20 7379 0609
Web site: http://www.eurospanonline.com

Library of Congress Cataloging-in-Publication Data

Enterprise architecture and integration : methods, implementation, and technologies / Wing Lam and Venky Shankararaman, editors.

   p. cm.

  Summary: "This book provides a detailed analysis of the important strategies for integrating IT systems into fields such as e-business and customer-relationship management. It supplies readers with a comprehensive survey of existing enterprise architecture and integration approaches, and presents case studies that illustrate best practices, describing innovative methods, tools, and architectures with which organizations can systematically achieve enterprise integration"--Provided by publisher.

  Includes bibliographical references and index.

  ISBN 978-1-59140-887-1 (hardcover) -- ISBN 978-1-59140-889-5 (ebook)

  1. Management information systems. 2. Business enterprises--Computer networks. 3. Information technology--Management. 4. Software architecture. I. Lam, Wing Hong. II. Shankararaman, Venky.

  HD30.213.E58 2007

  658.4'038011--dc22

                  2007007279

British Cataloguing in Publication Data
A Cataloguing in Publication record for this book is available from the British Library.

All work contributed to this book is new, previously-unpublished material. The views expressed in this book are those of the authors, but not necessarily of the publisher.

# Table of Contents

## Section I
## Business Requirements and Organizational Modeling

## Section II
## Business Process Management

## Section III
## Integration Methods and Tools

## Section IV
## Enterprise Integration Case Studies

# Detailed Table of Contents

### Section I
### Business Requirements and Organizational Modeling

**Chapter I**

This chapter examines the evolution of enterprise integration (EI) and its growing importance amongst organizations. The chapter describes how organizational divisions and silos have a limiting effect on business transformation. It gives an overview of how enterprise integration overcomes these limitations, both from the technical and business perspectives. The challenges associated with enterprise integration adoption are also discussed.

**Chapter II**

The effectiveness of information systems is closely related to the degree of integration between different applications. This chapter examines five critical success factors for enterprise application integration (EAI), namely, minimal project expense, optimal reuse of components, complexity reduction, optimal coupling of applications, and minimal cost of EAI infrastructure. The concept of agility is proposed as a confluence of these factors, which is defined as the extent to which a system is able to react to change. To improve agility, each factor must be managed and measured separately.

## Chapter III

This chapter explores how social and organizational aspects can affect ERP (enterprise resource planning) projects and their ability to integrate different parts of the enterprise. The chapter draws from a successful case of a multinational ERP implementation in a large international organization. The sociological theory of the actor network is used to analyse relationships and conflicts in the case. The chapter concludes with the recommendation that one should examine the roles of all actors with the power to affect not only the implementation project, but also the system being implemented.

## Chapter IV

This chapter discusses experiences in the acquisition of software-intensive systems prior to establishing a contract. One significant challenge in this respect is the identification of functional requirements. Drawing from two case studies of large-scale military applications, the authors contend that business process modeling is an effective way to explicitly define requirements while creating visibility and consensus among different stakeholders of major IT projects.

## Section II
## Business Process Management

## Chapter V

Business processes are so complex that an engineering approach is often required to design and construct them. This chapter presents an approach to constructing enterprise processes based on the integration of component processes. In this approach, process representations at the logical as well as physical levels are depicted in a hierarchy, and a graphical tool is used to support enterprise process construction.

## Chapter VI

In the past, B2B (business to business) has generally been adopted only by large organizations who can afford the technology and business investment associated with it. This chapter explores, through a case study, the use of XML (extensible markup language) and Web services in conjunction with an integration broker to provide a B2B solution. The authors argue that existing B2B solutions, largely based around EDI (electronic data interchange), present a high entry barrier to smaller organizations, and that the use of XML and Web services provide an alternative lower cost B2B solution that may be more suitable.

Today, the information systems development life cycle is seeing increasing effort and emphasis being placed upon front-end analysis rather than back-end development. This chapter provides an overview of business process management, which is becoming an integral aspect of system analysis. The chapter suggests that the growing maturity of business process management standards and technologies will drive the transition to more service-oriented IT architectures in the future.

# Chapter VIII

This chapter describes the importance of metamodels, particularly in relation to knowledge-intensive service industries. The key guiding principle is to define the process model at the logical level, free from any technical implementation. Business process integration is then a matter of achieving the best possible overall physical engine to implement that process model from available legacy applications, applied investment opportunity, and expert development resources.

## Section III
## Integration Methods and Tools

This chapter presents a methodology for the creation of EAI solutions based on the concept of software product-line engineering. In this approach, a designer builds not one system, but a family of systems, taking into account all the possible different integration contexts in which the system might be used now and in the future. This overcomes one of the weaknesses with existing enterprise application integration tools that provide little methodological support.

# Chapter X

Despite developments in supply chain management, there is relatively little in terms of tool support for supply chain design and development. This chapter demonstrates a visual tool for automatic supply chain integration. By clicking in the appropriate tables, functions, and fields, users can specify the data needed for integration. Given these design inputs, integration code can then be automatically generated in the form of Web services.

# Chapter XI

As data sources become more distributed, it is important for software applications to be able to access data in a transparent manner despite the fact that the data are residing physically in different places. This chapter presents an integration framework called DAVINCI aimed at providing mobile workers with mobile software applications to query and update information coming from different and heterogeneous databases. The chapter describes the trials developed using the DAVINCI architecture in different European countries.

This chapter describes the basic notions of intelligent agents and multiagent systems, and proposes possible types of their application to enterprise integration. The agent-based approaches to enterprise application integration are considered from three points of view: (a) It means using an agent as a wrapper of an application or service execution, (b) it means constructing a multiagent organization within which agents are interacting and providing emergent solutions to enterprise problems, and (c) it means using the agent as an intelligent handler of heterogeneous data resources in an open environment.

This chapter describes an attempt to introduce semantics to workflow-based composition. A composition framework is presented based on a hybrid solution that merges the benefits of the practicality of use and adoption popularity of workflow-based composition, with the advantage of using semantic descriptions to aid both service developers and composers in the composition process and facilitate the dynamic integration of Web services into it.

## Section IV
## Enterprise Integration Case Studies

# Chapter XIV

This chapter examines common EI challenges and outlines approaches for combining EI and process improvement based on the process improvement experiences of banks in several different countries. Common EI-related process improvement challenges are poor usability within the user desktop environment, a lack of network-based services, and data collection and management limitations. How EI affects each of these areas is addressed, highlighting specific examples of how these issues present themselves in system environments.

# Chapter XV

This chapter explores the gradual evolution of SOA (service-oriented architecture) through various phases and highlights some of the approaches and best practices that have evolved out of real-world implementations in regional retail banks. Starting from a step-by-step approach to embrace SOA, the chapter details some typical challenges that creep up as the usage of SOA platforms becomes more and more mature. Also, certain tips and techniques that will help institutions maximize the benefits of enterprise-wide SOA are discussed.

This chapter describes a case study of application integration in a Korean bank. The case examines the integration technology employed, and the adoption of the technology on a pilot project. The chapter highlights some of the managerial implications of application integration and discusses the broader, organizational impact of application integration. Importantly, application integration efforts should be justified by a sound business case and carefully introduced into the organization in an incremental fashion.

This chapter uses the example of a retail business information system to illustrate the concept of service-oriented development and integration in a number of different scenarios. The chapter shows how using a service-oriented architecture allows businesses to build on top of legacy applications or construct new applications in order to take advantage of the power of Web services.

# Chapter XVIII

RFID (radio frequency identification) is gathering increasing interest from many organizations. This chapter discusses the adoption and potential usages of RFID in the case of enterprise integration projects such as supply chain management. Through electronic tagging, RFID can enable stocks to be monitored in real time to a level that was previously not practical or cost effective. RFID can therefore be considered an important component of an enterprise integration solution.

# Preface

It is clear that we now live in a world where one expects everything to be connected, or for want of a better word, integrated. For example, when I get onto the Web and do my Internet banking, I expect to be integrated directly with my bank to see exactly how much money I have. Similarly, when I want to book air tickets online, I expect to be integrated into the airline reservation system to see what flights are available, and whether they are fully booked or not. Businesses too are demanding this kind of close integration. A supermarket, for example, that wants to match demand and supply needs to integrate its own stock control system with that of its suppliers so that goods being taken off the shelf are quickly replenished.

Increasingly then, there is a need for organizations to provide services in an integrated fashion. This, however, is easier said than done. Integration cannot be achieved unless the information technology used by the organization is also integrated. Unfortunately, organizations have had a long history of creating systems that operate in isolation from other systems. From conception, many of these systems have been designed to address the specific requirements in a particular functional area such as accounting, or personnel and stock control. This functional orientation, however, has tended to reinforce departmental silos within the organization, resulting in an IT architecture characterized by numerous "islands of applications" that remain quite separate from each other (Sawhney, 2001).

Of course, integration is not an entirely new problem. Many organizations have undertaken specific projects in the past where they had to integrate a number of different systems. The traditional approach to integration, often referred to as point-to-point integration (Linthicum, 2001), has tended to involve crafting custom interfaces between two systems, for example, System A and System B. However, when System A needs to share information with System C, another set of interfaces needs to be created between System A and System C. Similarly, when System B needs to communicate with System C, then another set of interfaces is created. In such an approach to integration, a new set of interfaces is created for each pair of systems that need to communicate. Unfortunately, such a piecemeal approach does not scale well when tens, even hundreds, of individual systems need to be integrated as is often the case in large organizations. Organizations that have attempted to use the point-to-point approach for large-scale integration have typically ended up with tangled webs of integration interfaces in which the high cost of maintaining such a large number of interfaces has become a burden on IT budgets.

In short, the challenge faced by organizations today is to find scalable and economical solutions to the problem of large-scale integration (Sharif, Elliman, Love, & Badii, 2004). This has given rise to the term enterprise integration (EI), which denotes the need to integrate a large number of systems that may be highly distributed across different parts of the organization.

## PAST SOLUTIONS

This is not to say that solutions to the challenge of large-scale integration have not been proposed in the past. In the early '90s, distributed objects and component architectures were mooted as the answer to the challenge of integration (Digre, 1998). This was typified by the Common Object Request Broker Architecture (CORBA), which provided an open standard for distributed systems to communicate (Vinoski, 1997). However, CORBA projects were perceived as technically very complex, requiring significant development effort (Henning, 2006). Furthermore, industry support and standardization efforts for CORBA proved problematic, leading to its slow demise. More recently, enterprise resource planning (ERP) solutions, such as SAP and Peoplesoft, which involve replacing existing systems with a suite of interconnected modular systems from a single vendor, were seen as the answer to systems integration (Davenport, 1998). However, organizations soon realized that no single ERP system could address all the requirements of an organization (Hong & Kim, 2002). In fact, an ERP system often needed to coexist with existing systems, therefore heightening, rather than removing, the need for integration (Themistocleous, Irani, & O'Keefe, 2001).

However, it is not just the scale of integration that is challenging. The integration of IT systems is also severely hampered by the technical complexities of integration (Lam, 2004). For one, a particular system may have been developed on a technology platform that is, at worse, incompatible with the technology platforms used by other systems. In the '70s and '80s, for example, many systems were developed on what is now considered legacy mainframe technology, while the '90s saw the adoption and proliferation of Internet and Web-based technologies. Therefore, most organizations have, over time, ended up with a portfolio of systems based on a diverse mix of technologies. It must also be mentioned that many legacy systems were originally conceived to serve as stand-alone systems with little intent for integration. Such systems, which represent a huge financial investment to the organization, tend to become operationally embroiled within the organization, which makes them difficult to replace with newer, more modern systems (Robertson, 1997).

## PROMISING DEVELOPMENTS

Recently, however, there have been some promising developments within the IT industry that, going forward, offer the potential for easing the technical challenge of integrating systems in the future. These developments center on the adoption of Web services (Cerami, 2002) as a standard for open communication over the Internet. In short, Web services enable systems to communicate to other systems over the Internet or, as the case may be, an intranet. For this to happen, systems must expose, as services, the functionality they wish to make available to other systems. A stock system, for example, may expose a "check current stock" service to other systems so that they can check, in real time, the current stock levels of a particular product. One can imagine how this might help buyers and suppliers coordinate the supply chain much more effectively. However there are several barriers, though insignificant, to the adoption of Web services. One of these barriers is the immaturity of the standards (Bloomberg & Schmelze, 2002). The standards relating to Web services are relatively new and continue to evolve at a rapid pace. For obvious reasons, organizations are often loath to make substantial technology investments relating to standards that may be quickly superseded. Other barriers include concerns over performance and reliability. Because Web services rely on the Web, performance and reliability cannot be guaranteed, nor are Web services generally suitable for high-volume transaction processing. For these reasons alone, Web services may not be a suitable technical solution to systems integration in mission-critical business processing. In addition, while it might be useful to design new systems with Web services in mind, existing systems may need to be substantially reengineered in order to conform to a Web services model. So, while Web services are certainly a promising technical solution to systems integration, they are by no means a complete one.

Another interesting development within the IT industry is that of the service-oriented architecture (SOA). SOA (He, 2003) is something that has piggybacked off the Web services bandwagon, but in the last 3 years or so, has gained a momentum of its own. In an SOA view of the world, systems expose their functionality as a set of services, typically as a set of Web services. It does not matter which technology platform the system sits on or what development language the system has been written in as the services are defined in a way that other systems can readily understand. In fact, other systems do not need to worry or necessarily care about how these services are actually implemented. These services can either be public or published in nature. If services are public, they are typically known to a limited set of other systems, such as in the case of a corporate intranet. If services are published, however, details of the services are registered in a directory from which other systems, including those external to the organization, can discover and use the services. In essence, in an SOA, there is a network of loosely coupled systems where a complex business function may be implemented by calling upon the services offered by other systems.

With SOA, the issue of integration is no longer problematic so long as every system conforms to the SOA model and is able to offer its functionality through a set of defined services. That, unfortunately, is the point where SOA suffers the same adoption problems as Web services, with which it is closely aligned. If organizations could start with a clean sheet again, they would probably develop their systems to conform to an SOA model. In reality, however, organizations need to manage, and live with, at least in the immediate and near future, the huge investments they have already made in existing legacy systems. So SOA and Web services are not the silver bullets that will resolve all IT woes, although some would like to believe otherwise. They clearly offer a pathway, or at a least a direction, for resolving many integration issues, but are not a solution that can be readily implemented by organizations today.

## ENTERPRISE APPLICATION INTEGRATION

Fortunately, what might serve as a more complete and holistic solution to enterprise integration are the enterprise application integration (EAI) tools being marketed by integration vendors such as Webmethods, TIBCO, IBM, Microsoft, Seebeyond, BEA, Mercator, and Vitria (McKeen & Smith, 2002). Such EAI tools did not appear overnight but evolved from the message-oriented middleware (MOM) tools that became popular as a means of providing high-volume, reliable communications between systems. In general, EAI tools have three main components. The first is an integration broker that serves as a hub for intersystem communication. The integration broker performs a number of functions such as multiformat translation, transaction management, monitoring, and auditing. The second is a set of adapters that enables different systems to interface with the integration broker. An adapter is essentially a gateway or wrapper that provides the means by which packaged applications (such as SAP), database applications (such as Oracle), legacy systems (such as mainframe), and custom applications (written in Java or another programming language) can connect to the integration broker (Brodie & Stonebraker, 1995). The third component is an underlying communications infrastructure, such as a reliable high-speed network, which enables systems to communicate with each other using a variety of different protocols.

Although EAI has, until now, occupied a rather niche market, the growing importance of enterprise integration can only mean that the size of the EAI market will expand. One can also observe a growing sophistication and maturity in EAI tools. One area of interest, for example, is that of business-process management (BPM). The reason for progress here is because the motivation behind many integration projects is to support business processes that span across different parts of an organization. An e-business transaction, for instance, may begin at the order-entry system, but such transactional information may be passed onto an account-management, payment, logistics, and then eventually a delivery-order system as part of broader business-process flow. Hence, the integration of these systems is driven by business-process needs. Some EAI tools are therefore beginning to include BPM tools that enable the modeling of business processes in a graphical format. These business-process models are subsequently linked to calls or operations that initiate processing within a system. As it turns out, few organizations have their business processes so rigorously defined and mapped out. The introduction of BPM tools therefore provides a timely and pertinent reason for organizations to revisit their business processes.

Another related area of growing sophistication in EAI tools is that of business activity monitoring (BAM). BAM is about monitoring the health of activities within a business process. If, for example, a business process fails for some reason, this will be picked up by the BAM tool and an appropriate alert can be raised. BAM can also be used to identify bottlenecks within a process by tracking throughput and assessing the rate at which the different activities within a business process are successfully completed. Clearly, BAM tools, and for that matter, BPM tools, are particularly well-suited to organizations that have business processes that are, or are inclined to be, heavily automated.

So, EAI tools are themselves evolving, and what one is beginning to see is a closer alignment between technical integration, which relates to how systems talk to each other, and business integration, which relates to why systems need to talk to each other. At the same time, business integration is being facilitated by the increasing standardization within the business integration space and convergence by EAI vendors in working toward these standards. Central to this effort is the business process modeling language (BPML 1.0) standard, developed under the auspices of the Business Process Management Initiative (http://www.BPMI.org), which provides a formal model for describing any executable end-to-end business process. In theory, if all organizations described their business processes in BPML, they would find it much easier to collaborate.

## STRATEGY

Aside from technology and process issues, the other important element of enterprise integration is the strategic element. Spending thousands of dollars on EAI tools and teams of individuals modeling business processes does not necessarily mean that organizations will solve their enterprise integration problems. One missing piece from all this, like any other major endeavor, is the importance of strategic direction and senior-management support (Lam, 2005). Enterprise integration is something that affects many different parts of the organization; the problem is not confined to one particular part of the organization. As such, success can only be achieved if the various departments buy into enterprise integration and share the same vision of how to achieve it. This, of course, is easier said than done, and there are several challenges. One of the challenges is the fact that, in a large organization, individual departments may be used to operating autonomously, with minimal interaction and engagement with each other. The notion of working with other departments, or at least coordinating their technology strategies, is something that may appear quite novel. At worst, the endeavor becomes a political battle, where certain divisions are seen to be vying for control, encroaching on the space occupied by others. This, of course, is not something that is peculiar to enterprise integration projects, but is seen any large project that involves different divisions within an organization working in new ways.

Importantly, each department must believe that there is a case for enterprise integration, either financially in terms of reducing the costs of systems integration, or from a business perspective in terms of enabling new business processes or enhancing business performance. Unfortunately, individual departments, by their nature, have a localized view of their integration needs. Getting the bigger picture of an organization's enterprise integration needs is a message that must be communicated to individual departments so that they understand the rationale for a strategic perspective. Of course, if one left each department to its own devices to develop its own solution to its own set of integration problems, there would be much reinvention of the wheel and wasted time and effort. A single strategic integration solution therefore makes much more sense, where an organization can address the integration requirements in different parts of the organization in a cost-effective and controlled manner. It certainly does not make sense for each department to purchase its own EAI tool, for example.

Another thing to bear in mind is that enterprise integration is not something that can take place overnight. Enterprise integration is a long-term initiative that, in some cases, may take years to complete depending upon the number of systems to be integrated and the complexity of the integration challenge. Organizations therefore need to think carefully about how to plan and roll out the enterprise integration initiative. One approach would be to identify the high-priority integration projects within the organization where the urgency or potential business returns from integration are greater. Such successful projects could then serve to reinforce the case for integration and perhaps even provide inspiration for further integration projects. Another more risk-adverse approach would be to identify pilot projects that could serve as the grounds for organizations to build up expertise and knowledge of enterprise integration before tackling larger and more substantial projects. Such a more cautious strategy might suit organizations with little prior experience with enterprise integration. It might also be wise to consider integration projects in parallel with other business improvement projects that, in turn, can help shape the integration project. A good example is business-process reengineering, where it does not make sense to automate a process that is intrinsically inefficient or problematic, but where an opportunity presents itself to make broader organizational changes. In fact, from an organizational perspective, information systems integration involves changes in business processes, and more broadly, a realignment of technology goals with business goals (Themistocleous et al., 2001).

To sum up, enterprise integration has become a strategic enabler for many of the business initiatives organizations are implementing or wish to embark on, whether it is supply chain management, customer relationship management, e-business, or simply more efficient ways of business processing. The traditional methods of systems integration have not proved to be scalable, so solutions are needed that can address both the scale and complexity of integration. Enterprise integration, however, is not just about technology integration, it is also about process and business integration, and so may involve a reconceptualization of how organizations work and do business.

## ORGANIZATION OF THE BOOK

This book is organized into four main sections, each of which has a number of chapters. The first section is entitled "Managing Enterprise Integration" and contains the following chapters.

Chapter I examines the evolution of enterprise integration and its growing importance amongst organizations. The chapter provides a good overview of the benefits of enterprise integration and some of the challenges associated with its adoption.

Chapter II looks at five critical success factors for enterprise application integration, namely, minimal project expense, optimal reuse of components, complexity reduction, optimal coupling of applications, and minimal cost of EAI infrastructure. The authors conclude that the success factors are closely integrated, and they develop from this a number of hypotheses.

Chapter III explores how social and organizational aspects can affect ERP projects and their ability to integrate different parts of the enterprise. The chapter draws from the successful case of a multinational ERP implementation in a large international organization and concludes with the recommendation that one should examine the roles of all actors with the power to affect not only the implementation project but also the system being implemented.

Chapter IV discusses experiences in the acquisition of software-intensive systems prior to establishing a contract. One significant challenge is the identification of functional requirements, and the authors contend that business-process modeling is an effective way to explicitly define requirements while creating visibility and consensus among different stakeholders of major IT projects.

The second section is entitled "Business-Process Management" and contains the following chapters.

Chapter V presents an approach to constructing enterprise processes based on the integration of component processes. In this approach, process representations at the logical as well as physical levels are depicted in a hierarchy, and a graphical tool is used to support enterprise process construction.

Chapter VI explores, through a case study, the use of XML (extensible markup language) and Web services in conjunction with an integration broker to provide a B2B (business-to-business) solution. The authors argue that existing B2B solutions, largely based around EDI (electronic data interchange), present a high entry barrier to smaller organizations, and that the use of XML and Web services provides an alternative lower cost B2B solution that may be more suitable.

Chapter VII provides an overview of business-process management. The chapter describes how systems development has changed from being largely implementation oriented to now being analysis oriented, and suggests that business-process management standards and technologies will drive the transition to more service-oriented IT architectures in the future.

Chapter VIII describes the importance of metamodels, particularly in relation to knowledge-intensive service industries. The key guiding principle is to define the process model at the logical level, free from any technical implementation. Business-process integration is then a matter of achieving the best possible overall physical engine to implement that process model from available legacy applications, applied investment opportunities, and expert development resources.

The third section is entitled "Integration Methods and Tools" and contains the following chapters.

Chapter IX presents a methodology for the creation of EAI solutions based on the concepts of software product-line engineering. The overall idea is to view the external applications with which a given system wants to integrate as a family of systems. In this way, the flexibility required by EAI applications can be assured.

Chapter X demonstrates a visual tool for automatic supply chain integration. By clicking on the appropriate tables, functions, and fields, users can specify the data needed for integration. Integration code can then be automatically generated using Web services.

Chapter XI presents an integration framework called DAVINCI aimed at providing mobile workers with mobile software applications to query and update information coming from different heterogeneous databases. The chapter describes the trials developed using the DAVINCI architecture in different European countries.

Chapter XII describes the basic notions of intelligent agents and multiagent systems, and proposes their possible applications to enterprise integration. Agent-based approaches to enterprise application integration are considered from three points of view: (a) using an agent as a wrapper of an application or service execution, (b) constructing a multiagent organization within which agents are interacting and providing emergent solutions to enterprise problems, and (c) using the agent as an intelligent handler of heterogeneous data resources in an open environment.

Chapter XIII describes an attempt to introduce semantics to workflow-based composition. A composition framework is presented based on a hybrid solution that merges the benefits of the practicality of use and adoption popularity of workflow-based composition with the advantage of using semantic descriptions to aid both service developers and composers in the composition process and facilitate the dynamic integration of Web services into it.

The fourth and final section is entitled "Enterprise-Integration Case Studies" and contains the following chapters.

Chapter XIV examines common EI challenges and outlines approaches for combining EI and process improvement based on the process improvement experiences of banks in several different countries. Common EI-related process improvement challenges are poor usability within the user desktop environment, a lack of network-based services, and data collection and management limitations. How EI affects each of these areas is addressed, highlighting specific examples of how these issues present themselves in system environments.

Chapter XV explores the gradual evolution of SOA through various phases and highlights some of the approaches and best practices that have evolved out of real-world implementations in regional retail banks. Starting from a step-by-step approach to embrace SOA, the chapter details some typical challenges that creep up as the usage of an SOA platform becomes more and more mature. Also, certain tips and techniques that will help maximize the benefits of enterprise-wide SOA are discussed.

Chapter XVI describes a case study of application integration in a Korean bank. The case examines the integration technology employed and the adoption of the technology in a pilot project. The chapter highlights some of the managerial implications of application integration and discusses the broader organizational impact of application integration.

Chapter XVII uses the example of a retail business information system to illustrate the concept of service-oriented development and integration in a number of different scenarios. The chapter shows how using a service-oriented architecture allows businesses to build on top of legacy applications or construct new applications in order to take advantage of the power of Web services.

Chapter XVIII discusses the adoption and potential usages of RFID (radio frequency identification) in the case of enterprise integration projects such as supply chain management. Through electronic tagging, RFID can enable stocks to be monitored at a level that was previously not practical or cost effective.

## REFERENCES

Bloomberg, J., & Schmelze, R. (2002). *The pros and cons of Web services* (ZapThink Report). Retrieved from http://www.zapthink.com/report.html?id=ZTR-WS102

Cerami, E. (2002). *Web services essentials.* Sebastopol, CA: O'Reilly.

Davenport, T. A. (1998). Putting the enterprise into the enterprise system. *Harvard Business Review, 76*(4), 121-131.

Digre, T. (1998). Business object component architecture. *IEEE Software, 15*(5), 60-69.

He, H. (2003). *What is service-oriented architecture.* Retrieved from http://www.xml.com/pub/a/ws/2003/09/30/soa.html

Henning, M. (2006). The rise and fall of CORBA. *ACM Queue, 4*(5). Retrieved from http://acmqueue.com/modules.php?name=Content&pa=showpage&pid=396&page=1

Hong, K. K., & Kim, Y. G. (2002). The critical success factors for ERP implementation: An organizational fit perspective. *Information & Management, 40*, 25-40.

Lam, W. (2004). Technical risk management for enterprise integration projects. *Communications of the Association for Information Systems, 13*, 290-315.

Lam, W. (2005). Exploring success factors in enterprise application integration: A case-driven analysis. *European Journal of Information Systems, 14*(2), 175-187.

Linthicum, D. (2001). *B2B application integration.* Reading, MA: Addison Wesley.

McKeen, J. D., & Smith, H. A. (2002). New developments in practice II: Enterprise application integration. *Communications of the Association for Information Systems, 8*, 451-466.

Robertson, P. (1997). Integrating legacy systems with modern corporate applications. *Communications of the ACM, 40*(5).

Sawhney, M. (2001). Don't homogenize, synchronize. *Harvard Business Review.*

Sharif, A. M., Elliman, T., Love, P. E. D., & Badii, A. (2004). Integrating the IS with the enterprise: Key EAI research challenges. *The Journal of Enterprise Information Management, 17*(2), 164-170.

Themistocleous, M., Irani, Z., & O'Keefe, R. (2001). ERP and application integration. *Business Process Management Journal, 7*(3).

Vinoski, S. (1997). CORBA: Integrating diverse applications within distributed heterogeneous environments. *IEEE Communications Magazine, 35*(2), 46-55.

# Acknowledgment

First and foremost, the editors would like to thank all the authors who contributed chapters to the book, without which this book would not have been possible. As writers ourselves, we know how difficult it is to write a substantial piece of work and appreciate the considerable efforts that the authors have made.

Many of the authors also served as reviewers, and so deserve a double thank you. Reviewing someone else's work is not a trivial task, but many authors who contributed to the book have shown themselves to be outstanding individuals with an in-depth understanding of the field. In addition, several other individuals served as reviewers only, and we thank them for their time and effort.

We would also like to thank our respective institutions, U21 Global and the Singapore Management University, for allowing us the space to develop this book. As academics and academic administrators, we have had to fit the editorial task of putting this book together with our other day-to-day duties. As such, the support from our respective institutions for this project is greatly appreciated.

Finally, thanks also go to the staff at IGI Global, whose timely reminders have been instrumental in completing what has been a long, and often tiring, project. We knew there was light at the end of the tunnel, but without their patience and assistance, we may never have reached the end.

*Wing Lam, Universitas 21 Global*
*Venky Shankararaman, Singapore Management University*
*March 2007*

# Section I
# Business Requirements and Organizational Modeling

# Chapter I
# Dissolving Organisational and Technological Silos:
## An Overview of Enterprise Integration Concepts

**Wing Lam**
*U21 Global, Singapore*

**Venky Shankararaman**
*Singapore Management University, Singapore*

## ABSTRACT

*Over the last few years, the importance of enterprise integration has grown significantly. As organizations expand, become more distributed, and rely more on technology, there is a need to ensure both business processes and technology systems are co-coordinated in a strategic rather than ad-hoc manner. In the early days of IT, integration was viewed largely as a technical activity. Today, there is a much stronger business focus. In fact, many of an organisation's strategic initiatives such as e-business, customer relationship management and supply chain management depend upon enterprise integration. There are four basic integration architectures, namely, batch integration, point-to-point integration, broker-based integration and business process integration. Each integration architecture represents varying levels of enterprise integration maturity. Enterprise integration is a significant undertaking, and carries with it significant technical and managerial challenges. Managing enterprise integration projects requires, among other things, a strategic framework, senior management support and risk management. A clear understanding of enterprise integration requirements is also crucial.*

## THE ENTERPRISE INTEGRATION CHALLENGE

### The Integration Challenge

Most organisations today are highly dependent upon information technology. In some organisations, IT provides a competitive edge, although it is generally accepted that most large organisations require a basic level of IT simply to function.

Over the years, organisations have invested heavily in IT and, as a consequence, accumulated large portfolios of IT systems comprising of hundreds, possibly even thousands, of separate IT applications. Historically, individual IT applications were designed as stand-alone systems addressing specific functional domains such as marketing, sales, personnel, billing, and manufacturing. As business needs evolved, however, it became necessary for individual IT applications to be integrated in order to support new business requirements, for example, the need for customer details to be automatically transferred from the sales system to the billing system.

While such integration raised organisational efficiency through increased automation, it was tactical rather than strategic in nature. Individual IT applications were integrated only as the need arose, and creating custom interfaces between individual IT applications was both expensive and time consuming. As a consequence of this piecemeal approach, organisations realized their IT architecture was riddled with a complex mass of custom interfaces often referred to as "spaghetti integration." Not only was maintaining such interfaces expensive, the need for increasing levels of integration sophistication have forced organisations to address their integration needs more strategically, giving rise to the growing interest in enterprise integration.

### What is Enterprise Integration?

We define enterprise integration as "[t]he strategic consideration of the process, methods, tools and technologies associated with achieving interoperability between IT applications both within and external to the enterprise to enable collaborative business processes."

Importantly, enterprise integration is not purely about technology integration. Enterprise integration also includes a consideration of business processes that cut across various IT applications, and so provides the overarching basis for technology integration. Enterprise integration is therefore an activity that:

- Is business driven rather than technology driven
- Coordinates business processes with and across different parts of the enterprise
- Involves multiple stakeholders
- Adopts a strategic rather than tactical or localized view

Enterprise integration is not some new technology fad that will come and go, but an essential feature of how IT solutions need to be designed in order to address today's business requirements.

Some of the generic kinds of business problems that enterprise integration can help solve include the following:

- **Information aggregation:** Aggregating, organising, and presenting information from multiple IT sources in one single view
- **Single point of data entry:** Replacing the need for manual and duplicate data entry into multiple IT applications with data entry into a single IT application
- **Process inefficiency:** Reducing the effort and time required to complete business processes and eliminating manual inefficiencies
- **Web channel integration:** Enabling Web-based customers and partners direct access

*Figure 1.*

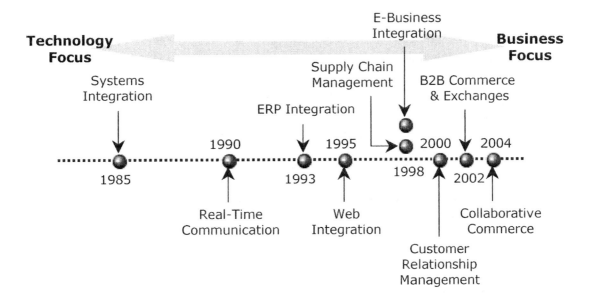

to the services provided by existing business systems

- **Supplier integration and supply chain optimization:** Enabling a supplier to integrate with a larger company's business process or an electronic marketplace

## From a Technology Focus to a Business Focus

Enterprise integration can be seen as an evolutionary concept, best viewed within the broader context of major IT trends and movements, as depicted in Figure 1.

In the '70s and '80s, a commonly used IT term was "systems integration." This had a distinct meaning and was often considered as the technical plumbing that went on behind the scenes to integrate different IT applications.

In the '90s, real-time integration became more prevalent as integration prior to that period was based largely on a scheduled, non-real-time basis. Real-time integration referred to the ability of IT systems to communicate with each other

in real time, thus providing immediate business processing and responsiveness. Technology-wise, middleware solutions based on messaging between applications and request brokers became popular.

Throughout the '90s, e-business integration and the integration of back-end legacy IT applications with modern Internet and Web-centric applications gained prominence. In the 2000s, we are seeing collaborative commerce, and businesses are now exploiting the Internet to transact with each other within a virtual value chain.

The perception of integration has therefore shifted over time from tactical, low-business-value technical plumbing to a strategic and high-value enabler for achieving business competitiveness.

## A CIO and CTO Priority

Enterprise integration is not an issue of concern solely for IT architects and designers. In fact, enterprise integration has consistently been high on the list of chief information officer (CIO) priorities. For example, in IDC's *Integration Drivers*

*Study* (2002), more than 80% of CIOs and CTOs responded that integration was either mandatory for addressing mission-critical activities or was a key enabler for meeting business-critical needs. Furthermore, Gartner Research estimates that 35% of all IT spending is for application integration. This reflects the trend toward the use of packaged IT applications offered by vendors that are integrated with an organisation's existing IT applications rather than developed from scratch.

## Assessing the Need for Enterprise Integration

To some organisations, it may be obvious that they are facing serious integration challenges and that there is a clear business case for enterprise integration. On the other hand, the need for enterprise integration may be less obvious. To quickly assess whether enterprise integration is worth investigating further, organisations should ask the following questions:

*Figure 2.*

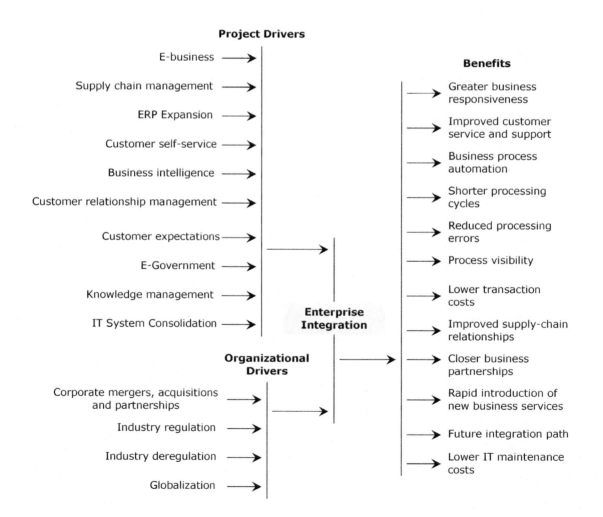

- Is there a requirement to collate information from several IT systems in real time to meet specific business objectives?
- Does the organisation have many IT systems that need to communicate with each other?
- Is there a need to key in and update the same details in separate IT systems?
- Is it difficult and time consuming to extract information from several IT systems?
- Does the organisation batch upload information from one IT system to another?
- Is the same information duplicated across many IT systems?
- Is there an issue with data inconsistency between IT systems?

If the answer is "yes" to several of the above questions, then it may well be worthwhile for the organisation to look further into enterprise integration.

## MOTIVATIONS FOR ENTERPRISE INTEGRATION

### Drivers for Enterprise Integration

Organisations are under constant pressure to remain competitive in an ever-changing business environment. Figure 2 shows some of the typical business drivers for enterprise integration, and the potential benefits that enterprise integration can bring.

The drivers for enterprise integration generally fall into one of two categories: project drivers and organisational drivers. Project drivers are those projects that cannot be realized or delivered without a significant amount of enterprise integration work. Examples of some typical project drivers are given in Box 1.

Organisational drivers are those that are brought about by organisational change or transition. (See Box 2.)

## Benefits of Enterprise Integration

Enterprise integration has a number of potential benefits to organisations. (See Box 3.)

## INTEGRATION STRATEGIES

### EAI, B2Bi, and Web Integration Projects

Most enterprise integration projects generally fall into one of three project categories, namely, enterprise application integration (EAI), business-to-business integration (B2Bi), and Web integration, as shown in Figure 3.

An EAI project is concerned with integrating the IT applications that reside within the organisation, for example, the integration of the customer-accounts IT system with the order-management IT system. A Web integration project is concerned with integrating an organisation's IT applications with Web applications to provide a Web channel. A B2B integration project, on the other hand, is concerned with integrating an organisation's IT system with those of its business partners or suppliers such as in an extended supply chain. EAI, Web integration, and B2B projects are compared in Table 1.

### Integration Timeliness: Asynchronous vs. Real Time

One of the main factors to consider when designing integration solutions is integration timeliness, or the manner in which IT systems communicate with one another. In real-time integration, IT systems communicate immediately with each other as and when required, and there is no time delay in the

*Box 1.*

| Project Drivers |
|---|
| **E-business** |
| • Integrating legacy IT systems containing core business functionality with front-end Web-based systems such as application servers and portals |
| |
| **Customer relationship management (CRM)** |
| • Building a single, consolidated view of the customer by integrating customer data that are distributed across different IT applications |
| • Integration of customer databases, and call-centre and Web-based customer service systems |
| |
| **Business intelligence** |
| • The collation of data from a variety of IT applications and data sources into a data warehouse |
| • The mining of data from multiple sources to identify patterns and trends |
| |
| **Customer self-service** |
| • Enabling customers to perform services traditionally performed by staff by integrating front-end customer service applications with back-end IT systems |
| • Integrating customer service IT applications running over multiple channels (Web, digital TV, mobile) |
| • Using customer service workflow events to trigger IT system services and processing |
| |
| **Expansion of ERP (enterprise resource planning) systems** |
| • The integration of ERP systems that provide core functionality with more specialised IT systems used within the organisation |
| |
| **Supply chain management** |
| • Integrating disparate IT systems for order management, manufacturing resource planning, scheduling, and logistics |
| • The exchange of real-time information between buyers and suppliers to optimize the supply chain and streamline workflow |
| |
| **Customer expectations** |
| • Meeting customer expectations for immediate access to real-time services and information that require the integration of IT systems |
| • Achieving straight-through processing (STP) for increased efficiency, particularly for those operating in the financial-services sector |
| |
| **Knowledge management** |
| • Real-time search and access to knowledge content that is distributed across multiple knowledge sources |
| • The management of knowledge assets within the enterprise |
| |
| **E-government** |
| • Integrating legacy back-end government systems with front-end Web-based systems |
| • Enabling intra- and intergovernment agency workflows and data exchange |

*Box 2.*

| Organisational Drivers |
|---|
| **Consolidation, mergers, and acquisitions**<br>•   Consolidating or merging duplicate or overlapping IT systems that perform similar functionality<br><br>**Globalisation**<br>•   The integration and consolidation of data from IT systems that reside in different countries and regions<br>•   The enablement of business processes and workflows that cut across multiple organisational and regional units<br><br>**Industry deregulation**<br>•   Forcing monolithic IT systems that supported a spectrum of business services to be broken up into smaller, more manageable IT systems that are loosely integrated<br><br>**Industry regulation**<br>•   Government and industry regulations that force systems to be integrated, for example, the United States Health Insurance Portability and Accountability Act of 1996 (HIPAA), which seeks to establish standardized mechanisms for electronic data interchange (EDI), security, and confidentiality of all health-care-related data |

*Box 3.*

| Benefits |
|---|
| **Greater business responsiveness**<br>It is easier to deliver real-time business services because the functionality and information of IT systems are immediately accessible.<br><br>**Customer support and service**<br>Customer service staff have a single, consistent view of the customer and immediate access to relevant customer data, thus allowing for a higher level of customer service to be delivered. Customers can also self-service, enabling them to take advantage of services anytime, anywhere.<br><br>**Business process automation**<br>Processes that were previously performed manually are now automated and streamlined.<br><br>**Process visibility**<br>Because business processes are automated, they can be more easily monitored and checked.<br><br>**Shorter processing cycles**<br>The time required to complete a business process is significantly reduced.<br><br>**Reduced processing errors**<br>Because less reliance is placed on manual processing, the scope for clerical errors is significantly reduced.<br><br>**Closer business partnerships**<br>The ability to exchange information with business partners in a dynamic and real-time fashion leads to closer and more effective business partnerships.<br><br>**Improved supply chain relationships**<br>The exchange of up-to-date supply chain information leads to more optimized supply chains between buyers and suppliers. Stock levels can be reduced and just-in-time (JIT) deliveries are possible.<br><br>**Future integration path**<br>With careful planning and design, the creation of a robust integration architecture will enable future IT systems to be integrated more easily.<br><br>**Rapid introduction of new business services**<br>New business services can be rapidly introduced because IT systems are integrated. |

*Figure 3.*

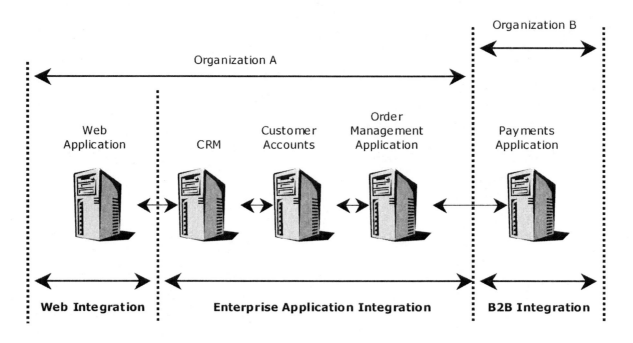

*Table 1.*

|  | **EAI** | **B2Bi** | **Web Integration** |
|---|---|---|---|
| **Scope of Integration** | Intra-organisation: IT systems within the organisation | Interorganisation: IT systems between different organisations | Intra-organisation: Existing IT systems with new Web-based front-end applications |
| **Typical Project Drivers** | Operational efficiency, customer relationship management, business-process automation | Supply chain management, B2B commerce | E-business, Web-channel services |
| **Key Challenges** | Business-process management, workflow | Document and data exchange standards | Online security, transaction management |

communication. Imagine, for example, a Web site that communicates with a back-end stock-price system to provide users with up-to-date information about stock prices. However, where there is a time delay in the communication, the integration is known as asynchronous. Imagine, for example, a Web site that allows a customer to reserve hotel rooms but can only confirm those room reservations 24 hours later.

Real-time integration typically relies on an event-driven architecture. In an event-driven architecture:

- Business events generate messages that are sent to all the appropriate IT systems
- Upon receipt of the message, the IT systems update information held locally
- A real-time (and typically sophisticated) messaging infrastructure is used to support communication between IT systems
- Information between IT systems is kept up to date

Many financial organisations, for example, deploy event-driven architecture to enable real-time trading and transaction processing. For such business applications, real-time processing is a necessity. Asynchronous integration, on the other hand, typically relies on an export-transfer-load (ETL) architecture. In an ETL architecture:

- Information from one IT system is exported, transferred, and loaded into another IT system
- The process is batch based and takes place at predetermined intervals
- Relatively simply query and file transfer routines are used to implement the ETL process
- Data may be inconsistent between IT systems for a period of time until the next ETL process takes place

Organisations need to decide whether their integration needs are for real-time integration or asynchronous integration. Given the business climate of short processing and faster turnaround cycles, there has been a general move toward real-time integration and the adoption of an event-driven architecture. However, for some organisations, the cost and feasibility associated with adopting an event-driven architecture may be prohibitive.

## Levels of Integration

Another way of looking at integration is in terms of the level at which integration occurs. Integration can occur at five levels, namely, presentation, data, application, service, and process integration, as shown in Figure 4.

Each successive level of integration represents what might be considered a more advanced form of integration.

- **Presentation integration:** The aggregation of data from multiple IT systems within a single view. A Web portal that aggregates and displays a customer's stock portfolio taken from one IT system and bank account balance from another IT system can be considered integration at the presentation level. However, there is no direct communication between the individual IT systems.
- **Data integration:** The synchronisation of data held in different databases. For example, if two different databases hold the same customer's address details, a change of address in one database should also by synchronised in the other database. If this synchronisation takes place only after a period of time, a certain amount of data freshness will be lost.
- **Application integration:** Where applications make some of the functionality directly accessible to other applications. Popular packaged applications, for example, such

*Figure 4.*

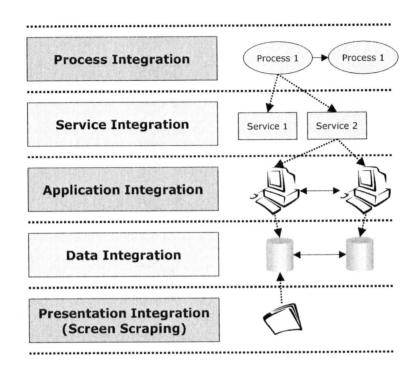

as SAP and PeopleSoft, often expose their functionality through well-defined application programming interfaces (APIs).

- **Service integration:** A common set of reusable services that are made available to other applications. For example, a service to obtain a customer's status may be created that can be called by any other application within the organisation. Such services are usually built from a set of specific application functions.

- **Process integration:** The definition of business-process or workflow models from which reusable services are called. Process integration is particularly relevant in collaborative contexts, such as B2B, where there business processes drive interactions and transactions between business partners.

To date, the integration efforts of most organisations have largely been focused on the presentation, data, and application integration levels. However, more organisations are now realizing the benefits and potential for integration at the service and process levels. This is evidenced by the interest in service-oriented architectures (SOAs), which we will discuss in more detail later on.

## CHALLENGES IN ENTERPRISE INTEGRATION

### Technical Challenges

Enterprise integration cannot be achieved overnight. Rather, it is a journey that may take an organisation many years to achieve. Some of the

*Box 4.*

| Technical Challenges |
|---|
| **Stand-alone design**<br>• Legacy IT systems that were originally designed to stand alone<br>• Legacy IT systems that lack a published API or set of external interfaces<br><br>**Internalised data models**<br>• Packaged applications with internalised data models that are not meant to be shared with other IT systems<br>• No data access or sharing of data from interfaces<br><br>**Heterogeneous technologies**<br>• The fact that different IT systems use different platforms, applications, and programming languages<br>• The presence of multiple distributed computing paradigms, for example, COM, EJB, and CORBA (common object request broker architecture)<br><br>**Legacy engineering**<br>• Legacy IT systems that use very old and possibly unsupported technology; original designers are no longer available and system documentation is sparse<br>• Compatibility issues between old legacy technologies and modern Web-based technology<br><br>**Lack of interfaces**<br>• Many IT systems that lack published interfaces or APIs that provide other applications access to functionality within the system<br>• Interfaces or APIs that are restricted in functionality, or mandate the use of a particular programming language such as C++<br><br>**Semantic mismatch**<br>• Semantic differences in the interpretation of data within individual IT systems that complicate the exchange of information<br>• Multiple business partners using proprietary business semantics and terminology<br><br>**Unclear business processes**<br>• Ill-defined and ad hoc business processes between organisational units<br><br>**Standards**<br>• Use of proprietary and organisation-specific standards for business data and documents<br>• Multiple business partners each with their own set of standards<br>• A lack of universally adopted standards, and new emerging interoperability standards such as XML (extensible markup language) and Web services that have yet to stabilize<br><br>**Security**<br>• Providing a consistent level of security across all integrated IT systems<br>• Providing users with single sign-on (SSO), and integrating authentication and authorization policies used by individual IT systems |

technical challenges that make enterprise integration difficult are described in Box 4.

It is because of these barriers that the technologies employed in enterprise integration are themselves both complex and diverse.

## Management Challenges

As well as the technical challenges, there are several management challenges. Some of the management challenges associated with enterprise integration are described in Box 5.

*Box 5.*

| Management Challenges |
|---|
| New collaborations<br>• Working across traditional organisational boundaries and silos<br>• A more intimate working relationship with external partners<br>• Overcoming differences in organisational policies, working practices, culture, and internal politics<br><br>Project scoping<br>• Satisfying the enterprise integration needs of multiple stakeholders<br>• Agreeing on a project scope that can be realistically achieved given limited resources<br>• Estimating project resources and timelines for a complex project<br><br>Continued stakeholder support<br>• Obtaining support from business stakeholders throughout the entire enterprise integration project life cycle, including detailed requirements analysis, business-process modeling, and user testing<br><br>Data ownership<br>• Resolving data ownership issues in situations where data are spread across different organisational entities<br>• Determining the policies regarding who has permission to change what data<br><br>Time constraints<br>• The need to deliver enterprise integration solutions rapidly in order to meet time-sensitive business requirements<br><br>Cost constraints<br>• The need to deliver effective, value-for-money enterprise integration solutions<br>• Operating within a set budget that offers little, if any, flexibility for cost overruns<br><br>Migration<br>• The need to keep existing business-critical IT applications running while the enterprise integration solution is being installed<br>• Executing migration plans that minimize downtime<br><br>Expertise<br>• Sourcing the necessary enterprise integration expertise, whether in house or externally |

# INTEGRATION ARCHITECTURES

## The Four Integration Architectures

There are four basic kinds of integration architecture: (a) batch integration, (b) point-to-point integration, (c) broker-based integration, and (d) business-process integration. Each kind of integration architecture represents differing levels of sophistication, as illustrated in Figure 5.
It is not always the case that an organisation should strive to achieve the most sophisticated integration architecture. The appropriateness of the integration architecture largely depends upon the business requirements for integration, and also on what can be achieved with the available technology. Each kind of integration architecture has its own capabilities and limitations. With a batch integration architecture, the most simple type of integration architecture:

•    IT applications communicate asynchronously with each other
•    ETL is used to transfer data from one IT application to another

*Figure 5.*

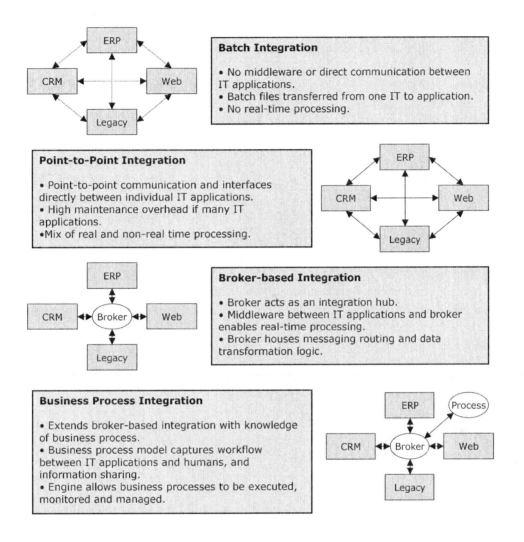

- There is no business requirement for real-time processing

A batch integration architecture is often suitable for organisations that perform high-volume back-end processing that is able to take place overnight, weekly, or on a scheduled basis.

In a point-to-point integration architecture:

- IT applications communicate with other IT applications through interfaces

- Communication supported by these interfaces may be real time or asynchronous
- The number of interfaces to be built rapidly grows as the number of IT applications increases

A point-to-point integration architecture is most suited when the number of IT applications to be integrated is small. However, as the number of IT applications grows, the number of interfaces to be maintained becomes problematic, and a broker-based integration architecture becomes more appropriate. In a broker-based integration architecture, an integration broker acts as a hub for connecting IT systems. A good analogy is to view an integration broker as a major airport that serves as a hub for connecting flights to smaller cities. A broker-based integration architecture can only be implemented through tools offered by companies such as TIBCO and WebMethods. In such tools, the main functions of the integration broker are as follows:

- **Message routing:** Routing data from one IT system to another, for example, routing banking transactions taken from the call centre to the legacy banking system
- **Schema translation**: Translating data given in one format or schema definition into data in another schema definition, for example, Siebel data fields mapped onto CICS records
- **Business-rules processing**: Applying business rules based on the content of the data or messages, for example, raising a warning flag if the account balance is less than zero
- **Message splitting and combining**: Splitting or combining messages
- **Transaction management:** Managing a sequence of messages as a single transaction

A broker-based integration architecture supports real-time integration. Such tools provide the basic integration broker and messaging infrastructure, but adapters must be purchased to interface with the kind of IT systems that reside within the organisation.

A business-process integration architecture builds upon the broker-based integration architecture by adding the concept of business processes that trigger the relevant messages to be sent to individual IT systems. For example, when a banking customer opens up a new account, the business process may involve doing a background check, creating an account if the

background check is positive, and then sending a notification to the customer. Each step in the business process involves communication with different IT systems.

## Why Business-Process Modeling?

Once encoded inside IT systems, business rules often become difficult to change. This prevents organisations from rapidly responding to new business circumstances, what is often considered a lack of IT agility. Greater business responsiveness can be achieved through more flexible IT architectures that allow business processes and business rules to be defined explicitly rather than hard-coded within the IT systems themselves. This has given rise to tools that can capture and model business processes, elements essential to the implementation of a business-process integration architecture.

## Service-Oriented Architecture

A new development in the area of integration architectures is that of SOA. In SOA, IT systems expose business functionality as services that can be accessed by other IT systems, both internal and external to the organisation. A set of standards, known collectively as Web services, enables services to be implemented in a standardized and consistent way, which allows them to be accessed via an intranet or the Internet.

In an SOA, organisations can treat services (whether provided internally or externally by another organisation) as a kind of building block from which they can compose new applications. Creating a service in the first place requires some initial development cost, but the cost is amortized when the service gets reused across multiple applications. In an SOA, IT systems are loosely coupled and, with Web services, use a common standardized protocol to communicate with each other. This makes the job of integration much easier and therefore potentially much cheaper as

*Figure 6.*

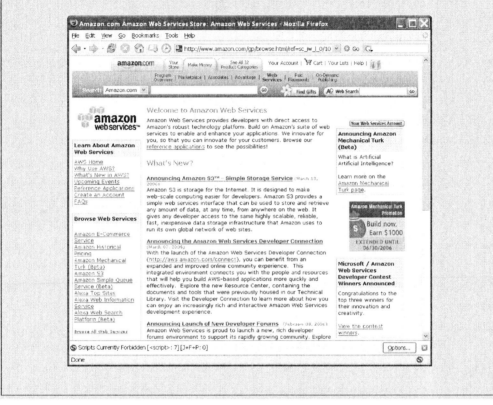

Amazon.com, best known for its success in selling books online, uses Web Services to make available certain business functionality to outside organisations. For example, an outside organisation can, through Web Services, search for books in Amazon's catalogue and display the results in its own website.

no special tools or proprietary technologies are required. The defining characteristics of SOA and how they are supported by Web services are shown in Table 2.

## Advantages of SOA

The potential advantages of SOA are as follows.

- **Shorter development cycles:** An organisation is able to rapidly create IT applications from the reuse and composition of existing services. The need to write new code from scratch is kept to a minimum.

- **Lower development costs:** The overall costs of IT are lowered due to the reuse of existing services.

- **Simplified systems integration:** IT systems that expose their functionality as a set of services are more easily integrated with other IT systems that are able to tap into those services.

- **Increased business agility:** An organisation can respond to changing business requirements more effectively because it is able

*Table 2.*

|  | **SOA Characteristics** | **Web Services Support** |
|---|---|---|
| **Service Exposure** | IT applications expose services that can be accessed programmatically by other applications or services. | Web services can act as a wrapper around legacy and existing applications, exposing their functionality as Web services that can be accessed by other IT systems. |
| **Distributed Services** | Services are distributed and may reside either within the organisation or outside the corporate firewall. | Web services can reside on servers within the organisation or anywhere on the Internet. |
| **Loose Coupling** | Services are loosely, rather than tightly, coupled. | A Web service represents a distinct piece of business functionality that is loosely coupled in nature. |
| **Service Ownership** | A set of services may not necessarily belong to a single owner. | A Web service may be owned by the organisation, its business partners, or its customers, or by external third-party service providers. |
| **Open Standards** | Services are described and accessed using open, rather than proprietary, standards. | Web services are described in WSDL (Web services description language), published via UDDI (universal description, discovery, and integration), and accessed over SOAP (simple open access protocol). |
| **Neutrality** | The services are platform independent, and vendor and language neutral. | Web services on a J2EE platform can communicate with Web services on the .NET platform and most other platforms. |
| **Service Creation** | New applications, services, or transactions can be created through the assembly, composition, and chaining of lower level services. | Web services can be called by other Web services, or strung together into a transaction. |
| **Transparency** | The underlying details of how a service is implemented (low-level transport, communications, platform, language, etc.) are hidden from the IT applications that use the service. | Web services hide back-end implementation and platform details from the IT applications that use them. |
| **Reuse** | The reuse of services is encouraged when possible. | Web services encapsulate business functionality into reusable services that can reused by many different IT applications. |

to create new services and IT applications more quickly.

- **Improved quality of service (QoS):** Customers and end users experience an improved level in the QoS because SOAs allow poorly performing services to be easily substituted with improved services without affecting other systems.

Until the advent of Web services, it was difficult to imagine how SOAs could be created and supported in a cost-effective manner using existing technologies. However, the huge support for Web services from major technology vendors, in terms of development tools and server software, provide organisations with the necessary building blocks for creating SOAs.

## MANAGING ENTERPRISE INTEGRATION PROJECTS

### Strategic Framework

Organisations need to take a strategic view of enterprise integration if they are to maximize their integration efforts and achieve overall success. This means understanding and coordinating the integration projects within the entire organisation at a strategic level. If integration issues are addressed solely at the tactical level of individual projects, the effort and cost of integration will be repeatedly incurred. Without a strategic view, multiple integration solutions may be developed each with their own associated development, implementation, maintenance, and infrastructure costs. Instead, adopting a strategic view of enterprise integration can be likened to managing multiple pieces of a large jigsaw. A strategic framework for enterprise integration is set out in Figure 7.

Crucially, enterprise integration is a major program of change that will affect the way the organisation does business. Understandably, there are several major elements within the strategic framework to consider:

• The organisation's business strategy, business directions, and consequently, the overall business case for enterprise integration

*Figure 7.*

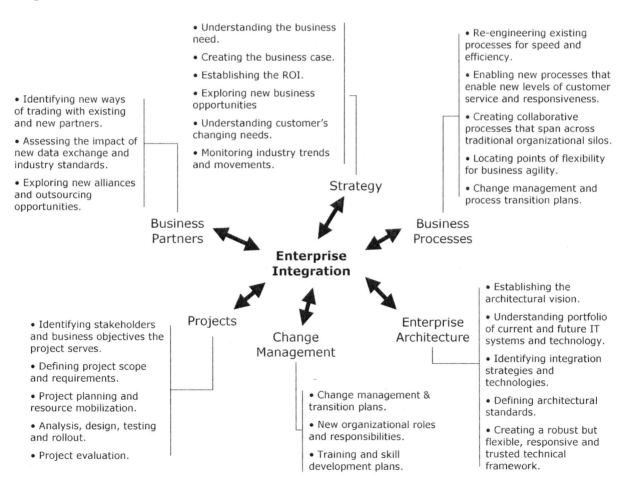

17

- The way in which the organisation engages and interacts with business partners
- Business processes within the organisation and how they span across units within the enterprise
- The enterprise IT architecture and portfolio of existing IT systems
- The management of change associated with the transition from nonintegration to integration
- The integration requirements for individual projects

An enterprise integration initiative therefore needs to be seen as a program that provides a coordinated and controlled platform for integration that is capable of sufficing the integration requirements of specific projects within the organisation.

## Project Risk

Estimates put forward by the Standish Group have suggested that 88% of enterprise integration projects fail (either failing to deliver within a time frame or within budget). This is an alarming statistic, even against the background of high project failure rates within the IT industry in general. Enterprise integration projects tend to be high risk for one or more of the following reasons:

- **Complex, cross-unit business processes:** The project involves the comprehension of complex business processes and business requirements that span across multiple organisations or organisational units.
- **Multiple stakeholder silos:** The project involves collaboration between multiple stakeholders who each represent an organisation or organisational unit that owns parts of the business process or IT systems that are pertinent to the business process.
- **Heterogeneous technologies:** The project involves the study of multiple IT systems

that employ a diverse mix of heterogeneous technologies ranging from older mainframe environments to modern Web-centric environments.

- **Legacy engineering:** The project involves work on or changes to legacy systems when those involved in the original design may no longer be available.
- **Novel technologies:** The project involves the use of integration technologies, such as messaging and integration brokers, that are novel to the organisation and IT professionals working within the organisation.
- **Lack of experience:** The organisation lacks experience in implementing enterprise integration solutions and therefore approaches the project without the benefit of lessons learned or previous working practice.
- **Shortfalls in specialist skills:** Individuals with the required skills set and expertise required to complete large enterprise integration projects are not readily available within the organisation or the market in general.
- **Lack of documented methodology and best practice:** Unlike for regular software development projects, there are few well-documented, tried, and tested methodologies for enterprise integration projects within the literature.

## Coordinating Multiple Projects

As argued, enterprise integration initiatives require the coordination of multiple projects. A layer of program management that coordinates these individual projects so that they address specific integration needs while avoiding duplicate work effort is essential. One model for program management is presented in Figure 8.

The program management model consists of four main phases:

*Figure 8.*

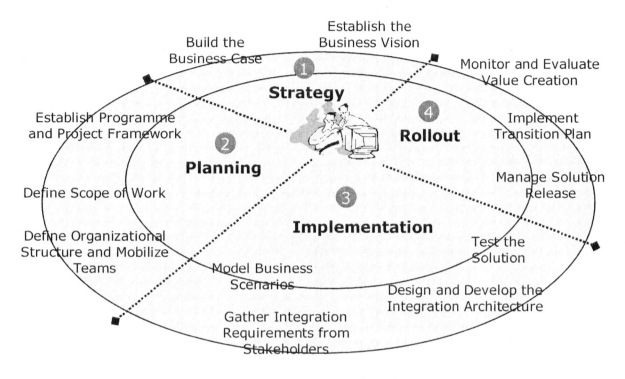

- **Strategy:** In this phase, the CIO, with support from senior business stakeholders, formulates the strategic need for enterprise integration though the creation of a business vision and business case.
- **Planning:** The enterprise integration business strategy is translated into a program of work. Individual enterprise integration projects are identified, prioritised, and planned.
- **Implementation:** The project manager, architects, and developers execute the enterprise integration project, working with stakeholders to gather requirements and implement the enterprise integration solution.
- **Rollout:** The enterprise integration solution is rolled out into the live environment.

The phases in an enterprise integration project are generally cyclic in nature because the value added, or not, from previous enterprise integration rollouts tend to influence the strategic directions of upcoming enterprise integration projects. The cycle of strategy, planning, implementation, and rollout is a pattern that applies to all enterprise integration projects, though the success of a project comes down to how well risks, both of a managerial and technical nature, are managed within each phase.

## UNDERSTANDING ENTERPRISE INTEGRATION REQUIREMENTS

### Enterprise Integration Requirements

Enterprise integration solutions can be large, complex, and costly. It is therefore important that enterprise integration requirements are understood before the organisation proceeds to design, develop, or procure an integration solution. A

*Figure 9.*

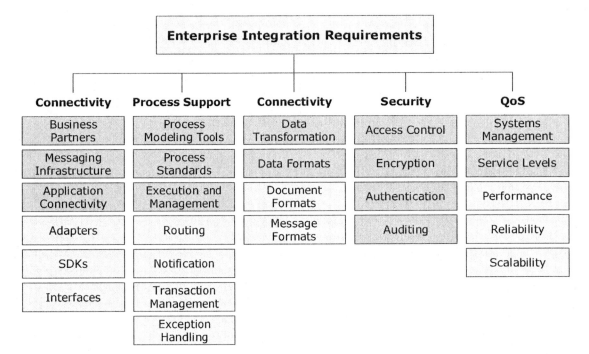

framework for gathering enterprise integration requirements is shown in Figure 9.

The framework identifies five major categories of integration requirements, key areas of requirements focus (the shaded boxes), and secondary areas of requirements focus that should follow on from discussions on the key areas. The five major categories of integration requirements are:

- Connectivity
- Process support
- Data exchange
- Security
- Quality of service

Most likely, an organisation will consider procuring an enterprise integration tool from a vendor such as TIBCO or WebMethods and look to tailor that tool to their specific requirements. In addition, cost and time-to-deliver constraints must also be considered as requirements that relate

to the project as a whole. Each major category of integration requirements is described in greater detail next.

## Connectivity Requirements

An enterprise integration solution may have the following connectivity requirements:

- Connectivity to business partners who are geographically dispersed and separated through several layers of corporate firewall
- Ability to add new business partners, such as new suppliers, and information about their connectivity profile (e.g., types of business interaction, data standards supported)
- Compatibility with the IT and communications infrastructure of business partners
- Minimal introduction of new IT and communications infrastructure, for example, dedicated lines such as with EDI

- Reliable messaging infrastructure for the exchange of messages between partners that works across and supports a variety of protocols, for example, HTTP (hypertext transfer protocol), HTTPS, FTP (file transfer protocol), and MIME (multipurpose Internet mail extensions)
- Support for, and connectivity across, multiple computing platforms including W2K, Windows XP, Unix, and Linux
- Off-the-shelf adapters that enable immediate connectivity to a wide range of packaged applications (e.g., SAP, Siebel) without any coding effort
- Adapters for legacy system connectivity
- Adapters for connectivity with database management systems such as Oracle and Informix
- Availability of software development kits (SDKs) or adapter development kits (ADKs) that allow adapters to be written in a widely supported language such as Java or C++ if they are not available off the shelf (as in the case of the bespoke applications)

## Process Support Requirements

An enterprise integration solution may have the following process support requirements:

- Graphical modeling tool that enables business processes to be defined diagrammatically, including the inputs and outputs of individual processes
- Scripting language that allows both simple and complex business rules to be defined, for example, "IF stock_level < stock_reorder_level THEN SEND reorder_message"
- Ability to link portions of the business-process model to either human actors or applications that are responsible for processing in that part of the model
- Repository for storing process models and different versions of the same process model

- Ability to change process models at a later time and link new actors and applications
- Environment for executing processes and for maintaining the state of a process
- Administration console that enables processes, and individual parts of each process, to be controlled (stopped and started)
- Monitoring tools that allow one to view the state of a process and collate statistics on process performance
- Transaction support and management, including transaction recovery and rollback
- Support for predefined or industry-standardized business processes, such as RossettaNet
- Alerts when process exceptions or time-outs occur

## Data Exchange Requirements

An enterprise integration solution may have the following data exchange requirements:

- Ability to handle files encoded in different formats, which may be proprietary, for example, iDOC (SAP format), .doc, or .rtf, or open, for example, XML
- Ability to transform files from one encoding to another for consumption by another application, for example, from XML to .doc
- Semantic mapping tools that enable data fields defined in one schema to be mapped onto data fields defined in another scheme
- Validation of data inputs and outputs, for example, to determine conformance to standards

## Security Requirements

An enterprise integration solution may have the following security requirements:

- Minimal violations or changes to existing corporate security setups, for example, firewalls
- Security administration console for defining users, roles, and access control properties
- Support for multiple authentication schemes, such as user names and passwords
- Support for digital certificates and pubic key infrastructure (PKI) standards
- Encryption of messages exchanged between business partners and applications
- Auditing, at multiple levels, from transaction to database operations
- Single sign-on

## Quality of Service Requirements

An enterprise integration solution may have the following security requirements:

- High performance and throughput
- Guaranteed delivery of messages between applications through the use of appropriate message storage and delivery architectures
- Overall high reliability, achievable through infrastructure redundancy and automated fail-over
- Administration console for the ease of systems management, for example, server start and closedown, memory thresholds, and database sizing
- Comprehensive suite of monitoring and performance tools

## CONCLUSION

Enterprise integration takes a strategic and coordinated view of an organisation's integration needs. In the past, these needs have been addressed on an ad hoc and piecemeal basis, resulting in a proliferation of interfaces and the high costs associated with their management and maintenance. This approach, however, has become unsustainable and a drain on IT resources. Enterprise integration is suited where an organisation's business requirements dictate the need for the real-time processing of complex business processes across different IT applications and parts of the enterprise. Enterprise integration solutions typically involve some form of integration broker that coordinates the flow of information from one IT application to another. Importantly, however, organisations should not underestimate the many technical and management challenges that need to be overcome in order to achieve enterprise integration success.

# Chapter II
# Success Factors and Performance Indicators for Enterprise Application Integration

**Alexander Schwinn**
*University of St. Gallen, Switzerland*

**Robert Winter**
*University of St. Gallen, Switzerland*

## ABSTRACT

*The effectiveness and efficiency of information systems are closely related to the degree of integration between applications. In order to support the management of application integration, five success factors are analyzed. For each success factor, appropriate performance indicators are proposed. Since the analysis indicates that the success factors are closely interrelated, these dependencies are discussed and hypotheses are derived.*

## INTRODUCTION

Design and management issues of information systems architecture are discussed from a practitioner's perspective (e.g., by Zachman, 1987) as well as from a scientific perspective (e.g., by Krcmar, 1990; Österle, Brenner, & Hilbers, 1992). Architecture models help us to understand and communicate enterprise architecture. They also support architecture design decisions.

Recently, some approaches have integrated the design and management of IS architecture with other architectures in an enterprise (e.g., Malhotra, 1996; Martin & Robertson, 2000; McDavid, 1999; Youngs, Redmond-Pyle, Spass, & Kahan, 1999). Some of these approaches focus on technologies, while others connect IS architecture to business requirements. This chapter addresses application architecture, one specific component of IS architecture. A company's application architecture describes applications (or application domains) and their relations (or interfaces) on a conceptual level (Winter, 2003b). Application architecture is designed and managed from a business rather than technical point of view. The design and management of application architecture aim at minimizing integration costs. For achieving this goal, development-time integration costs as well as run-time integration costs have to be considered.

After this introduction, conceptual considerations on the optimal level of application integration are used to identify general success factors. A broad literature review helps to identify specific success factors for application integration. Then, for every success factor, respective performance indicators are proposed. As some of the success factors seem to be closely interrelated, their interdependencies are examined qualitatively next. Finally, this analysis results in a set of hypotheses for successful application integration that have to be validated quantitatively in further research.

## APPLICATION INTEGRATION

In contrast to their technical interpretation as containers of software artifacts (e.g., modules and/or data structures), applications represent tightly interrelated aggregates of functionalities from a business perspective. While tight couplings between certain functionalities lead to their aggregation into the same application construct, loose couplings are represented by interfaces between applications. The number of application constructs depends on the definition of tight coupling. If a small number of (monolithic) applications are created in application design, only a few interfaces have to be implemented. As a consequence, costs for running and maintaining interfaces are low, while the total costs for running and maintaining applications are high due to more difficult change management and higher complexity. If many small applications are created in application design, much more interfaces are needed, which implies higher operations and maintenance costs. On the

other hand, the total application development and maintenance costs are significantly lower due to less application complexity. The question is how to find an optimal balance between the number of interfaces and the number of applications in order to reduce the total costs of operations and maintenance. These comprise (a) costs for developing, maintaining, and running applications, and (b) costs for developing, maintaining, and running interfaces. Figure 1 (Winter, 2006) illustrates this trade-off. Due to network effects, we expect a nonlinear growth of the costs for applications and interfaces.

In real-life situations, the optimal degree of integration cannot be determined analytically because the costs are not constant and often cannot be assigned directly to certain applications or interfaces. Therefore, instruments are needed that control and manage the evolution of an application architecture toward an approximated optimal degree of integration. An evolutionary approach (i.e., a bundle of IS projects that improve the degree of integration successively) is needed because normally a revolutionary redesign of application architecture is not feasible due to immense costs. In order to measure the contribution of proposed projects toward the degree of integration, it is necessary to define objectives and derive performance indicators. In the next section, success factors for application integration are analyzed.

## SUCCESS FACTORS FOR APPLICATION INTEGRATION

Numerous approaches to application integration can be found in the literature, many of them in the field of enterprise application integration (EAI). We analyzed not only scientific contributions, but also practitioner papers regarding the success factors mentioned. Table 1 summarizes the results. The following success factors were mentioned most often:

*Figure 1. Application vs. interface costs tradeoff (Winter, 2006)*

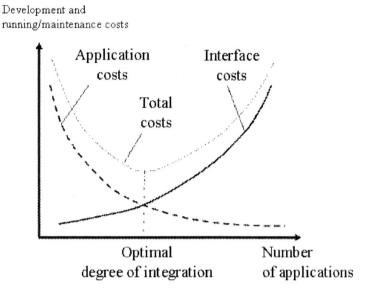

- Minimal project expenses (time and costs) for integrating applications into the present application architecture
- Optimal reuse of software components and minimal functional redundancy
- Reduction of complexity within the present application architecture
- Optimal coupling of applications (not tighter than needed, not looser than necessary)
- Minimal costs for and number of infrastructure components (e.g., middleware components like message broker or object request broker, ORB)

Assuming that all these factors affect information systems performance, we propose one central figure that is analyzed in the upcoming sections: the agility of an information system. Agility is defined as the ability to react to upcoming changes effectively and efficiently (Ambrose & Morello, 2004). The main goal of application architecture (re)design is to increase information systems' agility. The following sections intend to identify indicators that influence this agility. Besides this proposed central success factor, many other criteria are conceivable, for example, higher customer satisfaction by application integration. In the following we concentrate on the agility of an information system, only.

Related work mostly proposes general rules for application integration, for example, achieving cost savings by reducing the number of interfaces when using a bus architecture. Quantitative measurements and the derivation of specific performance indicators are usually not considered.

Figure 2 illustrates the identified success factors, the focus success factor—agility of the information system—and the assumed interdependencies between these success factors. In the following subsections, we describe and discuss

*Table 1. Success factors for application integration in related work*

| Approach (vertical) / Success factor (horizontal) | Minimal expenses of integration projects | Optimal reuse | Reduction of complexity | Optimal level of coupling | Minimal costs for infrastructure |
|---|---|---|---|---|---|
| **Scientific Contributions** | | | | | |
| (Linthicum, 2000) | | X | X | X | |
| (Zahavi, 2000) | X | X | X | | X |
| (Kaib, 2002) | X | X | X | X | X |
| (Ruh, Maginnis and Brown, 2001) | X | X | X | X | X |
| (Cummins, 2002) | X | X | X | X | X |
| (Fridgen and Heinrich, 2004) | | | X | | X |
| (Themistocleous and Irani, 2001) | X | X | | X | X |
| **Practitioner Approaches** | | | | | |
| (Liske, 2003) | X | | X | | |
| (Moll, 2003) | X | | X | | X |
| (Kuster and Schneider, 2003) | X | | X | X | X |
| (Bath, 2003) | X | X | X | X | |
| (Endries, 2003) | | | X | | X |
| (Gröger, 2003) | X | X | X | | X |
| (Knecht, 2003) | | X | | X | X |
| (Hofer, 2003) | X | X | X | | |
| (Friederich, 2003) | | X | X | | X |
| (Aust, 2003) | X | | X | X | X |

the identified success factors in more detail and propose appropriate performance indicators. Later in the chapter, positive and negative interdependencies are analyzed in detail.

## Information Systems' Agility

The agility of an information system expresses the ability to react to upcoming new or changed requirements (e.g., a new business function has to be supported). These requirements can be of technical nature (e.g., exchanging an application due to expired maintenance contracts), or they are triggered by business decisions (e.g., outsourcing decisions for certain business processes). Among the many factors that influence the agility of an information system, application architecture is of outstanding importance (Winter, 2003a). Until the end of the 1980s, software development was dominated by creating monolithic applications in nearly all business areas—independent from the core competences of the company. A wide "information systemization" of all business processes was the main intention. After this first phase, the trend of implementing standard software packages (COTS [commercial off the shelf], e.g., SAP R/3) came up. These packages provide a broad range of functionalities. The rapid growth and business importance of the Internet triggered the addition of many more applications. In this phase, time to market was much more important than a sustainable and cost-efficient application architecture. This led to redundancy as many functionalities were reimplemented, not integrated.

As a consequence of different application development phases with changing design goals, companies now struggle with an application architecture where functional silos, standardized packages, and nonintegrated Internet applications coexist. In order to increase consistency and reduce operations costs, most companies run evolutionary application redesign programs.

In addition, new business requirements trigger new applications that lead to even more integration efforts. A survey by the Gartner Group shows that about 40% of IT budgets are spent on implementing and maintaining interfaces (Krallmann, 2003).

The direct influence of the application architecture on the agility of the information system cannot be specified easily. A measurable coherence between the complexity of the interapplication relations and agility would be a precondition.

In general, however, agility is measurable. The idea is to measure all extensions and changes to the information systems in a certain period (C) and compare this to the total expenses needed for the extensions and changes (E). If two periods are compared, it can be checked whether extensions and changes have been implemented more

*Figure 2. Application integration success factors and their interdependencies*

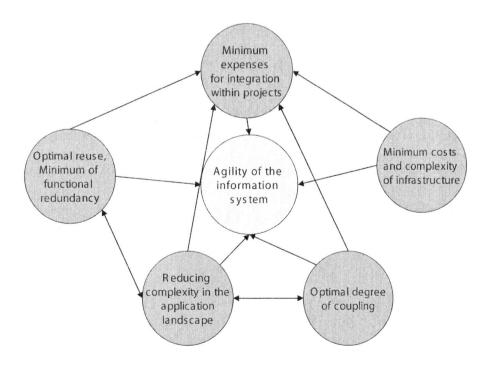

efficiently. The comparison, however, cannot be carried out on a project-by-project basis because architecture management is a long-term effort on an aggregate scale: "You can't 'cost-justify' architecture" (Zachman, 2001).

As the figures C and E cannot usually be provided directly from IT controlling, we propose figures for each success factor in the following.

## Application Architecture Complexity

Historically grown application architectures comprising hundreds of applications (Linthicum, 2000) cannot be managed as a whole. The complexity is too high and the dependencies are too multifaceted. Therefore, it is necessary to control the complexity by knowingly disintegrating the application architecture. In our context, complex means that it is difficult to describe the behavior of the whole application architecture in any language even if we have detailed knowledge about each single application. To reduce the complexity, the application architecture can be spilt up into smaller defined components (building blocks). Among these components we propose loose coupling, and within the components tight coupling. Loose coupling reduces dependencies among the components. This means that changes in one component will not affect the other component (Linthicum).

One way to disintegrate the application architecture is to define application domains that comprise a defined set of applications (e.g., the applications of one business unit; Hagen & Schwinn, 2006). The number of application domains should be small to keep the advantage of lower complexity. In a concrete case in the financial-service sector, the number of domains is about 20 (Hagen & Schwinn, 2006). It is important that applications within a domain can be modified without direct effects to other domains.

The most important figure is the degree of disintegration. To measure this figure, the number of loosely coupled controlled links (i.e., links that are directly controlled by architecture management) between application domains is counted. This figure has to be put in relation to the uncontrolled links between the domains. The quotient represents the level of disintegration.

To measure these figures, existing tools like source-code analyzers or application repository managers can be used.

## Degree of Coupling

General rules for the degree of coupling are not useful because each application relation is different. Intuitively, tight coupling is appropriate if two applications implement functionalities that belong to the same business process. If two applications implement functionalities of different business processes, loose coupling would be appropriate. The degree of coupling has direct influence on the agility of the information system: Tighter coupling necessarily will result in excess expenses for implementing new requirements or changing existing ones. If applications are coupled too loosely, run-time overhead may arise and additional middleware components for integration might be needed.

For each application relation, an appropriate level of coupling has to be chosen. Since a common methodology with objective criteria does not exist, it is difficult to derive measurable figures. A potential indicator could be the expenses for implementing changes in dependent applications due to modifications of an independent application. High expenses indicate too-tight coupling. On the other hand, the run-time overhead has to be measured. A high run-time overhead would indicate too-loose coupling. However, it is difficult to exactly determine the run-time overhead. It usually cannot be measured directly because it is hidden in other maintenance or infrastructure costs. Even if the run-time overhead could be measured, the interpretation of measured values is difficult as no benchmarks exist.

As a consequence, a random sampling of applications and the measurement of modification costs induced by context changes seem to be the only way to approximate the degree of coupling.

## Reuse and Functional Redundancy

The success factor of optimal reuse claims that every function is only implemented once by an application. If a function only has to be developed and maintained once, lower development and maintenance costs should be achievable. Furthermore, reuse supports the consistency, quality, and flexibility of applications (Cummins, 2002). To achieve maximum reuse, powerful middleware is needed to deliver the centrally implemented functionality to as many other applications as possible that are running on different platforms. In the design process, a framework is needed to ensure the future reusability of a component (design for reuse). One important aspect is the granularity of the function. If only large monolithic software components are developed, the potential for reuse is high because broad functionality or parts of it can be reused. On the other hand, dependencies are created as the frequency of changes is higher and release cycles are shorter. If the components are too modular, the benefit of reuse is lower, and run-time and maintenance overhead increases because many small functions have to be reused, not just one.

Another aspect that should be considered when designing reusable software components is the level of specialization. If only very specialized components are developed, the potential for reuse is low because only few applications or users need this very specialized function. If the components too general, the potential for reuse should generally be higher, but the benefit for the next user is low as only a very general service can be utilized. Furthermore, additional business logic has to be implemented by the next user, which leads to redundancy again. An example could be a service that checks the account balance of a customer of a bank. One function or service could be GetBalanceCustomerX(), which is designated to one special customer X and therefore is very special. A very general service would be QueryDatabase(DatabaseName, Query). The potential for reuse of this service would be very high as many applications have to access databases. Considering our example, the benefit of this service is pretty low as we still have to adapt the service so that it returns the account balance of customer X. Obviously, a service like GetBalance(Customer) would satisfy most needs. This example illustrates that it is very important to consider the level of specialization when designing new reusable functions or services.

For measuring reuse, one important figure is the average reuse per function, which means how many applications actually use a certain function. To measure this figure, repositories are necessary that document the utilization of functions by applications. If there is a central server (e.g., a CORBA [Common Object Request Broker Architecture] server), it could be analyzed which applications are using which service.

Another important indicator for the quality of the application architecture is the number and growth of public interfaces. A high amount and quick growth could indicate redundancy as we believe all functionality should be covered by reusable functions at a time. However, fast growth could also be the result of introducing new technologies (e.g., service-oriented architecture). If so, the figure indicates the user acceptance of the new technology.

## Integration Project Expenses

It is problematic to determine integration costs on a general level because integration effort is not only dependent on business requirements (e.g., timeliness) and technology support, but also on time. For example, the first project that uses a CORBA infrastructure has to pay for the infrastructure while the following projects can reuse it, thereby receiving indirect sponsorship by the initial project. Furthermore, an isolated application that only has some relations to other applications usually has lower integration costs than an application that needs information and functions of many other applications (e.g., a portal application). As a consequence of these problems, we do not consider single projects, but entire project portfolios over a certain period to measure integration expenses.

Implications of the quality of integration aspects within the application architecture can only be drawn if the integration problem and the expenses are normalized. The integration costs depend on many factors (e.g., number of interfaces, number of business units involved, quality of existing documentation, etc.) that are hard to determine. As it is very hard to measure the integration complexity, we propose an indicator that compares the entirety of integration efforts: We sum up all integration costs and divide them by the overall integration complexity within a certain period (e.g., 1 year). If we compare two periods by dividing the quotient from the first year by the quotient of the second year, the result should be smaller than 1. That means that we have implemented more integration complexity with lower expenses.

The only thing we can derive from this figure is the cost efficiency. However, without benchmarks, we cannot determine useful target values.

## Costs and Complexity of the Integration Infrastructure

The number of deployed integration technologies or tools within a company has direct influence on the IT expenditures. As a consequence, the number of utilized tools has an (indirect) influence on IT project costs. If only a few tools are used, they can be supported professionally. Higher numbers of tools lead to uncertainties as developers have to decide which tool is most appropriate

for specific requirements. On the other hand, a basic set of technologies is necessary to implement the requirements efficiently and to avoid work-arounds by simulating one technology by means of another (e.g., using a message broker to implement a service-oriented architecture).

Possible figures for measuring this factor are infrastructure costs, the number of deployed technologies and tools, standardization, or the degree of fulfillment of project requirements by standard technologies and tools.

## SUCCESS FACTOR INTERDEPENDENCIES

It seems likely that the identified success factors affect each other. Some of the success factors are complementary to each other, while others are competing. In the following, all possible relations between success factors are analyzed. The derived hypotheses form a basis for further quantitative research.

### Interdependencies Between Project Expenses and Reuse (Figure 3)

Project expenses can be kept to a minimum if a large number of components are reusable. On the other hand, project expenses are usually higher if there are no reusable components and the project has to implement new components. Implementing a reusable component is expensive because efforts are needed that would not be needed if a single-use component is implemented (Boehm et al., 2000; Ruh et al., 2001), mainly due to quality-assurance reasons.

As we do not want to minimize short-term single-project expenses, but instead want to minimize the total expenditures for the entire project portfolio over a long period, the influence of maximizing reusability should be positive. The savings should be higher than the additional expenses for developing reusable components.

*Figure 3.*

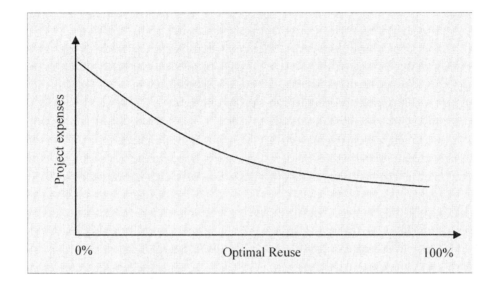

*Figure 4.*

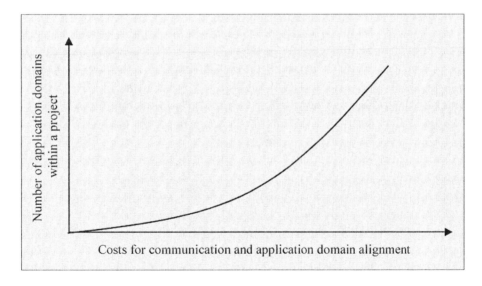

The earnings only can be realized when reusable components are actually reused. The more often a component is reused, the higher the savings are expected to be. However, the project costs will always have an initial effort to handle reusable components.

**Interdependencies Between Project Expenses and Complexity (Figure 4)**

To minimize complexity, disintegration of the application architecture (e.g., by separating manageable application domains) has been proposed above. If development projects affect one application domain only, no specific complexity

*Figure 5.*

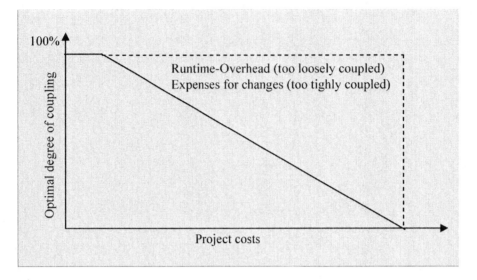

*Figure 6.*

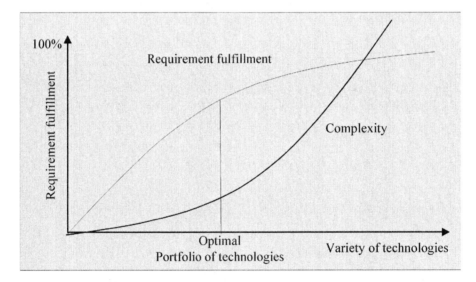

influence is present. If a development project affects more than one application domain, expenses are expected to increase. The more application domains are involved in a development project, the higher the communication costs are expected to be. Due to network effects, we expect nonlinear growth.

**Interdependencies Between Project Expenses and Degree of Coupling (Figure 5)**

Optimal coupling claims to minimize dependencies among applications and to avoid the run-time overhead. The more appropriate the degree of coupling between applications, the less development and run-time overhead may be expected (Linthicum, 2000). Both success factors are complementary to each other.

*Figure 7.*

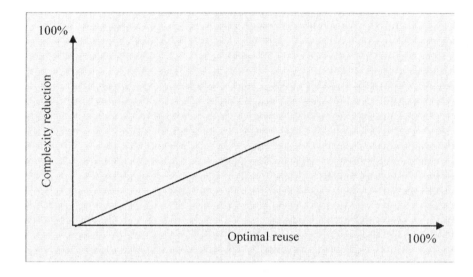

## Interdependencies Between Project Expenses and Complexity of Infrastructure (Figure 6)

To reduce the costs and complexity of infrastructure, the number of deployed infrastructure components has to be restricted. This leads to a limitation of applicable technologies within a project (e.g., only CORBA services are supported, and Web service technology is not used). The limitation of technologies incurs lower expenses for infrastructure, but they might not be able to meet project requirements perfectly. Therefore, additional project costs might be incurred to meet these requirements. Due to network effects, we expect nonlinear growth of the complexity as more and more technologies have to be compared against each other regarding whether they are appropriate within a project.

## Interdependencies Between Reuse and Complexity (Figure 7)

Like for development projects, communication and alignment problems between application domains occur when implementing reusable components (used by different application domains). Choosing an appropriate granularity and generality of reusable components is essential. In general, implementing reusable components has a positive influence on the complexity (Ruh et al., 2001). However, an optimal reuse does not imply a minimized complexity.

## Interdependencies Between Reuse and Coupling (Figure 8)

To realize a high potential of reuse, applications should be loosely coupled. A high potential of reuse means that one component may be used by many applications. If all applications are coupled tightly, the expenses for changing one component would be excessive. Tightly coupled components should therefore not be designed for reuse. The coupling of applications is necessary to reuse components (Kaib, 2002).

## Interdependencies Between Reuse Costs and Complexity of Infrastructure (Figure 9)

If reusable components are developed independently from any technologies, these success factors do not influence each other. If, however, reusable components are developed using a certain technology, the number of technologies within a

*Figure 8.*

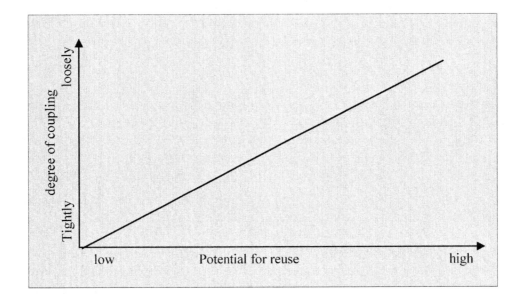

*Figure 9.*

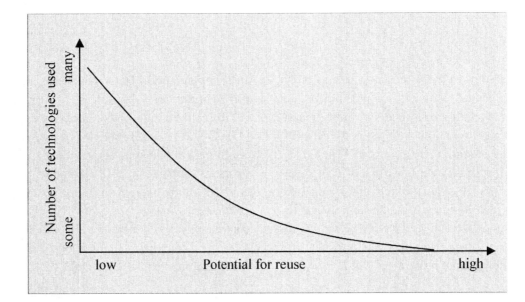

*Figure 10.*

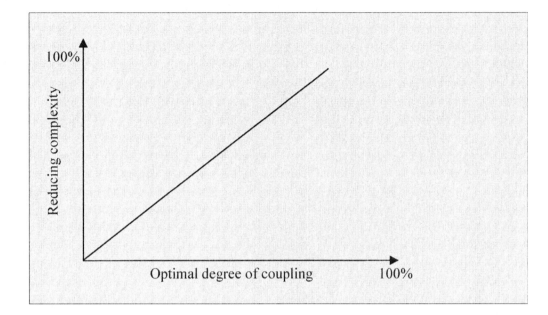

*Figure 11.*

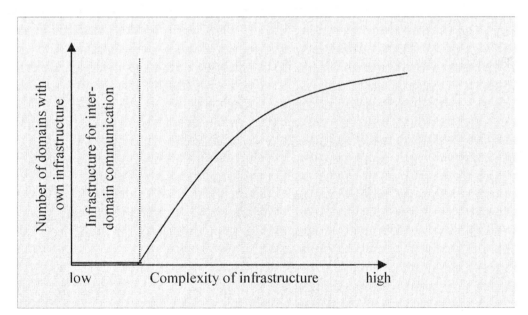

*Figure 12.*

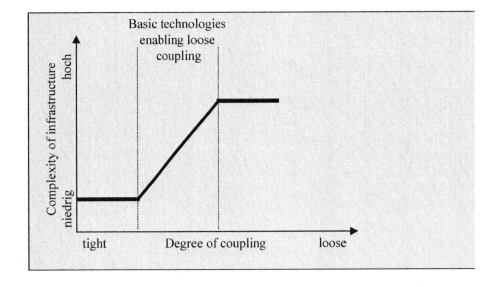

company obviously affects the reusability: For each reusable component, it has to be decided which technologies should be used to develop the component (e.g., developing a CORBA service and/or a Web service). The larger the number of technologies, the more difficult these decisions are. Hence, a positive influence among reusability and costs of infrastructure can be stated: The less technologies deployed, the higher the potential for reuse (Kaib, 2002). Due to network effects, we expect nonlinear growth as it gets harder and harder to make a decision (about which technology to use to develop a component).

**Interdependencies Between Complexity and Coupling (Figure 10)**

Forming many application domains leads to many loosely coupled applications: Within application domains, a tight coupling dominates, but multidomain applications are loosely coupled. If complexity is reduced, the influence on the degree of coupling is positive.

**Interdependencies Between Complexity Costs and Complexity of Infrastructure (Figure 11)**

If the infrastructure is managed centrally (i.e., independently from application domains), both success factors do not influence each other. If the infrastructure is managed for every application domain separately, the expenses and the complexity grow nonlinearly with every managed application domain. Moreover, a centrally managed infrastructure is needed anyway for supporting interdomain communication.

**Interdependencies Between Coupling Costs and Complexity of Infrastructure (Figure 12)**

If all applications are coupled tightly, infrastructure costs tend to be low as the applications do not have to communicate via middleware. If, in contrast, applications are coupled loosely, the distance between two applications has to be bridged. As a consequence, middleware is needed, for example, for transport, transformation, routing, and so forth. Since usually some loosely coupled applications exist in every application architecture, a standard set of middleware is needed anyway. If a set of middleware components is already available, the infrastructure costs and complexity do not grow further.

## CONCLUSION

Based on a literature review and an analysis of current practices in companies, five success factors for application integration have been analyzed: application architecture complexity, degree of coupling, reuse and functional redundancy, integration project expenses and costs, and the complexity of the integration infrastructure. For each success factor, performance indicators have been proposed that allow such success factors to be measured and ultimately managed. Furthermore, the interdependencies among the success factors have been analyzed qualitatively.

The significance of the proposed success factors and their influence on the agility of an information system have been thoroughly discussed. However, neither the completeness of the proposed system of success factors nor the hypotheses for their interdependencies have been validated quantitatively. Developing hypotheses in a top-down manner, but based on a similar system of success factors for application integration, two recent studies by Klesse, Wortmann, and Schelp (2005) and Klesse, Wortmann, and Winter (2005) analyze dependencies quantitatively. Future work will therefore focus on integrating the bottom-up, literature-based application integration management approach presented in this chapter with these quantitative studies.

## REFERENCES

Ambrose, C., & Morello, D. (2004). *Designing the agile organization: Design principles and practices.* Gartner Group.

Aust, H. (2003). *Einführung von EAI bei der PostFinance.* Proceedings of the St. Galler Anwenderforum, St. Gallen, Switzerland.

Bath, U. (2003). *Web Services als Teil einer serviceorientierten Architektur.* Proceedings of the EAI Forum Schweiz, Regensdorf, Switzerland.

Boehm, B., Abts, C., Brown, A. W., Chulani, S., Clark, B. K., Horowitz, E., et al. (2000). *Software cost estimation with Cocomo II.* NJ: Prentice Hall.

Cummins, F. A. (2002). *Enterprise integration.* New York: John Wiley & Sons.

Endries, T. (2003). *Schenker AG: EAI.* Proceedings of Integration Management Day, St. Gallen, Switzerland.

Fridgen, M., & Heinrich, B. (2004). *Investitionen in die unternehmensweite Anwendungssystemintegration: Der Einfluss der Kundenzentrierung auf die Gestaltung der Anwendungslandschaft* (Working paper). Augsburg, Germany.

Friederich, M. (2003). *Zusammenspiel verschiedener Integrationstechnologien und Werkzeuge bei der Züricher Kantonalbank.* Proceedings of the EAI Forum Schweiz, Regensdorf, Switzerland.

Gröger, S. (2003). *Enterprise application integration in the financial services industry.* Proceedings of Integration Management Day, St. Gallen, Switzerland.

Hagen, C., & Schwinn, A. (2006). Measured Integration: Metriken für die Integrationsarchitektur. In J. Schelp & R. Winter (Eds.), *Integrationsmanagement* (pp. 268-292). Berlin, Germany: Springer.

Hofer, A. (2003). *Projekt SBB CUS: EAI ermöglicht eine Erhöhung der Kundeninformationsqualität im öffentlichen Verkehr.* Proceedings of the St. Galler Anwenderforum, St. Gallen, Switzerland.

Kaib, M. (2002). *Enterprise Application Integration: Grundlagen, Integrationsprodukte, Anwendungsbeispiele.* Wiesbaden, Germany: DUV.

Klesse, M., Wortmann, F., & Schelp, J. (2005). Erfolgsfaktoren der Applikationsintegration. *Wirtschaftsinformatik, 47*(4), 259-267.

Klesse, M., Wortmann, F., & Winter, R. (2005). *Success factors of application integration: An exploratory analysis* (Working paper). St. Gallen, Switzerland: University of St. Gallen.

Knecht, R. (2003). *Application architecture framework UBS-WMBB.* Proceedings of Integration Management Day, St. Gallen, Switzerland.

Krallmann, H. (2003). Transformation einer industriell geprägten Unternehmensstruktur zur einer service-orientierten Organisation. *Proceedings des Symposiums des Instituts für Wirtschaftsinformatik "Herausforderungen der Wirtschaftsinformatik in der Informationsgesellschaft"* (pp. 1-12).

Krcmar, H. (1990). Bedeutung und Ziele von Informationssystemarchitekturen. *Wirtschaftsinformatik, 32*(5), 395-402.

Kuster, S., & Schneider, M. (2003). *Banking Bus EAI-plattform der Raiffeisengruppe Schweiz.* Proceedings of Integration Management Day, St. Gallen, Switzerland.

Linthicum, D. S. (2000). *Enterprise application integration.* Reading, MA: Addison-Wesley.

Liske, C. (2003). *Advanced supply chain collaboration enabled bei EAI.* Proceedings of the St. Galler Anwenderforum, St. Gallen, Switzerland.

Malhotra, Y. (1996). *Enterprise architecture: An overview.* @BRINT Research Institute. Retrieved from http://www.brint.com/papers/enterarch.htm

Martin, R., & Robertson, E. (2000). *A formal enterprise architecture framework to support multi-model analysis.* Proceedings of the Fifth CAiSE/IFIP8.1 International Workshop on Evaluation of Modeling Methods in Systems Analysis and Design, Stockholm, Sweden.

McDavid, D. W. (1999). A standard for business architecture description. *IBM Systems Journal, 38*(1), 12-31.

Moll, T. (2003). *Firmenübergreifendes EAI-Netzwerk: Integrierte Umsetzung von Geschäftsprozessen über einen Marktplatz als EAI-Hub.* Proceedings of Integration Management Day, St. Gallen, Switzerland.

Österle, H., Brenner, W., & Hilbers, K. (1992). *Unternehmensführung und Informationssystem: Der Ansatz des St. Galler Informationssystemmanagements.* Stuttgart, Germany: Teubner.

Ruh, W. A., Maginnis, F. X., & Brown, W. J. (2001). *Enterprise application integration.* New York: John Wiley & Sons Inc.

Themistocleous, M., & Irani, Z. (2001). Benchmarking the benefits and barriers of application integration. *Journal of Benchmarking, 8*(4), 317-331.

Winter, R. (2003a). *An architecture model for supporting application integration decisions.* Proceedings of 11th European Conference on Information Systems (ECIS), Naples, Italy.

Winter, R. (2003b). Modelle, Techniken und Werkzeuge im Business Engineering. In H. Österle & R. Winter (Eds.), *Business Engineering: Auf dem Weg zum Unternehmen des Informationszeitalters* (pp. 87-118). Berlin, Germany: Springer.

Winter, R. (2006). Ein Modell zur Visualisierung der Anwendungslandschaft als Grundlage der Informationssystem-Architekturplanung. In J. Schelp & R. Winter (Eds.), *Integrationsmanagement* (pp. 1-29). Berlin, Germany: Springer.

Youngs, R., Redmond-Pyle, D., Spass, P., & Kahan, E. (1999). A standard for architecture description. *IBM Systems Journal, 38*(1), 32-50.

Zachman, J. A. (1987). A framework for information systems architecture. *IBM Systems Journal, 26*(3), 276-292.

Zachman, J. A. (2001). You can't "cost-justify" architecture. *DataToKnowledge Newsletter (Business Rule Solutions LLC), 29*, 3.

Zahavi, R. (2000). *Enterprise application integration with CORBA.* New York: John Wiley & Sons.

# Chapter III
# ... and the Social Matters

**Amany R. Elbanna**
*The London School of Economics and Political Science, UK*

## ABSTRACT

*This chapter introduces one of the social problems that could affect the integration of the implementation team and highlights its effect on the integration capability of the ERP system. It adopts the sociological theory of the actor network and develops the notion of organisational "othering" to make sense of the raised intragroup conflict during implementation and to trace its effects on the final solution. The chapter contributes to current understanding of the organisational complexity of ERP implementation and the growing research on politics and intragroup conflicts.*

## INTRODUCTION

Enterprise resource planning (ERP) systems have received increasing attention from businesses, the press, and academia over the last few years. It is presented to the market as one of the large integrated packaged software. Its market boomed and is expected to continue to grow, from $14.8 billion in 1998 (Foremski, 1998) to $26.7 billion in 2004, with estimates of $36 billion by the end of 2008 (IDC, 2004). The top five vendors are SAP, PeopleSoft, Oracle, Microsoft, and Sage.[1] SAP is the market leader with a share of over 30% of the ERP market (AMR Research, 2004; SAP AG, 2004). It nearly became the standard for doing business in many industries and the de facto solution to replace the dispersed legacy systems within organisations. Organisations face tremendous difficulties in implementing ERP systems to the extent that their efforts were described by many authors as "the journey of a prisoner escaping from an island prison" (Ross & Vitale, 2000), "mission critical" (Davenport, 2000), and "the corporate equivalent of root-canal work" ("SAP's Rising in New York," 1998).

The complexity of ERP implementation is multidimensional with technical, organisational, and social facets (Dong, 2000; Holland, Light, & Gibson, 1999; Krumbholz, Galliers, Coulianos, & Maiden, 2000; Markus & Tanis, 2000). Previous research asserts that the social and organisational complexity has stronger effect on the implementation project performance than the mere technical dimension (Xia & Lee, 2004).

There is a growing body of research on the technical side of the implementation including research that focuses on the integration between ERP and other disparate systems that coexist with it (Themistocleous, Irani, & O'Keefe, 2001). The softer organisational and social issues involved in the implementation have not received much attention despite authors' assertions that they represent the major hurdle at the front of ERP implementation (Cadili & Whitley, 2005; Markus, Axline, Petrie, & Tanis, 2000; Mendel, 1999; Norris, 1998).

This chapter explores one of the social and organisational aspects that could affect the ERP implementation project, causing delays and jeopardising its integration capability and potentials. It draws from a successful case of a multinational ERP implementation in a large international organisation. Through the application of the actor-network theory's (ANT) notion of translation and the introduction of the concept of organisational "othering," it unravels the problematic nature of intragroup involvements in the implementation and how the organisation's existing social logic could affect negatively the implementation project.

The following section outlines the research's informed theory and concepts. It briefly reviews ANT and in particular the notion of translation in addition to developing and introducing the new concept of organisational othering. This is followed by a brief section on the research setting and methodology. The final two sections are dedicated to the analysis of the case study's details, and the conclusion and implications of the findings.

## THEORETICAL BACKGROUND

### Actor Network Theory

ANT was developed over the years in the field of science and technology studies (STS) through the collaborative work of many scholars (Bijker & Law, 1997; Latour, 1987; Law, 1992). ANT is occupied with unraveling the way societies come to accomplish certain goals (Latour, 1988). It maintains a distinct view of society since it views it as a network of human and nonhuman actors. Since the social is nothing but chains of associations between human and nonhuman actors, the theory keeps an analytically symmetrical view of both of the social constituents (human and nonhuman). It gained considerable attention in the IS field, and many IS scholars have applied it in their work (Bloomfield, Coombs, Knights, & Littler, 1997; Monteiro, 2000b; Walsham, 2001). ANT views technology as a product of active negotiation and network building where society actively inscribes on the technology a certain "programme of actions" (Monteiro, 2000a). It also views technology as what holds society together and renders it durable and relatively irreversible (Latour, 1991).

Translation is the dynamic by which an actor recruits others into its project. It is a continuous process and "never a complete accomplishment" (Callon, 1986). By and large, it describes how actors are bent, enrolled, enlisted, and mobilised in any of the others' plots (Latour, 1999). The word itself keeps its linguistic sense: It means that one version translates every other (Latour, 1987). It does not mean a shift from one vocabulary to another but "it means displacement, drift, invention, mediation, the creation of a link that did not exist before and that to some degree modifies two elements or agents" (Latour, 1999, p. 179). It also has a "geometric meaning" that is moving from one place to the other. Translating interests means at once offering new interpretations of these interests and channeling people in different directions (Latour, 1987). The translation or recruitment of entities toward a certain network could take place through implementing several strategies. All would lead the actors in whatever they do and whatever they are interested in to help the network builders to pursue their interests.

Each network consists of more actors and intermediaries. At the same time, a network could be collapsed to represent a node in a wider network. An actor, hence, is not only a member of his or her own local network, but this network is part of a wider global network. Intermediaries define the relationship between the local and global network.

ANT does not expect the network to hold forever since it has no inertia. The network holds only as long as the network builders involved invest their efforts to lock actors in a certain translation and prevent them from joining in any other competing ones. It is irreversible if the translation is able to suppress any other competitive translation. On the other hand, a network could reverse in front of a stronger competitive translation that pulls the actors away from the previous one (Callon, 1991).

## Organisational Othering

The notion of othering is implicitly embedded in ANT as the theory stresses differences and creates a distinction between "them" and "us" through its attempts to translate and recruit actors to a certain network, together with associated attempts to distance or weaken the relationships between these actors and other networks. Othering is part of the competition between networks that is needed to create sustainable boundaries, space, and distance between the actors and other networks.

The notion of othering itself is adopted from anthropology and politics. It was developed to understand how colonisers exercise power over the colonised by naming, labeling, stereotyping, and generally othering its subjects in order to divide and rule. The colonial ruler produces the colonised as a fixed reality that is at once an other and yet entirely knowable and visible (Bhabha, 1986).

Othering, then, refers to the articulation of differences. Yet, ANT is rather inclusive and recognises that the constituents of networks have,

and act on, different and sometimes competing interests. One of the strategies that can distance actors from networks is therefore to exclude, label, name, and blacklist some groups in order to marginalise them. Othering and differing also serves as a tool by which the identity of a group is assured against other groups. So, by othering and focusing on the differences between us and them, a group stresses its identity, creating "symbolic boundaries" around it to keep it pure and keep intruders, foreigners, and all kinds of others away (Hall, 1997).

In these ways, others are identified and outcast as they are different from oneself. Hall (1997) argues that, from many different directions and within many different disciplines, the question of difference and otherness plays an increasingly important role. He explains that difference is ambivalent because it can play both positive and negative roles. It is important for the production of meaning, the formation of language and culture, for social identities, and for a subjective sense of the self; at the same time, it is threatening and a site of danger, negative feelings, splitting, hostility, and aggression toward the other.

Organisational othering is about the ways in which some groups are perceived and categorised by others. It is to differentiate and to facilitate acting upon others. The labeling or stereotyping of others is carved and institutionalised over time, and it is often employed by a relatively powerful group as a means of defining other less powerful communities, mainly to stress the identity of the powerful group over others (Beall, 1997). It is a kind of encapsulation and fixation that is moved around. Hence, othering is to identify stereotyped notions about different entities and enact upon them accordingly.

It raises the question of how this othering could possibly be overcome in implementing ERP systems since the ERP systems' logic is more about integration, transparency, and coordination.

## RESEARCH SETTING AND METH-ODOLOGY

### Methodology

This research belongs to the qualitative school of research in information systems (Kaplan & Maxwell, 1994). It adopts an interpretive case-study approach since it explores the social aspects related to the ERP logic of integration. Interpretive case studies, by and large, are believed to be an appropriate approach for enquiries aimed at exploring the nature of phenomena (Klein & Myers, 1998; Orlikowski & Baroudi, 1991; Walsham, 1995). It does not aim to prescribe direct solutions but rather to highlight issues and invite reflections and discussions.

Data collection took place between August 2000 and March 2001 in a sound large international food and drinks organisation, anonymously Drinko, as part of a larger research project to study the implementation of ERP in various organisations. The system implementation project lasted for 4 years and consisted of implementing five modules of SAP in two major business units (BUs) of the group located in two different European countries. The researcher was allowed access to the prestigious organisation in the last three phases of the project. The data collection was based on interviews and document reviews. All quotes are from the field. When more than one source agrees on a particular view, only one quote is presented without referring to the source.

### The Case Study

Drinko is a global food and drinks group that owns many production, packaging, and sales sites, each of which represents a company or group of companies that operates locally. Drinko has major production operation in the United Kingdom and another European country (disguised as EUB). The case will focus only on the business units of the UK group (EUK) and the EUB group. This includes over 25 business units.

In 1998, Drinko announced the initiation of a "major Drinko-wide initiative unprecedented in scale and cost" (CEO [chief executive officer] in the project inauguration speech) to implement a single system based on SAP technology. The project took over 3 years and cost over £40 million. The project was later narrowed down to focus on two major business units, namely the EUK business unit and the EUB business unit. The EUB business unit was, in general, sensitive toward whatever came from EUK and had a long history of rejecting any sort of control coming from either the EUK business unit or the corporate centre.

## ANALYSIS OF THE CASE STUDY

### Organisational Othering

Drinko EUB was for a long time othered within Drinko. It was portrayed and stereotyped as old-fashioned, with complacent staff who were using the same procedures and concepts for over 20 years, who had typically worked for the company for 15 years or more, and who had no intention of changing, advancing, or modernising their "historic style" (as expressed by many EUK interviewees). In contrast, EUK perceived its own staff as being dynamic, modern, "capable of doing things," and able to face successfully the aggressive competition in the market.

Although EUK perceived EUB as being resistant and stubborn by rejecting any sort of idea coming from EUK, and although EUB was behind EUK in terms of business practices, structure, communication, and style, EUK had accepted for a long time a fluid relationship with EUB that allowed EUB to be distanced and separate as long as EUB was "bringing the cash back" (interviews with the executive manager and change manager).

From EUB's perspective, on the other hand, EUK unreasonably wanted to dominate, rule, and control EUB. It did not find a reason for EUK's perceived superiority and so always tried to assert its identity and the fact that EUB was the powerful part by providing the cash for the company, believing that Drinko would have collapsed without its hard work. EUB felt that despite it providing a valuable intermediary for the EUK network, it was not correctly valued and positioned within the organisation. Hence, it liked to stress how much it supported the company's whole network.

## How to Integrate the Other

In 1996, Drinko's top management became concerned with the cash flow (intermediary) that EUB provided and feared that the increased competition may decline it further over time. It therefore felt it needed to interfere in the "cash cow's" (EUB's) internal network to change it toward becoming more efficient and capable of meeting the increasing competition. However, management knew that EUB would be very sensitive toward whatever came from EUK and that no change programme would be accepted from there. This led Drinko's senior managers to identify a good opportunity to align EUB with EUK in the notion of having an integrated system, which would deal with Y2K (Year 2000) issues. They saw that facing Y2K compatibility issues offered a convincing excuse to implement the integrated system that would interfere with EUB and connect it operationally to EUK. The Y2K argument was a particularly powerful one for EUB since the sheer number of legacy systems it had to deal with was a serious problem.

For this reason, the corporate-wide SAP system was presented to EUB top management as a way to solve the threatening situation and the danger of a disastrous system collapse due to Y2K compatibility issues. This problematised the system for EUB top management as a survival solution. It also cut off the route to any other possible IT

solution to the problem as it convinced EUB that any other approach would be not only costly, but also high risk considering the large number of legacy systems in place in EUB. In doing this, Drinko's top management set the integrated SAP system as an obligatory passage point if EUB was to overcome its Y2K crisis.

However, Drinko's senior managers were concerned that this would not be enough to pull the EUB internal network toward EUK because EUB was "typically suspicious" of the EUK. They were sure that the invisible network that the EUB top management represented would render itself visible and problematic as the project became effective. Hence, they decided to proceed with recruiting more actors from the EUB network, namely, the "others'" location of EUB, and adopted it as the location of the project. In doing so, and by expressing publicly that the choice of location offers "significant resources" (CEO speech) because of the size of Drinko's operations in EUB and the "available capabilities" (project director speech) that EUB could provide, they appeared to follow EUB's explicit interests in bending the whole EUK network toward them, including asking EUK team members to fly out to work from EUB (excerpts from the corporate announcement).

## Network Reverse and Project Delay

EUK staff felt that they had been enrolled in EUB's internal network and that EUB was dominating the project. They first complained that the buildings were "old...like all the buildings [in EUB]." However, the buildings were not actually that old, but had traditional corridors and closed offices that were different to the open-plan layouts in EUK buildings.

The "old-fashioned" EUB offices and long corridors were viewed by EUK staff as constituting part of EUB's associations and networks, and hence part of the EUB identity that they strongly opposed and othered. This othering perspective viewed the building's internal layout as reflecting a hierarchical, slow, undynamic way of working, "which is a common practice in EUB" (interviews with a module manager, a change manager, and different team members). Team members from the EUK refused to enter EUB's network by, for example, sharing office rooms with each other to try to translate the buildings in their way. As an EUK manager expressed it, "Whenever [we] find a large room, they fit more than one person together to allow for 'informal ways' of working." Although EUB's buildings were criticised for enforcing a formal hierarchical relationship, the EUB business processes were later criticised as being too personal and informal, which reflects the contradictions and opposing opinions from the EUK side that stemmed from viewing EUB as the "other."

EUK staff continued to reflect their othering perception on everything else in the EUB network. For instance, as the project manager was from EUB, the EUK staff did not "see a point" in being recruited into his network and to exchange the agreed-upon intermediaries: schedules, milestones, and progress against targets. For this reason, the project office lost track of some teams because it continued to have outdated information on their progress. The project manager's invisible internal network was made visible when his aligned technical tools, such as project management software and Gantt charts, were not allowed to operate because the data they had was out of touch with what was happening on the ground.

EUK complained several times to Drinko's CEO about the uncooperative attitude of EUB, pointing to it as the reason for delays and a potential threat to the system integration capabilities. EUK tried to translate the top management's interests and channel them toward streamlining the teams and enhancing their cooperation. The top management commissioned a consultancy firm (Independent Consulting) to investigate the issue. Its report, confidentially submitted to the CEO, revealed that both EUB-dominated teams and the project office preferred to share the same building, while EUK teams and staff chose from the start to be in the other building in EUB. It mentioned that "the two buildings are taking on individual characters [characteristics] and alignment which might result in gaps appearing in the overall solution [system]" (consultants' report). This was also supported by the EUK process owners, who found it difficult to "conquer the in-built prejudices and impacts of their location in designing and communicating a "shared vision" with EUB. They preferred to work from their EUK offices and kept blaming EUB for not cooperating and overcoming "prejudices."

The top management thought another little detour for EUB was needed to move it away from direct involvement in the project teams, while keeping it broadly locked into the project network. Pulling the location out would have been quite risky as it represented EUB's "actorship" in the project and was strongly associated with all actors in the EUB network. If the EUB locations were removed from the SAP project network, the associated EUB network would probably have followed by withdrawing from the project network. Drinko top management therefore found that marginalising EUB required the creation of an invisible detour: taking EUB away from the centre to the periphery of the project without it realising the displacement. Hence, the top management asked Independent Consulting to give its input to this issue.

After long discussions with Independent Consulting, a joint report was compiled and issued under the consultants' authorship to convince EUB and justify the change. This explained the need for the project to have "a business pull" rather than "the programme push" that had been taking place, and justified the change by mentioning that "it is not unusual to change during a programme" and that the change proposed would be a way of going forward with the SAP project (consultants' report and different presentations, a change manager interview, a consultant interview). The changes that were then suggested in effect marginalised EUB actors while continuing to lock them into the project network. For example, the project's new structure moved the managing director of EUB from the active post of sponsoring the sales and operations planning team to a more ceremonial post of being a member of the steering committee. The sponsors of the new teams were all located in EUK. To ensure the locking in of EUB business units, the new "release owner" post was created for each stream, with three owners representing the companies within the project's scope: EUB, EUK (including the corporate centre), and Europe sales. This ensured the EUB release owner was responsible only for the businesses processes in EUB, and the rest was left for EUK release owners.

These changes guaranteed that EUB staff would not be in an effective powerful position, although they would remain actors in the network, thereby ensured their loyalty to the project. Most of the newly appointed actors worked from EUK offices without any formal announcement of a location change for the programme. Consequently, the project returned back to the EUK, putting an end to the EUK staff's feeling that it was being dominated by EUB in terms of staff and location and the significant delay in the project schedule.

## The Organisational Othering Inscribed into the System

The long othering of EUB was reflected in the SAP system configuration. SAP recommends and supports having one service centre for the whole organisation, which is considered to be a source of cost cutting and efficiency. However, the sensitivity of the EUK-EUB relationship and EUK's full realisation that its othering of EUB was well-known on the EUB side, Drinko top management had to take extra precautions. They believed that Drinko should have a single service centre,

although the question of its location turned out to be problematic. EUK staff believed that it had to be in the EUK. They problematised the issue to top management by arguing that EUB did not have the competences to operate it and the only staff who know how to do that were in the EUK. Hence, they argued that any single shared services centre should be located in EUK, not EUB.

The top management did not want any explicit manifestation of otherness and marginalisation at this point, so they did not want to "take away everything" from EUB territory, which "would jeopardise the whole thing in such a critical time" when the SAP configuration was being determined. Thus, top management decided to compromise the system and configured it awkwardly to have two shared services, one in EUB and the other in EUK, in order to ensure that an integrated SAP system would still assure the sovereignty of EUB and would not affect its right to be left alone, as before.

At the same time, Drinko's top management made it clear that this was a temporary solution and that it intended to move to a single shared services centre somewhere else in the future, with the time and location being determined later after the implementation. By the end of 2001, and after the implementation, the company decided to undo this "odd structure" and take away the two shared services from the two countries to amalgamate them in one shared service located in a third European country. In so doing, it hoped to avoid any controversy concerning who will "boss" whom.

## DISCUSSION AND CONCLUSION

Organisational integration is one of the key aims of the organisations implementing ERP and is usually taken for granted as a technical capability of the system that could be straightforwardly delivered. This chapter focuses on organisational othering and how it could hinder this essential function of the system. It emphasises the importance of understanding the social logic that dominates the organisation when managing ERP implementation projects. It doubts the notion that ERP systems straightforwardly enable organisations to coordinate across geographically dispersed locations (Davenport, 1998) and suggests that the implementation of ERP requires organisations to do so. Along the same line, ERP does not redefine previously known organisational boundaries (Brehm, Heinzl, & Markus, 2001; Foremski, 1998), yet it strongly requires organisations to bridge their social, behavioural, and structural barriers and prejudices.

Through the notion of organisational othering, the chapter reveals the historical and long-standing social logic of othering embedded in the business units' relationships in Drinko and reveals that othering can be reproduced and inscribed in the system, resulting in unusual and costly configuration. The reproduction of organisational othering and the subsequent creation of a separate structure for service centres despite the implementation of the integrated ERP system agrees with Quattrone and Hopper's (2005) observation that "ERP reproduced existing structure," yet Drinko's focus was on the effect of this on management control and not on understanding how this happened. This explains what researchers sensitively observed that organisation-wide integration is not always feasible and could be problematic, and that organisations, in many cases, reach a country-specific customisation of the system within the ERP framework (Markus, Tanis, & Fenema, 2000).

The team involved in ERP implementation tends to be large and heterogeneous, which makes it a complex entity to manage (Kay, 1996; Ward & Peppard, 1996). It involves staff from different organisational units. Together, these people lead the decision making concerning how the organisation's processes will be mapped or reconfigured to take advantage of the integrative functionality embedded in the ERP system (Sawyer, 2001b). This study reveals yet another aspect of the complexity of managing these teams stemming from institutionalised prejudices and othering. It demonstrates that careful monitoring and managing of the political sensitivity of the teams involved can overcome some of the conflicts and avoid the configuration of isolated modules and isolated business units. In this regard, it agrees with the remark that ERP project management is deemed to fail if it does not account for the business politics that comprise the framework within which the ERP project takes place (Besson & Rowe, 2001).

Studies on intragroup conflicts during IS development tend to divide the nature of the conflict into relationship- and task-related conflicts and focus on the task-related conflict and its effect on performance (Besson, 1999; Besson & Rowe, 2001; Sawyer, 2001a). This study reveals that arbitrary categorisation of the nature of conflicts and the decision of ERP researchers' to focus on one aspect or another a priori (Besson & Rowe) could cause one to miss the complexity of the encountered conflict. The application of ANT provides a vehicle to avoid any a priori categorisation and leave it as an empirical matter for the actors in the field to decide upon. The concept of organisational othering helps to unravel the historical roots and fixed perceptions (relationship conflict) behind the surfaced task-related conflicts. This concept adds a new dimension to the politics of implementing IS and the intragroup conflicts involved based on the combination of organisational history, culture, and institutionalised marginalisation of some groups within the organisation. The incorporation of nonhumans as actors in the organisational politics and the revealing of their role in the conflict and its resolution contribute to the ongoing discussion on the politics of IS implementation (Brooke & Maguire, 1998; Cavaye & Christiansen, 1996; Doolin, 1999; Markus, 1983).

For the practice of implementing ERP systems, the findings invite practitioners to reconsider the view that the technical integration capability of

ERP can be straightforwardly materialised, crossing the previous organisational boundaries and leading to a successful cooperation between previously isolated groups. Instead, they should be open to examining the roles of all actors, which have the power to affect not only the implementation project but also the system being implemented. Organisational othering should be accounted for, monitored, and managed before it gets inscribed into the ERP system, which could result in reproducing organisational boundaries. At the same time, practitioners can seize the opportunity of ERP implementation to loosen the established organisational barriers and tackle prejudices between different organisational groups.

## REFERENCES

AMR Research. (2004). *ERP market.* Retrieved March 21, 2005, from http://www.amrresearch.com

Beall, J. (1997). Valuing difference and working with diversity. In J. Beall (Ed.), *A city for all: Valuing difference and working with diversity* (pp. 2-37). London: Zed Books Ltd.

Besson, P. (1999). Les ERP a l'epreuve de l'organisaton. *Systemes D'Information et Management, 4*(4), 21-52.

Besson, P., & Rowe, F. (2001). ERP project dynamics and enacted dialogue: Perceived understanding, perceived leeway, and the nature of task-related conflicts. *The Data Base for Advances in Information Systems, 32*(4), 47-66.

Bhabha, H. K. (1986). The other question: Difference, discrimination and the discourse of colonialism. In F. Barker, P. Hulme, M. Iversen, & D. Loxley (Eds.), *Literature, politics and theory* (pp. 148-172). Methuen & Co. Ltd.

Bijker, W. E., & Law, J. (Eds.). (1997). *Shaping technology/building society: Studies in socio-technical change* (2nd ed.). Cambridge, MA: The MIT Press.

Bloomfield, B. P., Coombs, R., Knights, D., & Littler, D. (Eds.). (1997). *Information technology and organizations: Strategies, networks, and integration.* Oxford University Press.

Brehm, L., Heinzl, A., & Markus, L. M. (2001). *Tailoring ERP systems: A spectrum of choices and their implications.* Paper presented at the 34th Hawaii Conference on Systems Sciences, HI.

Brooke, C., & Maguire, S. (1998). Systems development: A restrictive practice? *International Journal of Information Management, 18*(3), 165-180.

Cadili, S., & Whitley, E. A. (2005). *On the interpretive flexibility of hosted ERP systems* (Working Paper Series No. 131). London: Department of Information Systems, The London School of Economics and Political Science.

Callon, M. (1986). Some elements of a sociology of translation: Domestication of the scallops and the fishermen of St Brieuc Bay. In J. Law (Ed.), *Power, action and belief: A new sociology of knowledge* (pp. 196-233). London: Routledge & Kegan Paul.

Cavaye, A., & Christiansen, J. (1996). Understanding IS implementation by estimating power of subunits. *European Journal of Information Systems, 5*, 222-232.

Davenport, T. H. (1998). Putting the enterprise into the enterprise system. *Harvard Business Review*, 121-131.

Davenport, T. H. (2000). *Mission critical: Realizing the promise of enterprise systems.* Boston: Harvard Business School Press.

Dong, L. (2000). *A model for enterprise systems implementation: Top management influences on implementation effectiveness.* Paper presented at the Americas Conference on Information Systems, Long Beach, CA.

Doolin, B. (1999). Sociotechnical networks and information management in health care. *Accounting, Management and Information Technology, 9*(2), 95-114.

Foremski, T. (1998, September 2). Enterprise resource planning: A way to open up new areas of business. *Financial Times*, p. 6.

Hall, S. (Ed.). (1997). *Representation: Cultural representations and signifying practices.* Sage Publications.

Holland, C. P., Light, B., & Gibson, N. (1999, June). *A critical success factors model for enterprise resource planning implementation.* Paper presented at the the Seventh European Conference on Information Systems, Copenhagen, Denmark.

IDC. (2004). *Worldwide ERP application market 2004-2008 forecast: First look at top 10 vendors.* Retrieved June 9, 2005, from http://www.IDC.com

Kaplan, B., & Maxwell, J. A. (1994). Qualitative research methods for evaluating computer information systems. In J. G. Anderson, C. E. Aydin, & S. J. Jay (Eds.), *Evaluating health care information systems: Methods and applications* (pp. 45-68). Thousand Oaks, CA: Sage.

Kay, E. (1996, February 15). Desperately seeking SAP support. *Datamation*, pp. 42-45.

Klein, H. K., & Myers, M. D. (1998). *A set of principles for conducting and evaluating interpretive field studies in information systems.* Retrieved December 11, 1998, from http://www.auckland.ac.nz/msis/isworld/mmyers/klien-myers.html

Krumbholz, M., Galliers, J., Coulianos, N., & Maiden, N. A. M. (2000). Implementing enterprise resource planning packages in different corporate and national cultures. *Journal of Information Technology, 15*, 267-279.

Latour, B. (1987). *Science in action: How to follow scientists and engineers through society.* Cambridge, MA: Harvard University Press.

Latour, B. (1988). *The pasteurization of France* (A. Sheridan & J. Law, Trans.). Harvard University Press.

Latour, B. (1991). Technology is society made durable. In J. Law (Ed.), *Sociology of monsters: Essays on power, technology and domination* (pp. 103-131). London: Routledge.

Latour, B. (1999). *Pandora's Hope: Essays on the reality of science studies.* Cambridge, MA: Harvard University Press.

Law, J. (1992). Notes on the theory of the actor-network: Ordering, strategy, and heterogeneity. *Systems Practice, 5*(4), 379-393.

Markus, M. L. (1983). Power, politics and MIS implementation. *Communication of the ACM, 26*(6), 430-444.

Markus, M. L., Axline, S., Petrie, D., & Tanis, C. (2000). Learning from adopters' experiences with ERP: Problems encountered and success achieved. *Journal of Information Technology, 15*, 245-265.

Markus, M. L., & Tanis, C. (2000). The enterprise system experience: From adoption to success. In R. W. Zmud (Ed.), *Framing the domains of IT research: Glimpsing the future through the past.* Cincinnati, OH: Pinnaflex Educational Resources, Inc.

Markus, M. L., Tanis, C., & Fenema, P. C. v. (2000). Multisite ERP implementations. *Communications of the ACM, 43*(4), 42-46.

Mendel, B. (1999). Overcoming ERP projects hurdles. *InfoWorld, 21*(29).

Monteiro, E. (2000a). Actor-network theory and information infrastructure. In C. U. a. o. Ciborra (Ed.), *From control to drift* (pp. 71-83). New York: Oxford University Press.

Monteiro, E. (2000b). Monsters: From systems to actor-networks. In K. Braa, C. Sorensen, & B. Dahlbom (Eds.), *Planet Internet*. Lund, Sweden: Studentlitteratur.

Norris, G. (1998). *SAP: An executive's comprehensive guide*. New York: J. Wiley.

Orlikowski, W. J., & Baroudi, J. J. (1991). Studying information technology in organizations: Research approaches and assumptions. *Information Systems Research, 2*(1), 1-28.

Quattrone, P., & Hopper, T. (2005). A "time-space odyssey": Management control systems in two multinational organisations. *Accounting, Organizations & Society, 30*(7/8), 735-764.

Ross, J. W., & Vitale, M. R. (2000). The ERP revolution: Surviving vs. thriving. *Information Systems Frontiers, 2*(2), 233-241.

SAP AG. (2004). *SAP annual report 2004*. Author.

SAP's rising in New York. (1998, August 1). *The Economist*.

Sawyer, S. (2001a). Effects of intra-group conflict on packaged software development team performance. *Information Systems Journal, 11*, 155-178.

Sawyer, S. (2001b). *Socio-technical structures in enterprise information systems implementation: Evidence from a five year study*. Paper presented at the IEEE EMS International Engineering Management Conference, Albany, NY.

Themistocleous, M., Irani, Z., & O'Keefe, R. M. (2001). ERP and application integration: Exploratory survey. *Business Process Management Journal, 7*(3), 195-204.

Walsham, G. (1995). The emergence of interpretivism in IS research. *Information Systems Research, 6*(4), 376-394.

Walsham, G. (2001). *Making a world of difference: IT in a global context*. John Wiley & Sons Ltd.

Ward, J., & Peppard, J. (1996). Reconciling the IT/business relationship: A troubled marriage in need of guidance. *Journal of Strategic Information Systems, 5*, 37-65.

Xia, W., & Lee, G. (2004). Grasping the complexity of IS development projects. *Communications of the ACM, 47*(5), 95-74.

# Chapter IV
# Precontract Challenges:
## Two Large System Acquisition Experiences

**Ayça Tarhan**
*The Bilgi Group Ltd., Turkey*

**Çiğdem Gencel**
*Middle East Technical University, Turkey*

**Onur Demirors**
*Middle East Technical University, Turkey*

## ABSTRACT

*The acquisition of software-intensive systems demands significant work on requirements prior to establishing the contract. Two significant challenges of the precontract phase are the identification of functional requirements and the determination of an approximate budget. In this chapter, experiences gained from the implementation of a business-process-based requirements-elicitation approach to two large innovative military applications are presented. The requirements-elicitation process proposes the determination of the requirements of a software-intensive system to be acquired from the business objectives and defined processes. In addition to the details related to the requirements-elicitation process, management practices for coordinating the efforts related to the acquisition process and determination of costs are also discussed.*

## INTRODUCTION

Acquisition cycles perceive requirements elicitation as the process to identify and understand needs and constraints of the customer. While generic approaches are successfully applied to acquire manufacturing goods and services, software acquisition provides unique challenges at this stage.

During the acquisition of large and innovative software systems, requirements elicitation entails more than obtaining and processing customer needs and involves major risks related to cost, quality, scope, and schedule. Specifically, the acquisition of contracted software demands significant work on functional requirements prior to establishing the contract.

In most cases, large and innovative software systems have numerous stakeholders with conflicting and frequently unidentified stakes. The attributes of the system to be acquired might require unique methodologies to be utilized, and the characteristics of the acquisition organization make a difference. Ideally, the acquirer needs to understand the concept, the domain as a whole, the technology to be utilized, and the technical and management constraints of the project. The current business processes should be explicit and potential new technologies should be prototyped. Only such an understanding creates a solid foundation for the challenges of defining the functional contract requirements as well as to make realistic estimates of size, effort, and cost.

Notations and tools developed primarily for business process modeling provide a natural tool set for the establishment and transformation of need from concept to system requirements. Organizational processes facilitate better understanding of the system as a whole, depict potential stakeholder conflicts, enable the determination of clear boundaries, and provide a natural medium of communication. Once the organizational process baseline is established, it can also be used as an enabler for technical processes. It can be used to estimate the size of the software system to be developed, which in turn is used to estimate the effort and related costs. It can be used as a source for software requirements specification. Both forward and backward traceability can be easily established and partially automated. It is also possible to automate the generation of specifications in natural language and software size estimation in Mark II function points. In other words, business process models, when used effectively, can be a basis to respond to the several challenges of precontract phases.

During the last 3 years, we have implemented an approach based on a business process model together with a unique set of notations and tools (Demirors, Gencel, & Tarhan, 2003, 2006) to guide two large military acquisition projects.

Our tasks involved business process modeling, requirements elicitation, size and cost estimation, and the preparation of technical documents for the acquisition of command, control, communications, computers, intelligence, surveillance, and reconnaissance (C4ISR) subsystems for Turkish Land Forces Command (TLFC). The outcomes of the implementations formed the major parts of the request for proposals (RFP) currently issued by TLFC.

In this chapter, we focus on the implementation of a business-process-based approach for requirements elicitation as well as on management practices for coordinating the efforts for RFP preparation. Based on the elicited requirements, the method used for estimating the size of the software products of the projects is also briefly discussed.

In the chapter, an overview of the literature on software-intensive system acquisition and business process modeling for requirements elicitation is presented. Then the descriptions of the cases are given. Next the chapter gives the summary descriptions of the acquisition planning processes and discusses the mapping between our approach and the C4ISR framework's descriptive views. The details of the two challenging processes of the precontract phase of software-intensive system acquisition, system requirements elicitation, and software size estimation are also discussed. Finally, we provide lessons learned and present conclusions.

## BACKGROUND STUDIES

There are several frameworks, models, and practices that guide agencies in acquiring software-intensive systems. These include the software acquisition capability maturity model (SA-CMM; Cooper & Fisher, 2002), the supplier agreement management process of capability maturity model integration (CMMI; CMMI Product Team, 2001), IEEE's (1998b) *Recommended Practice for*

*Software Acquisition*, and the C4ISR architecture framework (Department of Defense [DoD] Architecture Working Group, 1997).

SA-CMM offers a framework for organizational improvement, and it focuses on building the acquisition-process capability of an organization. It defines five levels of software-acquisition process maturity.

The CMMI supplier agreement management process area is targeted to manage the acquisition of products from suppliers for which there exists a formal agreement. It has a context for system and software acquisition, and remains more generic when compared to other models. It includes the practices for determining the type of acquisition that will be used for the products to be acquired, selecting suppliers, establishing and maintaining agreements with suppliers, executing the supplier agreement, accepting delivery of the acquired products, and transitioning acquired products to the project.

The Software Engineering Institute's (SEI) models and practices specified above describe what characteristics the acquisition process should possess, and do not mandate how the acquisition process should be implemented or who should perform an action. In other words, neither of them guides as an acquisition life cycle. IEEE's (1998b) *Recommended Practice for Software Acquisition* offers such a life cycle for a typical software-acquisition process, which primarily includes planning, contracting, product implementation, product acceptance, and follow-on phases. IEEE defines and relates one or more steps to each of these phases, as given in Table 1. It should be noted that these steps might overlap or occur in a different sequence, depending upon the organizational needs.

Another model, which was primarily developed and has been used for military acquisitions at the U.S. Department of Defense, is the C4ISR

*Table 1. Software-acquisition phase milestones (IEEE, 1998b)*

| Phase | Phase Initiation Milestone | Phase Completion Milestone | Steps in Software-Acquisition Process |
|---|---|---|---|
| Planning | Develop the idea | Release the RFP | 1) Planning organizational strategy<br>2) Implementing organization's process<br>3) Determining the software requirements |
| Contracting | RFP is released | Sign the contract | 4) Identifying potential suppliers<br>5) Preparing contract requirements<br>6) Evaluating proposals and selecting the supplier |
| Product Implementation | Contract is signed | Receive the software product | 7) Managing supplier performance |
| Product Acceptance | Software product is received | Accept the software product | 8) Accepting the software |
| Follow On | Software product is accepted | Software product is no longer in use | 9) Using the software |

Architecture Framework (DoD Architecture Working Group, 1997). This framework provides the rules, guidance, and product descriptions for developing and presenting architecture descriptions of the systems to be acquired. The aim of this effort is to ensure building interoperable and cost-effective military systems.

The DoD Architecture Working Group (1997) defines an architecture description in the C4ISR architecture framework as "a representation, as of a current or future point in time, of a defined domain in terms of its component parts, what those parts do, how those parts relate to each other, and the rules and constraints under which the parts function" (p. 2-1). There are three major views that logically combine to describe architecture: the operational, systems, and technical views. Since each view provides different perspectives on the same architecture, DoD suggested using an "integrated" description that would provide more useful information to the planning, programming, and budgeting process and to the acquisition process. Figure 1 shows the linkages that describe the interrelationships among the three architecture views.

One of the primary reasons underlying the definition of these models is to develop a mechanism to appropriately define the requirements of a software-intensive system, which will be taken as a basis by both acquirer and supplier organizations throughout the acquisition. Defining software-intensive system requirements is not straightforward; it requires extensive understanding of the domain and representing domain knowledge formally and visually, not getting lost in the domain and skipping any point, as well as enabling the ease of understanding and health of validation. There are several approaches and related techniques used to gather domain knowledge, including functional approaches such as data flow diagrams (DFDs), scenario-based approaches such as use cases, and business-process-oriented approaches such as business process modeling.

Business process modeling is a technique used to understand, depict, and reorganize business processes of an organization. Davenport (1993) defines a process as "a specific ordering of work activities across time and place, with a beginning, an end, and clearly defined inputs and outputs: a structure for action" (p. 5). A business process

*Figure 1. Fundamental linkages among the views (DoD Architecture Working Group, 1997)*

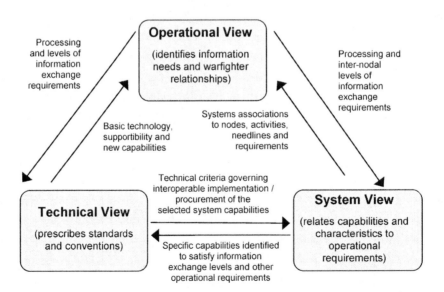

defines how an organization achieves its purpose including its vision and goals.

The analysis and design of business processes are the major tasks of business process reengineering (Hammer & Champy, 2001; A. W. Scheer, 1994). Business process modeling has been implemented by a large number of organizations as part of business process reengineering during the last decades (Davenport, 1993; Hammer & Champy). It is claimed that it adds the most value when the application environment is complex and multidimensional, and many people are directly involved in using the system (Yourdon, 2000).

The analysis of business processes requires understanding the current business processes of the organization before designing the new ones. Especially in complex organizations, there is no way to migrate to a new process without understanding the current one. Existing business process models enable participants to develop a common understanding of the existing state, and are used to establish a baseline for subsequent business process innovation actions and programs. The design of business processes is performed before implementation of the system in order to translate the requirements of the business process innovation into proposed system requirements. Existing business processes are the foundation of this kind of study for finding out the weak points. When designing the new system, it is worthwhile to try to describe the new processes based on the business objectives to respond to business requirements (Cummnis, 2002). The target process models serve as the foundation for implementing the new business processes and defining the requirements for new information technology systems.

As a result, business process modeling brings the following advantages to the requirements elicitation of software-intensive systems.

- Brings broader view to business domain
- Creates a common language between analysts and customers or users, leading to an increase in customer and user participation in the requirements-elicitation process
- Increases efficiency in determining deficiencies in the current business processes and modeling the target business processes
- Helps in identifying business requirements as the knowledge is captured at a more abstract, business-need level than many other specification approaches (Demirörs, Demirörs, Tanik, Cooke, Gates, & Krämer, 1996; Wiegers, 1999).

## DESCRIPTION OF THE CASES

We have implemented the planning phase of the acquisition life cycle in the context of acquiring two large innovative military applications for TLFC. TLFC is a part of the Turkish Armed Forces and consists of four armies, one training and doctrine command, and one logistic command.

The projects targeted RFP preparation for two different but closely related C4ISR subsystems for TLFC. The Middle East Technical University (METU) Project Office, as depicted in Figure 2, counseled the TLFC Project Office for preparing the technical contract of the system to be acquired.

Throughout the chapter, we code these projects as A and B. Due to security constraints, names and descriptions of the organizational processes are not provided. We briefly define the characteristics of the projects to provide insight on the implementations of the acquisition planning process. The domains of Project A and Project B are different but complementary with a high degree of data exchange requirements. There are four other C4ISR subsystem projects that are planned to be integrated with these two. Therefore, not only the functional requirements of each subsystem domain, but also the integration requirements of Project A and Project B with these projects had to be considered.

*Table 2. Information on case projects*

| Project | Duration (months) | Effort (person-months) | Number of METU Staff | Estimated Size (FP) |
|---------|-------------------|------------------------|----------------------|---------------------|
| A | 8 | 18 | 11 | 10,092 |
| B | 13 | 26.5 | 9 | 25,454 |

*Figure 2. Projects organization*

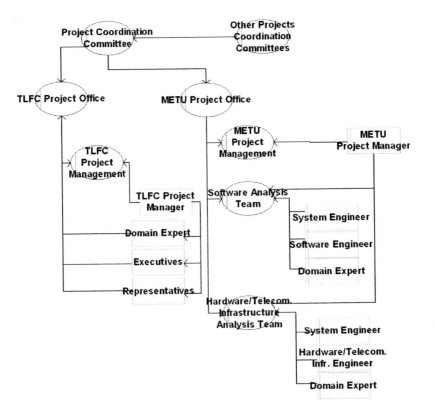

Both projects required taking a system viewpoint to map domain requirements to hardware, software, and telecommunication components. Estimated sizes for the software components of these systems as well as the number of staff involved and the duration and total effort utilized for the acquisition planning process by the METU Project Office are given in Table 2. The effort utilized by the TLFC Project Office is not included as the collected data were not consistent.

Both projects required various resources to be utilized including human resources, process modeling notations and tools, and domain-specific materials such as books and guidelines. The characteristics of the resources utilized in Project A and Project B and their organizations were similar since not only the purpose of both of the projects was to acquire C4ISR subsystems, but the customer was TLFC in both cases as well.

Human resources included the Project Coordination Committee, the staff of METU and TLFC project offices, domain experts, and the current executives and representatives of the military units where the system would be used. The organization of the project staff is shown in Figure 2.

The Project Coordination Committee coordinated and monitored the tasks of the METU and TLFC project offices and was in coordination with the committees of other C4ISR subsystem projects of TLFC in order to depict the system interfaces.

The METU Project Office counseled TLFC for preparing the technical contract of the system to be acquired within the boundary of the project and included a project manager and software, hardware, and telecommunication analysis teams. Analysis teams modeled the business processes and specified the software, hardware, and telecommunication requirements.

The TLFC Project Office executed the processes of the project and included a project manager, externally involved domain experts who have domain knowledge, executives, and current representatives of the military units who would use the system to be acquired.

In Project A, the project staff consisted of seven part-time persons from the METU Project Office, four graduate students of METU who have military background, and five part-time persons from the TLFC Project Office. In addition, two domain experts and four representatives of the organizational units where the system would be used joined the validation activities.

In Project B, the project staff consisted of nine part-time persons from the METU Project Office and nine part-time persons from the TLFC

Project Office, who are also domain experts. Not all of the TLFC Project Office staff participated in the workgroup meetings at the same time. They participated in an interchangeable manner. In addition, seven more domain experts, who are not members of the TLFC Project Office, and two representatives of the organizational units where the system would be used joined the validation activities.

Other resources we utilized in the projects included process modeling notations and tools, and domain books and documents. We proposed in Demirörs et al. (2003) that the candidate tool for business process modeling should support definitions for process, process flow, input and output, input and output flow, role, and responsibility entities at minimum. Specifically in these projects, the architecture of integrated information system (ARIS) concept and ARIS tool set (A. G. Scheer, 2003) were used as the modeling tool. ARIS is frequently used by consultants and companies in creating, analyzing, and evaluating organizational processes for business process reengineering.

While modeling the business processes, organizational charts, function trees, extended event-driven process chain (eEPC) diagrams, access diagrams, and network topology diagrams were used as basic process modeling notations. The organizational chart reflects the organizational units (as task performers) and their interrelationships, depending on the selected structuring criteria. Business processes consist of complex functions that are divided into subfunctions, and basic functions represent the lowest level in semantic function-tree diagrams. The procedure of a business process is described as a logic chain of events by means of an event-driven process chain. Events trigger functions and are the result of functions. By arranging a combination of events and functions in a sequence, eEPCs are created. We created 210 distinct diagrams in Project A and 295 in Project B to model existing business processes of different levels of organizational units by using the eEPC notation.

In Project B, in order to generate user-level functional system requirements with the KAOS tool (Su, 2004; Turetken, Su, & Demirors, 2004), we derived and used a special notation while modeling target business processes. This notation differed from Project A's in the way that color codes and specific rules for the naming of functions, inputs, and outputs were used.

Hardware components and their high-level relations for each organizational unit were modeled by using the access-diagram notations. The assignment of software components to hardware components, and the domain users of these hardware components were also modeled by using the same type of diagram notations. Network-topology diagram notations were utilized to represent the system architecture.

Military books and instructions were among the basic resources, especially when domain experts had uncertainties and disagreements related to the concept of the processes. Throughout the requirements-elicitation processes of Project A and Project B, 15 and 9 military books and guidelines related to the domain were utilized, respectively.

## ACQUISITION PLANNING PROCESSES

The processes we implemented for acquisition planning are shown in Figure 3. Summary descriptions for the processes of the acquisition planning phase are as follows.

- **Planning and managing acquisition planning project:** A project management plan was prepared at the start of the project in order to describe the activities, responsibilities, schedule, and effort as related to acquisition planning. Throughout the projects, the performances of the projects were tracked and accordingly the plans were updated.

- **Eliciting system requirements:** We performed a business-process-based requirements-elicitation approach to define software-intensive system requirements (Demirörs et al., 2003). We determined user-level functional requirements for software components of the systems, nonfunctional system requirements, commercial off-the-

*Figure 3. Acquisition planning phase*

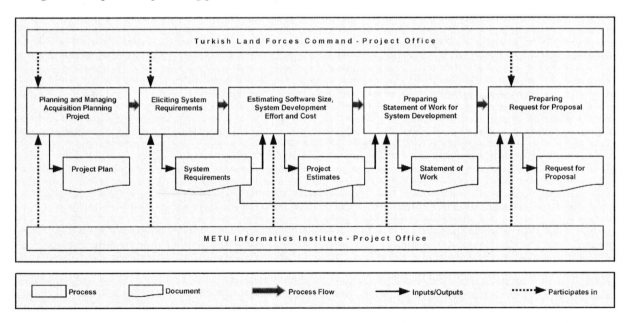

shelf (COTS) product requirements, and hardware and telecommunication infrastructure requirements for both systems.

- **Estimating software size, and system development effort and cost:** We estimated the sizes of the software components of the systems based on the functional software requirements elicited in the previous step and using the Mark II function points analysis method (Demirörs & Gencel, 2004). Effort and cost for the system development were also estimated by using software size estimates.

- **Preparing statement of work for system development:** We described system and software development life cycles, which are to be applied by the supplier organizations, together with engineering process steps and their outputs. We used IEEE's system and software engineering standards (IEEE, 1998a, 1998c) and recommended practices as a reference in describing the templates for process outputs.

- **Preparing RFP:** We gathered the system requirements, system development estimates, and statement of work. Then we integrated these with the acquisition regulations of the TLFC in the form of an RFP. We included the acquisition schedule, management practices, and deliverables; quality requirements for system and software development and management processes; quality assurance requirements for the deliverables; and qualifications of the system and software development and management staff in the RFP to be issued for the system development.

In this study, we focused on the details of the two challenging processes of the precontract phase of software-intensive system acquisition: system requirements elicitation and software size estimation processes.

The Requirements-Elicitation Process

The requirements-elicitation process based on business process modeling we implemented is depicted in Figure 4. Since the projects were military, the C4ISR architecture framework (DoD Architecture Working Group, 1997) also influenced us while defining this process.

The process includes four technical subprocesses, namely, concept exploration, analysis, and modeling of current business processes; modeling of the target system; system requirements definition; and the quality assurance activities of verification and validation.

The C4ISR architecture framework provides the rules, guidance, and product descriptions for developing and presenting architecture descriptions of the systems to be acquired. As discussed in the background studies, there are three major views that logically combine to describe the architecture: the operational, systems, and technical views. In this process, we utilized an integrated description. Table 3 demonstrates the mapping between the C4ISR architecture views and the requirements-elicitation process we defined.

The practices were performed by the collaboration of the customer and contractor project offices. The detailed steps of the requirements-elicitation process and, if any, the differences in the subprocesses of Project A and Project B are explained in the following subsections.

## Concept Exploration

This process was performed to get knowledge about the domain and review previous documents. All the material, including military procedures, forms and guidelines, and related class notes, were gathered. Those resources helped in the requirements-elicitation process several times, such as in describing business processes and developing target conceptual and detailed design in the direction of the business goals. The following activities were performed to execute this process.

*Figure 4. Requirements-elicitation process*

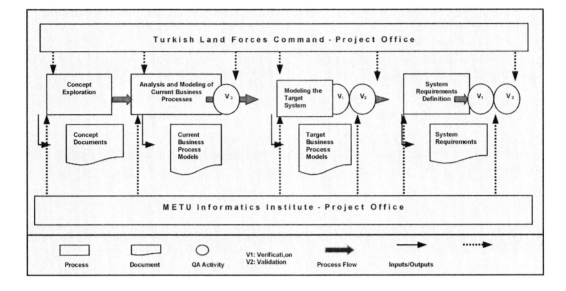

- Concept documents for the projects had been prepared prior to requirements-elici•tation contracts. The METU Project Office reviewed the concept documents in order to have a general understanding of the domain. The documents had mostly addressed current hardware and telecommunication infrastructure at an introductory level, so it was just used as a means to start domain analysis.

- The TLFC Project Office attended several orientation meetings together with the METU Project Office for the staff to get to know each other, and it provided short presentations about the domain and answered the questions of the METU Project Office. During the life cycle, orientation meetings for the domain experts and representatives of the organizational units who will use the system were also held.

- Written resources such as books, instructions, forms, and reports were asked for by the METU Project Office and provided by the TLFC Project Office when available.

## Analysis and Modeling of Current Business Processes

This process was performed to understand the current business processes with their business flows, inputs, outputs, and responsible bodies, as shown in Figure 5. Current business processes were modeled using the ARIS tool set, which provided the foundation for modeling the target business processes. The activities performed to execute this process are discussed in the following paragraphs.

- **Identifying organizational units of business domain:** People to be interviewed were determined by the METU Project Office together with the TLFC Project Office. Then, a schedule was made to hold these interviews. As a result of these interviews, the organizational units of the business domain were identified, and organization charts were generated. In addition, the key stakeholders were identified and the stakeholder representatives, who would join the workgroups for

*Table 3. Mapping between requirements-elicitation process and C4ISR architecture views*

| Subprocess Name | Operational View | System View | Technical View |
|---|---|---|---|
| Concept exploration | √ | √ | √ |
| Analysis and modeling of current business processes | √ | | |
| Modeling of the target system | | | |
| *Reviewing and enhancing current business processes* | √ | | |
| *Updating the data dictionary* | √ | | |
| *Identifying business processes that need IT support* | √ | | √ |
| *Identifying software components to provide IT support* | | √ | |
| *Assigning software components to business processes that need IT support* | | √ | |
| *Identifying hardware components* | | √ | √ |
| *Assigning software components to hardware components* | | √ | |
| *Identifying data transmission requirements* | √ | | |
| *Identifying telecommunication infrastructure* | | √ | √ |
| *Identifying system architecture* | | √ | √ |
| System requirements definition | √ | √ | √ |

determining the key business processes as well as analyzing and modeling the current business processes, were determined.

- **Identifying key business processes of organizational units:** This study was the foundation for more detailed business process analysis. Key business processes were determined after a study of the workgroup. Further modeling activities were based on these identified key business processes. Then, the models of the key business processes were produced.

- **Decomposing key business processes into business subprocesses:** The workgroup

analyzed the key business processes, and further decomposed them into business subprocesses. Decomposition was performed up to several levels when required until the lowest level subprocesses did not involve any nesting and had a simple flow of execution from beginning to the end.

- **Modeling lowest level business subprocesses:** Each lowest level business subprocess model was constructed in terms of processes, flow of processes, inputs and outputs of the processes, and the responsible bodies of the processes, which are all supported by eEPC notation. Uncertainties

*Figure 5. Analysis and modeling of current business processes*

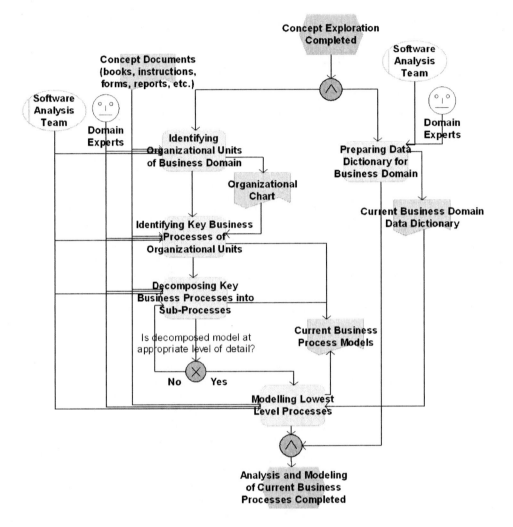

and disagreements about the execution of the processes were discussed by domain experts and stakeholder representatives, and resolved by referring to domain books and instructions where available. Ease of understanding of the eEPC notation resulted in extensive contribution and feedback from the customer. Feedback, including change requests taken at anytime, was reflected to the models by the staff of the METU Project Office.

• **Creating data dictionary for business domain:** Concepts and entities of the business domain that were introduced during current business process modeling were described in a data dictionary that formed a basis for developing a common language between the TLFC Project Office and the METU Project Office. The entities put into the dictionary included forms, reports, and instructions related to the business domain.

## Modeling of the Target System

This process was performed to describe the IT-oriented target system, as depicted in Figure 6. In modeling process enhancements, defining hardware and software components, and constructing target business process models, we also used the ARIS tool set. Target business process models provided the foundation for defining system requirements. The activities performed to execute this process are discussed in the following paragraphs.

- **Reviewing and enhancing current business processes:** Current business process models were revisited from the beginning for possible enhancements, or even redesign, considering the business goals and needs. The weaknesses, bottlenecks, and deficiencies of the current business processes were determined and removed wherever possible. However, enhancements were limited due to strict policies and regulations of the military. In addition, the TLFC Project Office gave up some enhancements because of the long approval cycles in the organization.
- **Updating the data dictionary:** The current data dictionary was updated to reflect the changes in the concepts and entities of the business domain, which appeared during the enhancement of current business process models and construction of target business process models. The data dictionary was extended to include new concepts and entities whenever required.
- **Identifying business processes that need IT support:** Enhanced business process models were reviewed to identify the business processes that need IT support. It was decided that all standard manual operations such as document recording and reporting were to be automated. Since the systems have strong interrelations with other systems, management operations were to be automated with a need for a workflow management system to provide coordination. Some processes required decisions on intelligence in IT support such as decision-support systems. The customer decided to evaluate such requirements based on price and benefits to be detailed and suggested by the supplier.
- **Identifying software components to provide IT support:** While identifying business processes that need IT support, we had a clear insight on which software components should cooperate in providing that support. Therefore, the software components of the system were determined, and high-level relationships among these components were set. The software components that are not specific to the domain were determined basically as the documents management system, reports management system, workflow management system, geographical information system, and messaging system. Since the projects had system contexts, hardware-interface and system-interface software components were also included in the set.
- **Assigning software components to business processes that need IT support:** During the two previous activities, we informally performed the mapping between business processes and software components. This activity was performed differently in Project A and Project B.

In Project A, the models of target business processes did not include the identification of the IS infrastructure within the business process models since these were reflected in the system breakdown structure (SBS) that was prepared simultaneously. At the end of this study, gaining satisfactory domain knowledge, we could associate the system architecture with the business process models, which enabled us to determine the requirements of the target system for each organizational unit separately at a later stage.

*Figure 6. Modeling of the target system*

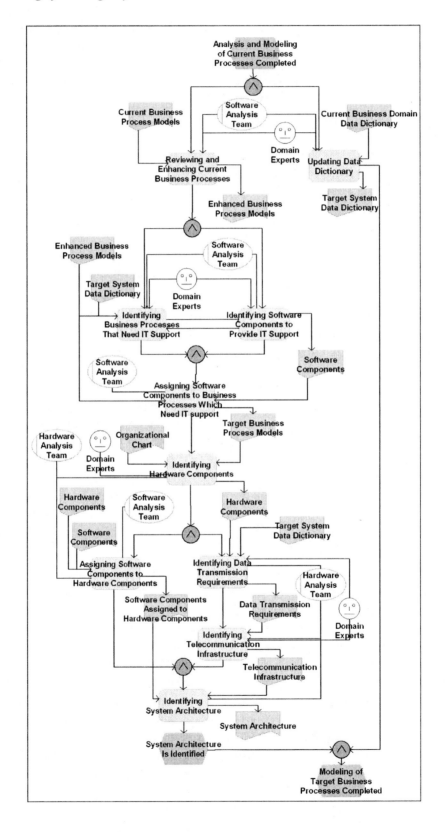

In Project B, we performed a careful analysis on current business process models to transform their organization-based structure into a system-based structure. This was performed to obtain a unique and generic set of target business process models for all organizational units rather than having several sets of target business process models for different organizational units, which overlap in functionality. Simultaneously, we formalized our insight on the relationships between business processes and supporting IT components by putting corresponding assignments on current business process models as to construct a unique set of system-based target business process models. We connected each business process needing IT support to the corresponding software to provide that support while constructing target business process models. This notation, which included business processes, software components, and their connections, formed a basis for defining user-level functional system requirements at a later stage.

- **Identifying hardware components:** Hardware components of the system were identified based on target business process models and the organization chart. Organizational units were analyzed to help us decide on hardware support. As a result of this activity, hardware components and their high-level relations were modeled by using access diagrams for each organizational unit. In addition, the existing and the needed hardware were also depicted.
- **Assigning software components to hardware components:** The software components to run on each hardware component were identified. An access diagram, showing the software components assigned to hardware components with the domain user types, was constructed for each organizational unit.
- **Identifying data transmission requirements:** By using the target business process

models, the data transmission requirements were determined based on the data exchange requirements between organizational units as well as between other C4ISR subsystems. The sizes and frequencies of transmissions determined the requirements related to bandwidth and speed.

- **Identifying telecommunication infrastructure:** The telecommunication infrastructures of the systems were determined using the access diagrams showing the hardware components, organizational units, and data transmission requirements. The existing telecommunication infrastructure, as well as new ones, and the related constraints were analyzed. Then, the customer evaluated the required bandwidth and speed requirements based on the performance of the transmission media that can be supplied and on the importance of the data to be transmitted. Accordingly, some decisions were made. Due to some limitations on basic technology supportability and organizational rules governing the implementation of system elements, it was decided that some processes of the organizational units were not to be automated. In addition, it was decided that some data were to be sent manually as in the current system.
- **Identifying system architecture:** This activity was simply the integration of the work done up to this point. Software components assigned to hardware components together with the telecommunication infrastructure constituted the system architecture. As a result of this activity, system architecture diagrams, showing all hardware components and software components at each organizational unit, and the telecommunication infrastructure among these units, were constructed. The system architecture diagram was constructed by using network-topology diagram type.

## System Requirements Definition

This process was performed to generate the target systems' software, hardware, and telecommunication requirements, as shown in Figure 7. The activities performed to execute this process are discussed below.

- **Preparing SBS:** Using the information on system architecture diagrams, an SBS including all IT components of the systems to the fifth level was prepared.
- **Defining user-level functional system requirements:** User-level functional requirements of the target system were elicited based on processes, flow of processes, inputs, outputs, and actors defined in the target business process models. As we mentioned previously, we generated the requirements manually from target business process models in Project A, and automatically from target business process models using a tool in Project B.

Since the requirements were generated manually in Project A, a small organizational unit of the domain (approximately 1/10 of the model system in terms of size) was selected as pilot, and its user-level functional requirements were defined as a trial by the whole workgroup. The experience of the pilot study was then used in planning the rest of the user-level functional-requirements-elicitation efforts. Three teams were formed to define the functional system requirements for each organizational unit and worked in parallel for the rest of the process, which benefited the workgroup in defining a standard way of working and in time saving. By using 210 business process models, the manual generation of the 10092 FP requirements document took 40 person-days. The functional system requirements filled about 400 pages, and there were about 1,380 basic requirements, each composed of three subrequirements on average (e.g., 1a, 1b, 1c).

In Project B, the requirements were generated automatically from target business process models using the KAOS tool (Su, 2004) developed specifically to generate requirements in natural language from target business process models. We derived and used a special notation while modeling target business processes, and generated a generic target system model that involves all the processes of each organizational unit. The tool runs on that notation to generate user-level functional system requirements with respect to system components assigned to each target process. After modeling target business processes using the notation's conventions at required detail, the tool benefited in time saving for requirements definition. The number of business process models was 295 and they are instantiated by the KAOS tool; in total, 1,270 user-level functional requirements were generated. The generation took 30 minutes. On the other hand, results showed that 517 generated requirements (40.7% of generated requirements) required corrections, and they were corrected in 3.5 person-days.

- **Defining COTS requirements:** COTS product requirements of the target systems were largely determined based on the requirements of the Turkish Armed Forces, and many of the regarding make and buy decisions and cost analyses (Aslan, 2002) were skipped due to this reason. The types of software components that were identified as COTS products included the operating system, database management system, office programs, document management system, report management system, workflow management system, and geographical information system.
- **Defining nonfunctional system requirements:** Nonfunctional system requirements were determined by analyzing implicit requirements of the target business process models as well as by discussing special

*Figure 7. System requirements definition*

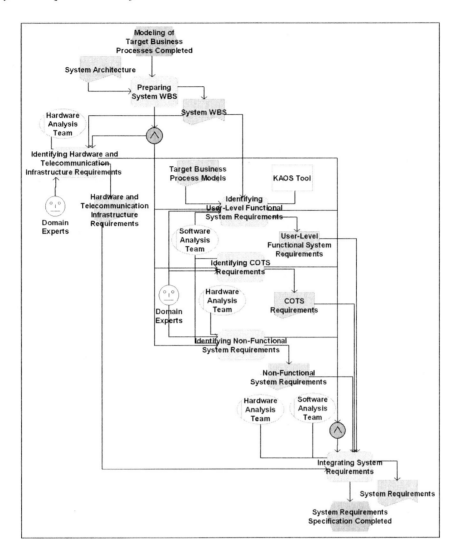

requirements and likely weaknesses, bottle-necks, and deficiencies of the systems with the customer. ISO/IEC 9126 (ISO/IEC, 1991) was used as a guideline while determining nonfunctional requirements. Function understandability, inherent availability, mean time between breakdowns, mean time to repair, and the failure resolution rate were defined and assigned target values for both systems. In addition, security requirements of the target systems were determined by analyzing access restrictions of the actors to processes, and process inputs and outputs. Subjects and objects of the model systems were identified and subject-object authorization matrices were constructed.

- **Defining hardware and telecommunications infrastructure requirements:** SBS, which included all IT components of the systems and system architecture diagrams, showing IT components at each organizational unit, and the telecommunication infrastructure among these units, was used as a basis for the detail requirements of these components. Since the applications were military, there were a lot of constraints put by rules and regulations while specifying the hardware and telecommunication requirements.

- **Integrating system requirements:** The requirements for all system components, including functional and nonfunctional system requirements, COTS requirements, and hardware and telecommunications infrastructure requirements, were integrated at this activity to construct overall system requirements. System requirements generated as the end product of the requirements-elicitation process were included in the RFP.

## Software Size Estimation

The size of the projects to be acquired was estimated by using the Mark II FP method (United Kingdom Software Metrics Association [UKSMA], 1998).

For both cases, as the user-level functional system requirements were generated from business process models with respect to different subsystems, the size of each subsystem was estimated and then summed to compute the sizes of the whole development projects.

The size estimates of the software components of the systems to be contracted for the development projects, the number of staff, and the effort utilized to make software size estimations are given in Table 4.

Since the estimation method and the procedure we followed were very similar in both of the projects, we will only discuss the size estimation procedure of Project B in this chapter.

In Project B, the user-level functional system requirements were generated from business process models with respect to 11 different subsystems as shown in Table 5. Original names of the subsystems and modules in the project are not given due to security constraints.

While making the software size estimation, some difficulties were faced as the Mark II FP method is designed to estimate size after the software requirements specification is complete. The first one is the differences in the abstraction levels of the system-level functional requirements. Therefore, some assumptions on the kind of transactions and the number of data entity types (DETs) had to be made while making the Mark II FP estimation. The percentage of such transactions was about 60% overall. The accuracy of the method decreases as the abstraction level of the requirements gets higher. Another difficulty was that, due to insufficient details of some requirements, the method could not be used to size these requirements at all. Thus, we had to use expert opinion to estimate their sizes. The percentage of the number of such requirements overall was 2.2%, and the percentage size of the subsystems involving these requirements in the whole project was found to be 14.1%.

*Table 4. Summary of the estimations, effort, and number of staff used for estimations*

| Project | Effort Needed to Make Size Estimation (person-hours) | Number of Staff (part time) | Estimated Size (MkII FP) |
|---------|---------|---------|---------|
| A | 54 | 1 | 10,092 |
| B | 131 | 4 | 25,454 |

*Table 5. Size estimates of subsystems by Mark II*

| Subsystem | Module | Unadjusted FP |
|---|---|---|
| A | A1 | 2,886.64 |
|  | A2 | 4,882.10 |
|  | A3 | 9,281.55 |
| B |  | 8.48 |
| C |  | 185.34 |
| D |  | 3,344.96 |
| E |  | 878.31 |
| F |  | 386.66 |
| G |  | 1,000.00 |
| H |  | 1,000.00 |
| I |  | 1,000.00 |
| J |  | 200.00 |
| K |  | 400.00 |
| **Total Project Size** |  | 25,454.04 |

## LESSONS LEARNED

Identifying domain experts and reaching them as scheduled were among the most critical tasks. Orientation meetings at the start of the projects, and regular progress meetings between the METU and TLFC project offices enabled effective communication throughout the projects. These meetings provided the opportunity to discuss conflicting issues, to notify demands on resources, and to replan the work under progress.

Modeling existing business processes took significant time and almost half of the total project effort, but other than being a baseline for requirements specification, it helped the stakeholders to identify the bottlenecks and the improvement points needed in the business processes.

The domains of Project A and Project B are different but complementary with a high degree of data exchange requirements. In addition, both Project A and Project B have numerous integration points with seven other C4ISR subsystem projects of the TLFC. Therefore, during the execution of Project B, we organized coordination meetings with other TLFC C4ISR subsystems' project committees. Although this task was difficult due to the required formalization in the organization, the business process models created an excellent baseline to discuss issues related to the integration of projects. The process models enabled the representatives of other C4ISR projects to visualize the whole picture in a short time and create an early consensus on how the data would be exchanged among these projects.

Currently, the systems for which we completed acquisition planning have entered into the development phase. The TLFC has decided to integrate the development of Projects A and B since corresponding systems are complementary and their requirements are redundant. The TLFC has suspended the release of the RFP, and decided to complete the system and software requirements specification stages on its own, specifically by assigning responsibility to one of its departments that develops in-house systems and software. The development group has used elicited system requirements as a basis for their planning as well as for requirements specification. They are currently generating use-case specifications and scenarios based on user-level functional system requirements.

Recently, the TLFC has contracted a new project to define integration requirements of all nine C4ISR subsystems including Systems A and B using this approach.

Another challenging task was the orientation of domain experts for business process modeling. Domain experts needed assistance to think in terms of business processes, and to identify and decompose the key business processes using specific notations. Almost all domain knowledge was documented in books, instructions, guidelines, or reports. The existence of written resources helped in understanding the context of the domain in detail, and speeded up the orientation of consultants to the domain. However, identifying and modeling business processes by the TLFC Project Office following these resources were stringent since the domain knowledge is captured in terms of business work products rather than business processes in these resources. In other words, there was confusion between preparing a domain document and executing the processes behind it. For example, documenting the sections of a domain report actually requires executing the steps of a business process that generates that report. This confusion between business work products and business processes slowed down the modeling from time to time, and frequently required elaboration of what business process modeling is.

In both projects, orientation was provided via meetings and frequent discussions. However, regular formal training sessions might have been better to save overall effort in such projects. Since the TLFC Project Office had been staffed by domain experts, we needed to assist them in the details of business process modeling and system requirements elicitation throughout the projects. This assistance was one of the most important indicators of success, and was therefore provided with great care as to proceed within context but not to restrict the expectations of the domain.

Modeling and analyzing current business processes provided considerable guidance for cre-

ating target business processes. Process-oriented analysis also enabled extensive understanding of the domain. The determination of IS-supported functions, new information flows, and changes in existing workflows were the results of the target business modeling. Indeed, current and target business modeling were performed together to some extent because bottlenecks, deficiencies, and problems were already captured in current business modeling. This activity was performed differently in Project A and Project B. In Project A, the models of target business processes did not include identification of the IS infrastructure within the business process models. Gaining satisfactory domain knowledge, we could associate the system architecture with the business process models and determine the requirements of the target system. In Project B, we constructed target business process models in order to generate user-level functional system requirements automatically.

During the requirements-elicitation process, we generated functional requirements manually from target business process models in Project A, and automatically from target business process models by using the KAOS Tool (Su, 2004) in Project B. For Project A, the size of which was estimated to be 10,092 Mark II FP, the manual generation of the requirements document took 2 person-months. After modeling target business processes using the notation's conventions at required detail, the KAOS tool generated the functional requirements of Project B, which was 25,454 Mark II FP in size, in 30 minutes. Thus, the planning of the target system modeling and functional-requirements-generation processes was made according to whether the requirements generation would be made manually or automatically.

Readers will notice that the ratio between the number of user-level functional system requirements and software size in Mark II FP in Project A is quite different from that in Project B. That is, 10,092 Mark II FP were estimated from 1,380 requirements in Project A, whereas 25,454 Mark II FP were estimated from 1,270 requirements in Project B. There were two reasons for this: The first one lies in the difference between the methods used while generating target business process models as a basis for requirements elicitation, and the second reason was the difference in abstraction levels of the generated requirements of the projects.

In Project A, we kept the organization-based structure of current business process models while constructing target business process models, and as a result, we had different sets of functional requirements for different organizational units. However, these sets had overlapping requirements among organizational units due to similarities in functionality, causing a greater number of requirements. Among the organizational units, 1,156 requirements of 1,380 were overlapping, which constitute about 85% of the total requirements; these overlaps in the requirements were handled while performing software size estimation for Project A. By using this experience gained from Project A, in Project B, we transformed the organization-based structure of current business process models into a system-based structure while constructing target business process models. As a result, we obtained a unique and generic set of target business process models, and therefore user-level functional system requirements that had no overlaps.

This resulted in a greater number of user-level functional requirements in Project A than in Project B, although the real number of functionalities required were not so. This means that if the requirements had been organized in a system-structure-based manner, Project A would have only 224 requirements.

Another point is that the abstraction levels of target business processes were higher in Project A than in Project B. If both projects had used a system-structure-based organization, the number of requirements would not have been comparable since the number of functionalities, which each requirement constitutes, would be very different.

During both projects, we maintained a project effort metric database. The team members entered data related to the type of work, activity name, and effort attributes into this database. This helped to reflect more realistic figures to the management plan as the projects progressed since we utilized this database to estimate the effort needed for modeling the remaining business processes.

## CONCLUSION

The establishment and transformation of need from concept to system requirements can be supported by means of notations and tools developed for business processes reengineering. The process-oriented approach not only helped the organization to see the big picture, but it enabled us to focus on the required improvements to be able to utilize the acquired system as well.

Although business process modeling requires significant effort, it brings various advantages. For example, the automatic generation of software functional requirements as well as the consistent application of software size estimation methodologies was possible in Project B: Both have great value, especially for large and innovative projects. In addition, we observed that business process modeling is an effective way to explicitly define the requirements of a software system while creating visibility and consensus among different stakeholders of other systems that are to be integrated with each other.

We implemented our approach in two systems to elicit the requirements of activity-based operations of two organizational units. However, not all technology solutions, such as highly graphical systems, embedded systems, and real-time systems, are inherently activity based. We have no observation on the usability of our approach for such systems.

There is no doubt that predevelopment processes need further research studies for the development of systematic approaches (Demirörs, Demirörs, & Tarhan, 2001). Specifically, extensive research on predevelopment methodologies including processes, notations, and heuristics as well as tools and techniques to support such methodologies are required. These methodologies should also link the development phases with the work products of the predevelopment phases.

# REFERENCES

Aslan, E. (2002). *A COTS-software requirements elicitation method from business process models.* Unpublished master's thesis, Department of Information Systems, Informatics Institute of the Middle East Technical University, Ankara, Turkey.

Capability Maturity Model Integration (CMMI) Product Team. (2001). *CMMI-SE/SW/IPPD, v.1.1. CMMI^SM for systems engineering, software engineering, and integrated product and process development: Staged representation* (Tech. Rep. No. CMU/SEI-2002-TR-004). Carnegie Mellon University, Software Engineering Institute.

Cooper, J., & Fisher, M. (2002). *Software acquisition capability maturity model (SA-CMM®) v.1.03* (Tech. Rep. No. CMU/SEI-2002-TR-010). Carnegie Mellon University, Software Engineering Institute.

Cummnis, F. A. (2002). *Enterprise integration: An architecture for enterprise application and systems integration.* John Wiley & Sons.

Davenport, T. (1993). *Process innovation: Reengineering work through information technology.* Ernst & Young.

Demirörs, O., Demirörs, E., Tanik, M. M., Cooke, D., Gates, A., & Krämer, B. (1996). Languages for the specification of software. *Journal of Systems Software, 32,* 269-308.

Demirörs, O., Demirörs, E., & Tarhan, A. (2001). Managing instructional software acquisition. *Software Process Improvement and Practice Journal, 6,* 189-203.

Demirörs, O., & Gencel, Ç. (2004). A comparison of size estimation techniques applied early in the life cycle. In *Lecture notes in computer science: Vol. Proceedings of the European Software Process Improvement Conference* (p. 184). Springer.

Demirörs, O., Gencel, Ç., & Tarhan, A. (2003). Utilizing business process models for requirements elicitation. *Proceedings of the 29^th Euromicro Conference* (pp. 409-412).

Demirörs, O., Gencel, Ç., & Tarhan, A. (2006). Challenges of acquisition planning. *Proceedings of the 32^nd Euromicro Conference on Software and Advanced Applications (EUROMICRO 2006)* (pp. 256-263). IEEE CS Press.

Department of Defense (DoD) Architecture Working Group. (1997). *C4ISR architecture framework, version 2.0.*

Department of Defense (DoD) General Services Administration. (2001). *Federal acquisition regulation.*

Hammer, M., & Champy, J. A. (2001). *Reengineering the corporation: A manifesto for business revolution.* Harperbusiness.

IEEE. (1998a). *IEEE software engineering standards.*

IEEE. (1998b). *IEEE Std. 1062: IEEE recommended practice for software acquisition.* New York.

IEEE. (1998c). *IEEE Std. 1220: IEEE standard for application and management of the system engineering process.*

ISO/IEC. (1991). *ISO/IEC 9126: Information technology. Software product evaluation: Quality characteristics and guidelines for their use.*

Scheer, A. G. (2003). *ARIS toolset, version 6.2.*

Scheer, A. W. (Ed.). (1994). *Business process engineering: Reference models for industrial enterprises.* Berlin, Germany: Springer.

Su, O. (2004). *Business process modeling based computer-aided software functional requirements generation.* Unpublished master's thesis, Department of Information Systems, Informatics Institute of the Middle East Technical University, Ankara, Turkey.

Turetken, O., Su, O., & Demirörs, O. (2004). Automating software requirements generation from business process models. *Proceedings of the First Conference on the Principles of Software Engineering.*

United Kingdom Software Metrics Association (UKSMA). (1998). *MkII function point analysis counting practices manual, v.1.3.1.*

Wiegers, K. E. (1999). *Software requirements.* Microsoft Press.

Yourdon, E. (2000). *Managing software requirements.* Addison Wesley Publishing Company.

# Section II
# Business Process Management

# Chapter V
# Process Integration Through Hierarchical Decomposition

**Ayesha Manzer**
*Middle East Technical University, Turkey*

**Ali Dogru**
*Middle East Technical University, Turkey*

## ABSTRACT

*A mechanism is presented to construct an enterprise process by integrating component processes. This approach supports the formation of value-added chains based on process models. Process representations in logical as well as physical levels are offered in a hierarchy as a graphical tool to support process integration. An enterprise represents a complex process that needs to be constructed within an engineering approach. Such construction is a matter of integration if component processes are made available by smaller enterprises. Process integration is supported by task-system formalism that may especially prove useful in the analysis of the preservation of process attributes during integration. Approaches using the decomposition of the problem definition have matured for various engineering disciplines in an effort to match definition pieces with previously built subsolutions. Among these disciplines, the techniques used in software engineering are exploited as this domain is closer to business process construction than the others.*

## INTRODUCTION

Businesses are complex organizations, therefore their formation requires some scientific guidance. Similar to the construction of any other structure, the construction of an enterprise is expected to benefit from engineering practices. Similar to the role of information technologies in shaping the way we conduct our operations, structuring the business will also require modern technologies. Developing rapidly, software engineering has become the closest discipline to enterprise engineering for the study of processes, the logical representation of any structure integration, and modeling. Similar leverage has been exploited in other disciplines, but software technologies have used them most recently and effectively. In this chapter, some mechanisms used in component-oriented software integration are adapted to enterprise engineering, namely, hierarchical decomposition and component integration.

We will benefit from engineering practices, and such rigorous techniques are crucial in achieving dependable outcomes. However, we also need simplicity and understandability, especially when it is a matter of preserving our intellectual control over the business (Laguna & Marklund, 2004) through a mind not necessarily that of an engineer's. It is a fact that complex artefacts require interdisciplinary teams. In other words, the big picture should not be neglected while low-level details and formalisms of engineering mathematics are accounted for. There is a racing condition here; we want to understand the enterprise easily, and yet the enterprise needs to be engineered rigorously.

Another complicating factor is the need for agility. Our organizations, like our production infrastructure, need to be flexible and dynamic to meet the rapidly changing market demands

(Madachy & Boehm, 2006). Rather than designing the process from scratch, it is always faster and more reliable to reuse what was invented and tested before. Our goal is to utilize a number of existing component processes in a smart integration for constructing the next enterprise-level process. Agility can be achieved through component-process replacement. Wherever a modification need arises in the system, one corresponding process can be replaced with another that might satisfy the need. This replaceable part may correspond to a single process or to a set of neighboring processes in the hierarchy model.

We treat enterprise engineering as a design problem. Being applicable to any design area, hierarchical decomposition has been proposed as the fundamental algorithm (Simon, 1969). Applied to business processes, this means an enterprise engineer can think of the organization as a huge process that can be decomposed into component processes. Assuming a top-down approach to designing a new process, as an initial step the

*Figure 1. Decomposition in a hierarchy*

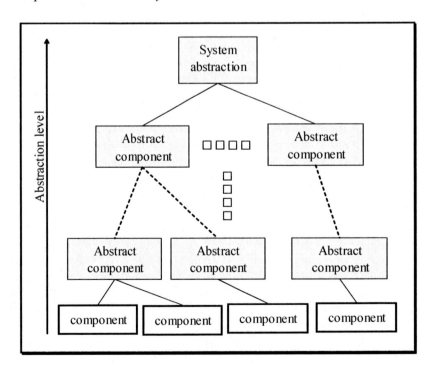

whole operation (system) can be modeled as an abstract process. Likewise, the immediate parts of the whole, the smaller processes, can be represented as logical entities in the decomposition. At the outset, we want to visualize the relative positions and perhaps interactions of the component processes in the system. Eventually, when sufficient decomposition is done, the decomposition model will come closer to real processes: Process models that represent real business processes will take place at the bottom of the hierarchy. Figure 1 depicts the decomposition model.

Component processes are represented as boxes that connect to others (integration) through links that carry information about material flow, resource allocation, and synchronization. A box can be modeled by any existing modeling technique that helps in ordering the activities, assigning resources to activities, and organizing material and product flows. In other words, a process in the system will be presented as an activity inside a process. This view not only allows for process abstraction, that is, hiding the information internal to a process and viewing it as a black box so that the big picture is easier to understand, but also, in cases of too many processes, it becomes possible to understand the system by looking at it one region at a time. The region of focus could correspond to the whole enterprise when it is at the top of the hierarchy, or to any other local context. In the following sections, information about the related mechanisms in software engineering and their adaptation to enterprise engineering will be presented with a process integration example. Also, a task-systems-based formal representation is included so that some process attributes can be analyzed for their preservation after integration.

## Similar Experience: Decomposing Software

Hierarchy and decomposition have always been important issues in software development. Re-cently, the emergence of software-component technologies has enabled the application of such mechanisms to practical and efficient engineering. Traditional approaches offer modeling techniques for design that present various cross sections of the system with respect to different concerns. The outcome is a set of models that would not converge to a real system unless the designers have the experience to combine the divergent information. Data, function, and other conceptual dimensions were accommodated by such models. For human cognition, structure is the best choice for a decomposition concern. Rather than other concepts, components, being structural units, facilitate the divide-and-conquer paradigm based on the structure dimension. Structure corresponds to a chunk of software code that can contain other constituents such as data, function, control, and so forth.

A purely component-oriented approach has been proposed (Dogru & Tanik, 2003) that suggests starting with a decomposition of the problem definition. The goal is to achieve a software application without having to develop a single line of new code. Rather, locating and integrating existing components is the suggested methodology. This revolutionary thought is not an unfamiliar one: Other engineering disciplines have already discovered an attitude that is against reinventing the wheel. This approach is also analogous to the outsourcing avenue to producing a complex product or service.

A designer decomposes the system definition into abstract units that, preferably, correspond to components that are anticipated to exist. A good decomposition will present units that completely correspond to existing components so that minimal modification or development is necessary. We want to develop software by integration rather than code writing. The software market is rapidly moving in this direction; component-ware and related methodologies are maturing day by day.

## Process Modeling

As a recent and rapidly developing field, software engineering has already experienced different process approaches for developing software. Exposure to such a variety in a short time, combined with the versatile infrastructure, has helped this discipline to offer modeling mechanisms that have transcended the software domain. A typical process model represents tasks and material or product flows in and out of these tasks, and resources can be modeled and assigned to tasks. Advanced tools allow simulations on these models to run and expose bottlenecks in the business conduct.

Humphrey (1990) described the process as "the set of tools, methods, and practices we use to produce a software product." We model the processes as an ordered set of tasks to conduct a manufacturing or service operation. The process concept soon became an important issue for enterprises to investigate. A new consulting field emerged in the form of business process reengineering (BPR; Hammer & Champy, 1994). A typical reengineering operation started with the modeling of the existing practices of an enterprise. The experienced consultants could then improvise a "to be" model for the enterprise that would most probably incorporate an important IT component. Finally, the prescription to convert the operation to the new process was coined. Substantial savings have resulted by applying BPR, and the topic enjoyed a sustained demand.

Workflow systems have also developed as a new terminology. Process modeling was a close concept, but the workflow concept has rapidly attained the meaning of "executable process model." Workflow engines are available today as components to integrate with business software. Workflow-engine-based hospital automation software, for example, will accept the definition of the process in a specific hospital through a graphical user interface. Without having to develop any further code, the system will adapt itself to this definition, and for every different hospital, a tailored software version will be created with ease, even without having to compile the code.

Recently, related concepts have evolved into the definition of businesses using newer languages and tools. Also, services available as software operations that can be employed over the Internet have evolved into smoothly operating Web services. Business process definition languages are currently becoming capable of working with Web services. As a result, the definition and enactment of a new business process are available to users of the correct software tool set.

The business process execution language (BPEL; Juric, Mathew, & Sarang, 2004) provides a language for the formal specification of business processes, and its newer form for Web services (BPEL4WS) also provides interaction protocols. In this way, business transactions are enabled through the interaction protocols of the Web services. This technology can facilitate process integration.

Business process modeling notation (BPMN, 2004) is another graphical modeling framework that targets technical as well as business-oriented users. With its abstract and formal mechanisms, this tool supports interaction among businesses and also among different modeling media.

## PROCESS DECOMPOSITION MODEL

We view a process as a set of tasks ordered to accomplish a goal (Curtis, Kellner, & Over, 1992). There have been various approaches to modeling processes (Acuna & Juristo, 2005). The fundamental ideas in our framework are applicable to a variety of modeling techniques. That idea is based on a recursive definition of a process: A process can be made of smaller processes. An activity is traditionally regarded as a unit-processing element inside a process. Actually, in our perspective, an activity is defined similar to a process. However,

*Figure 2. Process modeling concepts in three dimensions*

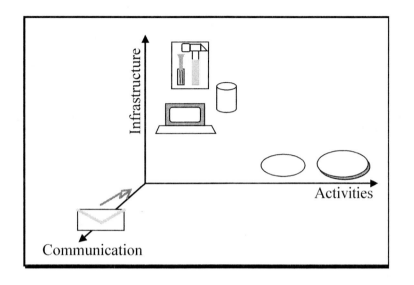

there is still a differentiation: If an activity is not going to decompose to further tasks, we do not call it a process. A process can be represented by a complex model of its internals, but in our system of processes, it can be drawn as a black box at logical levels in the hierarchy. Visualization of the interconnection of processes is important, which reveals the synchronization and product or material flows. To accommodate at least one view to process modeling, we will present the visual process modeling language (VPML; Dutton, 1993) representation, which offers a generic foundation. The three dimensions of modeling elements on which VPML is also based are depicted in Figure 2.

The activity dimension is the main one, where activities are basically situated. Unit actions that make a process are usually modeled as activities. Sometimes these activities are composite; they are explained through lower level activities. The infrastructure dimension actually stands for material and products, besides resources that alone are traditionally more representative of the infrastructure concept. Nevertheless, activities produce products that could be either the final product or

intermediate products. Finally, communication could be interpreted as any kind of information flow between activities and personnel, but, as a most important concept, this dimension is also responsible for the ordering of the tasks. If synchronization is required to signal the beginning or the ending of the activities, it will be carried out through communication among the activities.

## Abstraction

Abstraction is a key concept in studying complex structures. Our initial response to complexity is to employ divide-and-conquer tactics. Such division, however, should be guided in a hierarchy, meaning that a group of the dividends should be represented by a generalization concept. A generalized unit as a logical entity, hiding many enclosed details, allows us to concentrate on its relative position inside the system. In other words, observing from a higher point of view, we can ignore the internal details, thus reducing the complexity in our cognition process to aid in the understanding of the whole rather than being required to view at the same time the internals of all parts: The whole

*Figure 3. Abstract and actual process representations*

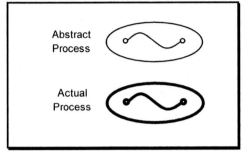

is prohibitively complex to study with all details considered simultaneously; it is relatively easier to understand a part alone. We can study a system at two levels: (a) the whole without details, and (b) parts with details, one at a time. Thus, abstraction, combined with a hierarchical decomposition of a complex structure, helps us to manage complex systems. Here our system is the combined process of an enterprise, and its parts are the lower level processes. To study the integration, we introduce abstract processes that represent more than one component process.

The system under construction is treated as an abstract process until the integration of the component processes is finished. A process engineer decomposes this whole process into lower level logical processes. Each newly identified abstract process is investigated: Is it specific enough for an existing process to correspond to it? If not, this abstraction will need to be divided further. When all the abstract processes are covered by one real process, the decomposition activity is terminated. For the processes to be integrated into the model, material flows and synchronization links should be finalized. Actually, such relations among processes can also be represented in the higher (abstract) levels. We can start with representing interactions in abstraction, whose details in terms of flows and synchronization will be defined later in the lower levels. Figure 3 shows the graphical representation of processes.

The opposite concept to abstraction is refinement. In the decomposition hierarchy, higher levels are more abstract and lower levels are more detailed. Enhancing the model toward lower levels corresponds to refinement. In a case where an existing set of processes were already acquired and it is desired to predict how a system would work that will be constructed through their future integration, a bottom-up hierarchy of abstractions can be built. Conversely, starting with the goal represented at the top, the hierarchy can be built top-down through decomposition and eventually a set of refined processes will be modeled. The proposed hierarchy model can be used both in the understanding of existing complex processes, and in the design of new complex processes. In this chapter, mainly the latter goal is exploited.

## Building the Hierarchy Model

Our goal is to propose a structure for complex process models. Structures are not required for the operation of a system. They are, however, crucial in our understanding of a complex system. Models in general are tools for understanding a complex system. What to include and what to exclude in a model are therefore important. At any point during the development, a designer should make careful decisions regarding the main concern for understandability and should determine the degree of detail accordingly.

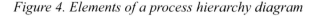

*Figure 4. Elements of a process hierarchy diagram*

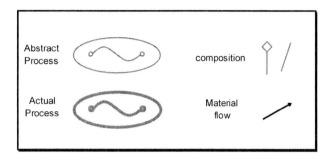

We have mentioned the representation of a process. Besides activities, their ordering (synchronization), assigning resources, and flow of material and products are involved. A super process of integrated processes is also a process, and therefore it should also address all the issues besides the activity dimension. Our main concern is the ability to design such a process. The main view promoted by our approach is decomposition where the activities are the important players. Nevertheless, other constituents of the model can be added to the hierarchy diagram. The only danger here is to populate the diagram with different concerns and reduce understandability. We suggest leaving the usual process modeling details to the internal modeling of our process nodes that are the component processes. If other concerns such as resource assignment can be appropriately represented at high levels, a copy of the hierarchy diagram can be maintained to study different concerns.

One important reminder is to establish the connections among the nodes during the decomposition: The time during which a higher level process is being divided into its parts is the best time to declare connections among the parts. Considerations about splitting a process will share a similar context with the interaction view concerning the split parts. Of course, connections can be declared after completing the whole system decomposition, but some information (context) may be forgotten after concentrating on different parts of the design. Figure 4 depicts the hierarchy diagram elements.

Actually, the mechanism exists in most process modeling tools: The composite-activity notion is usually included. A composite activity is made of other activities and a detailed process model can be drawn for it. However, the concept of composite activities as an implicit representation of hierarchy is not very indicative of the complex process structure. Such a structure is better visualized when displayed as a tree.

## AN EXAMPLE PROCESS CONSTRUCTION

In this section, a decomposition of a business process is demonstrated through an example. English Text Doctor (Tanik, 2001) as an existing electronic enterprise is decomposed into logical processes and then integration is demonstrated. This organization offers help in the proofreading and correction of written documents. There is a customer interaction part of the business that receives new texts and, when edited, returns them. Customers are billed based on the number of pages. The other part of the business is like the back office: A list of proofreaders is maintained and the proofreaders receive jobs based on a circular order; documents are ordered and assigned to proofreaders, and finalized documents are collected and directed to the customer. There are criteria for evaluating the proofreaders; timely delivery, for example, is a critical issue in the cost of the service.

This organization seems to have two main processes: one for customer interaction and the other for the actual service production. The enterprise engineer can start with the top-level decomposition of the operation as customer and office processes. Predicting further component processes to consider later to do the actual jobs, these top-level subprocesses are determined to be of abstract type. Figure 5 shows the initial decomposition outcome. Other relations among the processes are ignored at this point. The most important tool offered to the enterprise engineer is the divide-and-conquer mechanism. It is important to organize the complex operations as clearly separated units.

At this point in the design, customers and the documents flowing to and from them could be entered in the model. The main concern is to visualize the process structure, and for this problem, presenting the customer and the documents are considered to be details at this phase. The

design continues with the effort to decompose the enterprise process into further components.

Next, the customer interactions process is perceived to comprise two important constituents: the receive and sell operations. We realize that the two newly introduced kinds of operations are not sequentially joined. They are under the customer interactions abstraction only logically because they both interact with the customer. Actually, receiving the documents and selling the processed documents, process-wise, are very different operations. This is the use of abstractions: They do not necessarily correspond exactly to the real processes but they help us organize the model (understand the system). In this way, although process-wise they are not related, two subprocesses are organized at some proximity in the structure because of a logical relation. At the actual processes level, the detail may prevent a designer from understanding the model. The opportunity to break down the structure based on a logical concern is regarded as better from a designer's point of view. Once constructed, the position of the component processes in the hierarchy diagram is immaterial for the enactment of the integrated process.

Continuing with the decomposition, the designer introduces the receive and sell processes. On the production side, there is a central process that monitors the delivery from the proofreaders.

*Figure 5. First-level decomposition of the process hierarchy*

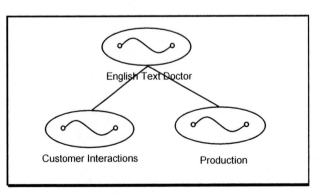

*Figure 6. Hierarchy diagram for English Text Doctor*

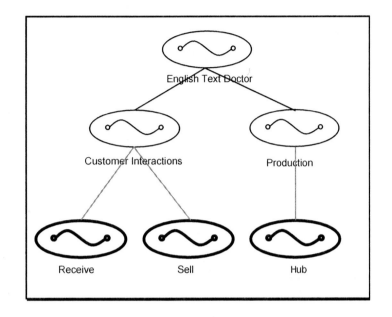

Actually, the incoming documents are tagged, categorized, and then dispatched to proofreaders. This central process is referred to as the hub process. This simple enterprise is finally represented through three actual processes that are organized under the abstract processes. Figure 6 shows the final decomposition for the processes. The fundamental operation is proofreading, which is actually only an activity, and it is suggested that it be allocated under the hub process.

The top-down progress of the design would naturally continue with forming the connections among the bottom-level (actual) processes. Now that we have configured the structure that is the part-whole relations, it is time to specify the refinement. It is also possible to depart from the hierarchy diagram and continue development with the actual processes determined so far. This route is easier for trivial processes, but any development activity in a complex enterprise would require the preserving of the big picture through the hierarchy diagram. It is possible to keep both views—the

hierarchy and a detailed process model—while refining the design to its maturity.

## Component Processes

Now that the skeleton of the system is defined, it is time to concentrate on the functional modules. This organization does not reveal any information about what the receive, sell, and hub processes should look like. Their process models are to be introduced next. Even if such models were existent, the hierarchy will still help in their integration and related fine-tuning. Figure 7 depicts the process for the receive side of the organization.

Following the regular flow of the operation, incoming documents are directed by the hub process. Figure 8 presents a model for the hub operations. Here, documents are sorted, assigned IDs, and dispatched to proofreading groups and then to the customers.

Finally, the sell side is where the edited document is directed to the customer. Figure 9 depicts a process model for the sell side.

*Figure 7. A process model for the buy-side process of the English Text Doctor organization (Tanik, 2001)*

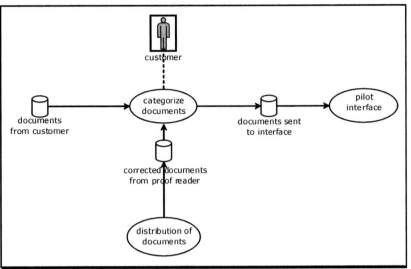

*Figure 8. A process model for the central hub for English Text Doctor*

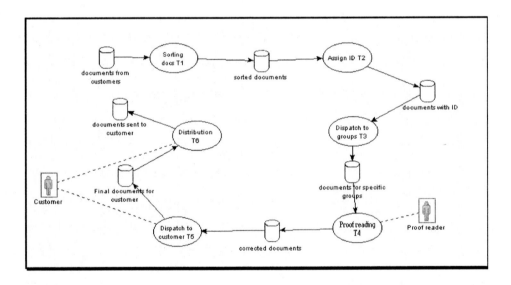

Now it is time to integrate the identified component processes. Some modification to the individual processes is possible, hence distorting the isolated views of the component processes in the hierarchy diagram. Still, the hierarchy view will remind us about where the components came from. Actually, the abstract versions of the com-

ponent processes are where the integration should start. The connection for these processes is more important than the internals of the processes. Such connections can be superimposed on the hierarchy diagram also. Next, the black-box views for the component processes will be considered and their abstracted versions and connections will be

*Figure 9. The sell-side process model for English Text Doctor*

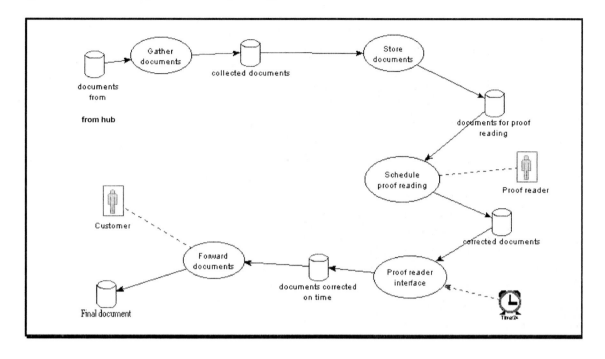

determined. Once all the processes are integrated through connections, it is also possible to present a combined process model for the system in detail. For the complex processes that are the focus of this chapter, such a combined model will rarely be used. The hierarchy diagram will be used as a map to direct the designers to the point of interest, and that special area can be studied in detail in the process models for the components. An intermediate model, that is, the black-box processes and their connections, can also be referred to. Figure 10 presents the black-box view for the three actual processes. Here, details are hidden, but the context of a process is presented: Input, output, and related resources are shown with the process.

The integration can be represented on the hierarchy diagram with less detail on the connections as shown in Figure 11. One point can be observed in Figure 11 about the different concerns between the hierarchy diagram and the actual integrated

process. The logical organization of the hierarchy, which assumes proximities for the processes that are related to the customer, does not provide the best layout in terms of the flow sequence of operations. Positioning the hub between those processes would have resulted in a better picture. It should be remembered here that the hierarchy diagram did its contribution in the determining of the component processes. It is natural to have different views in any modeling for visualizing different concerns.

It is also possible to draw connections among the abstract processes in the hierarchy. This is useful if an early idea is desired about the integration before proceeding toward details. Again, it all depends on what needs to be studied on the model. Different copies of the hierarchy diagram can be maintained that present different views, even with partial views. A diagram that only contains the abstract levels, for example, is possible.

*Figure 10. Actual processes and their connections*

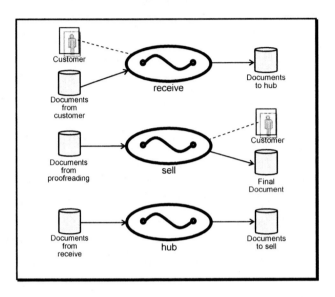

*Figure 11. Integration through connections in the hierarchy diagram*

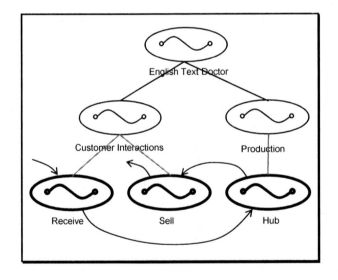

This much work seems to be sufficient in the design of an integrated process. However, the actual process model is also required due to the detailed modifications that will appear neither in the hierarchy diagram nor in the black-box integration diagram. The black-box concept means hiding the details of a unit if its connections are enough to study a system of boxes. In this case, a box corresponds to a process. An integration view that ignores where the processes came from (hierarchy diagram) and the internal details (process model for the component processes) is presented

*Figure 12. Integration of component processes*

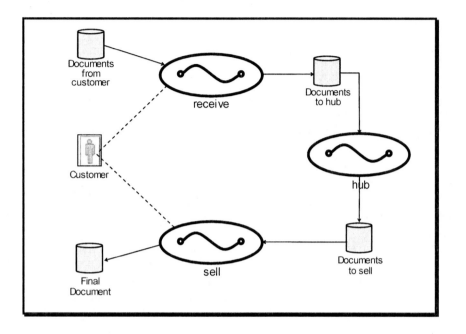

*Figure 13. Integrated super process in detail*

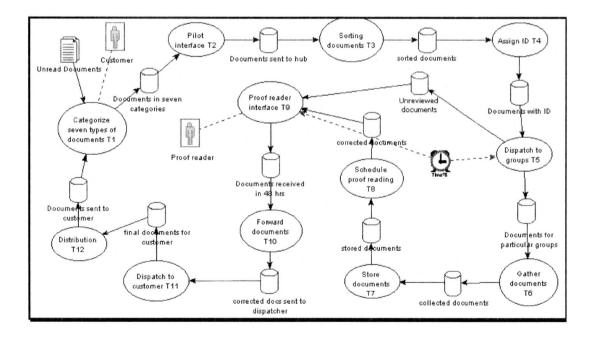

in Figure 12. This view can be constructed by studying the component processes in their abstract or black-box forms and their connections as shown in Figure 12.

One point can quickly catch the eye: The customer icon represented twice in Figure 10 is only drawn once in Figure 12. This is an example of the general statement that an integrated process is not simply a gathering of component processes; some modification may be necessary. Finally, the combined process model for the integrated processes is shown in detail in Figure 13.

## INTEGRATION GROUNDWORK

It is advantageous to support a graphical integration environment with further tools for investigating various concerns. A process engineer can decompose a complex process logically, but then locating existing processes and integrating them will have peculiar difficulties. There may

be technical or theoretical hurdles in composing the super process. Verification tools can detect such problems. Also, the preservation of some attributes of the processes is investigated for the integration. Given the component processes with their existing attributes, a forecasting capability will indicate the attributes for the super process to construct.

For a more instrumental analysis of such issues in integration, more formal representations will be required (Havey, 2005). This section proposes modeling processes and their attributes in task systems (Delcambre & Tanik, 1998) that have also historically provided a theoretical basis for operating systems. The preservation of some process attributes can be investigated through this formal representation (Manzer, 2002). The selected demonstrative set of process attributes comprises efficiency, repeatability, manageability, and improvability. These attributes are indirectly mapped to task features such as determinacy, deadlocks, mutual exclusion, and improvability.

*Figure 14. Mapping process attributes to task features*

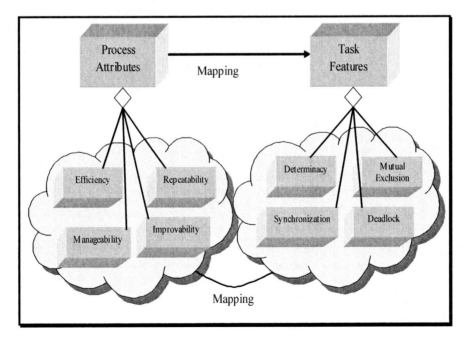

Figure 14 depicts the mapping of process attributes to task features where a process corresponds to one task in the formal system. Such a tool, while being more adept for rigorous analysis, can be utilized for higher level concerns with more subjective evaluation. These decision-supporting measures for integration can be a starting point for the otherwise totally unsupported task of integration. Based on the decisions, specific component processes will be acquired, or it will be possible to assess the infeasibility of the goal.

In the following sections, some process parameters are investigated and an assessment of their variance due to integration is discussed. It should be reminded that a mapping from the listed process attributes to the addressed task features cannot be a perfect one, nor can the interpretations of the model for high-level process attributes be perfect. This approach equips the decision maker with a tool that will gain efficiency with experience.

## Task Systems

According to Coffman and Denning (1973), concurrently executing processes can interact in a number of undesirable ways if resources are improperly shared or if there is a faulty communication of results (signals) between processes. Generally, the problem is one of timing.

To provide a better understanding of processes, task systems were introduced that model a process as $C = (\tau, <\cdot)$, where $\tau$ is a set of tasks and $<\cdot$ represents the partial order (precedence relation) of the tasks. If T is a task, the initiation of T is denoted by $\overline{T}$ and the termination of T is $T_-$. A superscript bar is used to denote starting up, and a subscript bar denotes closing down. A task with no successor is a terminal task, and one with no predecessor is an initial task. If task T is neither a successor nor a predecessor of T', then T and T' are independent. This section will not provide a description of task systems in depth. Some of its constituents will be included in the example process for an analysis of its composition.

The following sections contain discussions about the task features (determinacy, deadlock, mutual exclusion, and synchronization). Since the ordering relations among the tasks in the example are known, together with the input-output resources, articulation can easily be developed. Also, comments are made about the process attributes for the integrated organization based on the task-features discussions.

## Determinacy

The super task system is determinate according to the following definition (Coffman & Denning, 1973): "Task system C is determinate if for any given initial state $s_0$, the resource-value sequences depend uniquely on the initial values in $s_0$."

In our example, the resource-value sequences depend uniquely on the initial values in the initial state. Every task in the super system is noninterfering as each task is either a predecessor or a successor of another task. Consequently, the super system is determinate. In the super system, the tasks of dispatching to groups and the proofreader interface are indeterminate as the dispatching task depends on the completion of the proofreading. It is stated that repeatability is partially dependent on determinacy (Manzer, 2002). It can be stated that this process is repeatable.

## Deadlocks

In the super system, the threat of a deadlock is prevented by preventing circular wait: The circular-wait condition is avoided for the proofreading task. There is a timer connection between the tasks of dispatching to groups and the proofreader interface. After the dispatcher sends the documents to the groups of proofreaders, 48 hours are allowed: If a proofreader does not return a document by that time, then the document is passed to the next proofreader in the same group. Since resource usage is known in advance, deadlock can be avoided. Still, there is a risk of deadlock if the number of incoming documents (that is, requested resources) is larger than the resources held (that is, the number of proofreaders in a specific group). In order to avoid deadlocks, the number of people in various groups must be greater than the number of incoming documents from customers. There is a high risk of experiencing an unsafe state if the number of documents becomes greater than the number of proofreaders. In order to manage the process, the manager must know the resources used in the process (Coffman & Denning, 1973). Since the number of used resources is known in advance, this process is manageable.

## Mutual Exclusion

The mutual-exclusion problem concerns the control of a reusable resource so that it is never in use by more than one task at a time. In the super system, the reusable resource is the documents sent to the proofreaders for correction. The system must confirm that the same document is not sent to more than one proofreader. Similarly, the document must not be sent to the proofreader if he or she is not available. It is obvious that mutual exclusion helps the traceability of resources. Also, it indirectly promotes the efficient use of resources. This implies the process attribute of efficiency. As has already been stated (Manzer & Dogru, 2002), efficiency can be related to the traceability of resources.

## Synchronization

Setting the time synchronizes the dispatching and proofreading tasks in the super system. When the document is sent to the proofreader, the task of dispatching to groups waits for 48 hours. If the corrected document is not received, then it is sent to the next available proofreader. By achieving the timing requirement in this process, performance of the process is improved. Hence, the process complies with the process attribute of improvability.

## CONCLUSION

Process modeling is an important task that organizations will increasingly refer to. We can assume that our world is in the information age simply by observing the growing demand for software and the fact that information technologies are being employed as an inevitable component of any organization. This improvement is enabled through engineering approaches; disciplined methodologies are very important for an efficient supply in an effort to address demand. Process notions can be regarded as positioned at the heart of the disciplined approaches in any field, most dramatically and quickly felt in the fields of engineering related to information. In addition, utilizing the Internet and other electronic means is gaining popularity in the forming or restructuring of organizations. We foresee an increasing demand for process modeling and electronic enterprise engineering (Bieber et al., 1997) or virtual enterprises.

The proposed techniques can find many areas for application. One application is assuming the Internet for locating component processes in an effort to build value-added chains (Manzer, 2002). Verification can also be conducted on the process model, which can be integrated using the Internet. This start may also find a smooth integration into software development projects. A model introduced in the manner suggested can act as a requirements tool for any software development to follow the enterprise design. Such development will be necessary in the enabling of newly founded organizations, and the super organization can also provide a specific value-added chain. We may witness the quick formation of such chains in an effort to exploit rapid changes in the market. Tools such as the one proposed in this chapter will aid in the design of these organizations. As a summary, we expect to see agile organizations that can quickly shape, adapt, and function. As enabling technologies, practical yet sound tools will be required for use in the formation of value chains and for verification of the outcome.

# REFERENCES

Acuna, S. T., & Juristo, N. (2005). *Software process modeling: International series in software engineering.* Springer.

Bieber, M., Bartolacci, M., Fierrnestad, J., Kurfess, F., Liu, Q., Nakayama, M., et al. (1997). Electronic enterprise engineering: An outline of an architecture. *Proceedings of the Workshop on Engineering of Computer-Based Systems,* Princeton, NJ.

Business Process Modeling Notation (BPMN). (2004, May 4). *Business process modeling notation (BPMN) information.* Retrieved April 5, 2006, from http://www.bpmn.org

Coffman, E. G., & Denning, P. J. (1973). *Operating systems theory.* Englewood Cliffs, NJ: Prentice-Hall.

Curtis, B., Kellner, M. I., & Over, J. (1992). Process modeling. *Communications of the ACM, 35*(9), 75-90.

Delcambre, S. N., & Tanik, M. M. (1998). Using task system templates to support process description and evolution. *Journal of System Integration, 8,* 83-111.

Dutton, J. E. (1993). Commonsense approach to process modeling. *IEEE Software, 10*(4), 56-64.

Hammer, M., & Champy, J. (2001). *Reengineering the corporation: A manifesto for business revolution.* New York: HarperCollins Publishers.

Havey, M. (2005). *Essential business process modeling.* O'Reilly Media Inc.

Humphrey, W. S. (1990). *Managing the software process.* Reading, MA: Addison-Wesley Publishing Company.

Juric, B. M., Mathew, & Sarang, P. (2004). *Business process execution language for Web services BPEL and BPEL4WS.* Birmingham, UK: Packt Publishing.

Laguna, M., & Marklund, J. (2004). *Business process modeling, simulation, and design.* Prentice Hall.

Madachy, R. J., & Boehm, B. W. (2006). *Software process modeling with system dynamics.* Wiley-IEEE Press.

Manzer, A. (2002, June). Towards formal foundations for value-added chains through core-competency integration over Internet. *The Sixth World Conference on Integrated Design and Process Technology.*

Manzer, A., & Dogru, A. (2002, April). Formal modeling for the composition of virtual enterprises. *Proceedings of IEEE TC-ECBS & IFIP WG10.1, International Conference on Engineering of Computer-Based Systems,* Lund, Sweden.

Tanik, U. (2001). *A framework for Internet enterprise engineering based on t-strategy under zero-time operations.* Unpublished master's thesis, Department of Electrical Engineering, University of Alabama at Birmingham.

## Chapter VI
# Using a Standards–Based Integration Platform for Improving B2B Transactions[1]

**Andrew P. Ciganek**
*Jackson State University, USA*

**Marc N. Haines**
*University of Wisconsin – Milwaukee, USA*

**William (Dave) Haseman**
*University of Wisconsin – Milwaukee, USA*

**Lin (Thomas) Ngo-Ye**
*University of Wisconsin – Milwaukee, USA*

## ABSTRACT

*This chapter presents a specific case in which a company explores the use of XML Web services in conjunction with an integration broker. Several business-to-business (B2B) approaches are developed to provide insight into the ability of modern integration technology to improve B2B interactions and to allow a broader group of businesses to participate. Furthermore, technical details and integration methodology principles are presented that can be used as points of reference for scenarios in which the same or a similar set of technologies is applied to advance the efficiency and effectiveness of B2B transactions. An evaluation of the B2B approaches is presented accompanied by a discussion of several key lessons. This chapter then ends with some concluding remarks.*

## INTRODUCTION

Many organizations have introduced electronic means to perform their business-to-business (B2B) transactions. These means include electronic data interchange (EDI) via value-added networks (VANs). Although EDI technology has existed for quite sometime and is mature, a large number of companies are still using manual phone- or fax-based processes to transfer business information (e.g., purchase orders). The literature suggests EDI is largely unattainable or at least impractical for many small or even medium-size enterprises (Iacovou, Benbasat, & Dexter, 1995; Teo, Wei, & Benbasat, 2003; Wigand, Steinfield, & Markus, 2005). Larger businesses, on the other hand, have widely adopted EDI and benefited from it, allowing them to grow at the expense of smaller organizations (Clemons & Row, 1988). Recently, however, the use of the Internet in conjunction with technologies such as the extensible markup language (XML) and XML Web services is being explored as a possible alternative to traditional EDI (Havenstein, 2005). As a result, businesses belonging to the "long tail" (Brynjolfsson, Yu, & Smith, 2003) that were unable or reluctant to adopt EDI may now have an opportunity to participate in electronic B2B transactions.

In this chapter, we present a specific case in which a company explores the use of XML Web services in conjunction with an integration broker to drastically reduce, if not eliminate, the need for placing orders by phone or fax. A B2B solution is developed that poses a much lower adoption threshold than previous B2B initiatives and provides greater flexibility.

The purpose of this chapter is to provide the reader with insights into the ability of modern integration technology to improve B2B interactions and to allow a broader group of businesses to participate. We first describe the business scenario and provide an overview of the technology employed in the solution. We then present three different approaches for placing purchase orders electronically that were implemented in a proof-of-concept system. The three approaches are compared and evaluated based on their advantages and disadvantages in the given scenario, as well as their relation to overall business strategy. Finally, we present important lessons learned in the process of implementing the solution and the options the company has moving forward with the implementation of a solution.

## BUSINESS SCENARIO

## The Company

Midwest Manufacturing Company (MMC)[2] is a manufacturer located in the Midwest of the United States of America that supplies customers with a diverse set of products on a global scale. Not only are its customers geographically distributed, but they also differ largely in their technological capabilities and the means with which they order MMC's products.

While many of MMC's customers, particularly larger organizations, employ EDI or use Web-based purchase forms, the majority of its customers still use a combination of fax and phone to conduct purchases. In particular, exchanges using EDI account for 35% of the orders that customers place. Another approach, an extranet where customers log on and place orders using a Web-based application, accounts for 12% of orders. Lastly, fax-based transactions account for the remaining 53% of orders that customers place.

The fax- and phone-based approach requires several steps in which the order information has to transition from one system to another system. This requires manual interventions and thus creates opportunities for errors. To avoid some of the problems inherent in the fax-based approach and reduce the cost of business-to-business transactions, MMC is seeking solutions to further streamline and automate the process with a flexible integration solution that would encourage more customers to use direct electronic means for purchase transactions.

Since MMC has already made a substantial investment implementing an SAP R/3 enterprise system, the new solution must leverage the functionality implemented in the enterprise system package. Therefore, MMC decided to explore developing a solution using the integration capabilities of the SAP NetWeaver platform to improve interaction with current customers and attract new ones. In this case, the new approach may not only be viewed as an attempt to reduce costs, but also an attempt to create a competitive advantage for MMC by facilitating easier interaction with a broader range of customers.

## The Order Process

The customers placing orders by fax are running disparate legacy enterprise applications that lack both application program interfaces (APIs) to facilitate integration and a means to directly communicate with MMC's SAP enterprise system. Therefore, the customer first enters an order into his or her own enterprise system. Next, the customer prints out the order in his or her native language (e.g., English, German, French, and others) and faxes it to MMC for further processing. After receiving this order by fax, staff at MMC manually translate and enter the order information into the SAP system. There are several limitations to this approach. Chief among them is the significant amount of effort involved with placing these fax orders for both the customer and MMC. Furthermore, there is a relatively high amount of errors in the orders placed by fax compared to orders placed via EDI or the Internet. Lastly, since there are manual steps involved in the process, there are increased lead times to the supply chain for these types of orders. Consequently, MMC would like to move the fax order process to an automated electronic process that is more cost effective and highly interoperable, that leverages its investment in SAP systems, and that presents a practical solution that MMC's customers can easily adopt.

The customers that currently send fax-based orders have so far resisted joining the EDI VAN and were also not signing on to the extranet solution to place orders despite discounts offered by MMC. This is partially attributed to a reluctance of customers to make significant investments to update their legacy systems to allow them to directly interface with MMC's SAP system. Consequently, an approach that improves the order process, reduces the cost of placing orders, and also provides a low-cost solution in terms of implementation and maintenance may present enough incentive for these customers to adopt it and move away from ordering via fax. Since the recent SAP NetWeaver platform, particularly the SAP Exchange Infrastructure (XI), supports open standards and enables interoperability with a variety of disparate systems, MMC decided to explore SAP NetWeaver for technical options for a solution to improve the order process.

## TECHNOLOGY OVERVIEW

At the center of the technical solution implemented at MMC is XI. XI is an integration broker and business process management engine that constitutes a key part of the SAP NetWeaver platform. The integration broker can leverage adapters to connect to a diverse set of systems using a variety of communication protocols. At its core, however, it uses XML and XML Web services to exchange and manipulate messages. Internally, data structures are defined using XML schema definitions (XSDs), message transformations can be performed using extensible stylesheet language transformation (XSLT), messages are exchanged in an extended SOAP[3] format (XI protocol), and business processes can be defined using the Web services business process execution language (WS-BPEL). To better understand the potential solutions in this scenario, we provide an overview of the key technologies: XML, Web services, and XI.

## Extensible Markup Language

According to the World Wide Web Consortium (W3C, 2005):

*Extensible markup language (XML) is a simple, very flexible text format derived from SGML (ISO 8879). Originally designed to meet the challenges of large-scale electronic publishing, XML is also playing an increasingly important role in the exchange of a wide variety of data on the Web and elsewhere.*

XML is commonly used as the canonical data format in integration solutions. XML itself, however, is a metalanguage and only prescribes a way to construct XML-based markup languages (i.e., XHTML) that serve a specific purpose, including markup languages associated with XML Web services (i.e., SOAP).

## XML Web Services

Even with XML, an important obstacle in the process of the integration of diverse systems was the lack of a common message exchange mechanism for distributed systems. While several technologies (e.g., DCOM, CORBA [common object request broker architecture]) have been developed in the past to connect distributed systems, they either were too focused on a specific technology platform and lacked industry-wide support or they were too complex to become a widely adopted and practical solution. XML Web services have the potential to overcome this obstacle. Web services are based on a set of established core standards that are platform independent and have broad industry support. The essential standards for Web services are SOAP, which describes the messaging protocol, and the Web service description language (WSDL), which describes the service interface. A third standard often associated with Web services is universal description, discovery, and integration (UDDI), which describes the interaction with and the structure of a registry for services. Additional Web-service-related standards have been or are still being developed that address orchestration (e.g., WS-BPEL), security (e.g., WS-Security), reliability (e.g., WS-Reliability and WS-Reliable Messaging), and other aspects of Web services that are important for the commercial applications of

Web services. While XML Web services define a standard mechanism to exchange messages, none of its standards describe the format of the actual payload that is exchanged within the message beyond requiring it to be formatted in XML. Therefore, for successful message exchanges, it is important that the involved parties agree on a payload format (McAfee, 2005). This format may be based on an industry-specific standard, such as ACORD (insurance industry) or HL7 (health care industry), or may be developed as a custom solution for a particular message exchange.[4]

## SAP NetWeaver and the Exchange Infrastructure

The SAP Exchange Infrastructure is part of a suite of products that form the SAP NetWeaver platform. XI serves as an integration broker and business process management engine, which facilitates building a service-oriented architecture based on XML Web services, or in SAP's terminology, an enterprise service architecture (ESA). XI can be leveraged for internal application-to-application data exchanges as well as in business-to-business scenarios. It also facilitates business process management across heterogeneous application components. Interoperability is provided either through the use of open standards (i.e., XML Web services) or specialized adapters, which are available for a wide variety of systems and support proprietary APIs, databases, and messaging systems, such as IBM MQ Series.

The two key components for building an integration solution in XI are the integration repository and the integration directory (see Figure 1). The integration repository contains data type definitions (XML schema), interface definitions (WSDL), process definitions (WS-BPEL), interface and message mappings (XSLT or Java code), and context objects. There is a set of tools that can be leveraged to manipulate these objects. All these definitions are abstract and do

*Figure 1. Building an integration solution in XI*

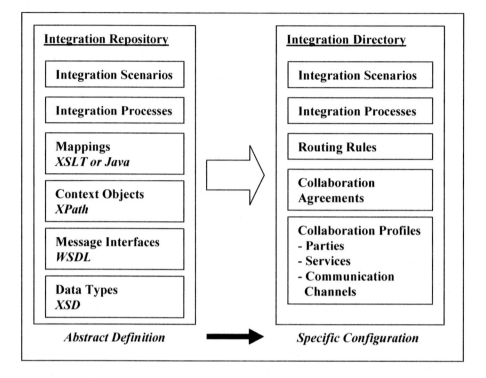

not depend on specific execution-environment configurations.

The integration directory contains integration scenarios and integration processes that leverage objects defined in the integration repository and links them to actual execution environments. Collaboration profiles with parties, services, and communication channels are configured, and routing rules and collaboration agreements are determined here. There is also a set of tools that facilitate manipulating the objects in the integration directory. The purpose of the integration directory is to adapt abstract integration content, as defined in the integration repository, to a specific system configuration in an organization.

The XI run time executes the scenarios and integration processes defined in the integration directory (see Figure 2). It consists of three layers: the adapter engine, which manages the adapters

that may be used to connect to other systems that require proprietary solutions; the integration engine, which handles message routing and transformation; and the business process engine, which executes cross-component business processes.

To interact with another SAP system, XI can directly communicate using the XI protocol or SOAP if the other SAP system is running on a recent version of the SAP Web application server (WAS). Otherwise, either the SAP RFC (remote function call) adapter or IDOC adapter has to be used.

## IMPLEMENTATION OF A PROOF-OF-CONCEPT SOLUTION

MMC decided to implement a proof-of-concept solution to get a better understanding of how XI and

*Figure 2. Executing an integration solution in XI*

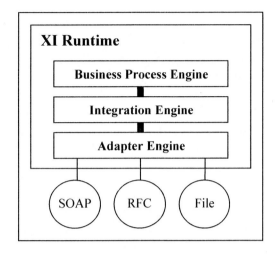

Web services can be leveraged to improve the sales order process. After investigating the technologies discussed previously, we decided to implement and compare three different approaches. The first two approaches involve the integration broker, while the third approach directly leverages the Web services capabilities of the WAS on which the SAP R/3 system is running.

The first approach uses XML Web services to expose the order functionality to the customer. Since the customer systems did not have Web services capabilities, we needed to develop a Web service adapter that could bridge the communication between the customer's legacy system and MMC's Web service. The integration broker handles the SOAP message exchange with the customer, message transformation, and the communication with the back-end SAP R/3 enterprise system.

The second approach uses the file transfer protocol (FTP) instead of Web services to communicate with the customer. Each customer is sending a data file (typically in a comma-separated value file format) to an FTP server located at MMC. The integration broker then detects the presence of a new order file, converts it, and passes it on to the back-end enterprise system using the same mechanisms as in the previous approach.

The third approach excludes the integration broker and involves developing a Web service directly on MMC's enterprise system. As in the first approach, we needed to install a Web service adapter at the customer's location to bridge the gap between the legacy applications and MMC's Web service. Now, however, the Web service adapter communicates directly with the SAP enterprise system.

The data that were used to design and test these three solutions were provided by MMC and included orders from three different customers, here referred to as Customer UK, Customer DE, and Customer FR. These customers were each located in a different country, used different native languages, and ran different legacy systems.

These customers are wholly owned subsidiaries of MMC and operate primarily as aggregators of orders for their respective countries. While each of their legacy systems was different, it was determined that each system had the capability to produce comma-separated files, albeit using different content formats. All three approaches used these files as the starting point for the electronic order process. The following is a more detailed description of each of the three approaches.

### Approach 1: Integration Broker and Web Services

This approach involved MMC's integration broker, in this case XI, and XML Web services to communicate with customers. Since the customers' current legacy systems do not directly support Web services, a Web service adapter had to be developed for each customer. The Web service adapters were implemented using Microsoft .NET. The following provides a step-by-step description of the message exchange between the customer and MMC using this approach (see Figure 3).

1.   The customer-specific Web service adapter detects and parses the flat files written by the legacy system and produces the data

structures necessary for calling MMC's order Web service.

2.   The adapter then sends a SOAP message to the appropriate Web service endpoint on the integration broker, as defined in the WSDL for MMC's order Web service.

3.   MMC's integration broker, SAP XI, receives the incoming SOAP message via its SOAP adapter. The SOAP adapter converts the incoming SOAP message into an XML-based SAP XI protocol message to match the outbound interface structure.

4.   Mappings in XI then transform the payload of the incoming message into the format matching the inbound interface structure, which reflects the data structures required by the receiving SAP R/3 system.

5.   The XI RFC adapter is then utilized to send the message to the SAP R/3 system.

6.   The SAP R/3 system processes the order using a custom function module developed for creating and committing a sales order.

*Figure 3. Approach 1: Integration broker and Web services*

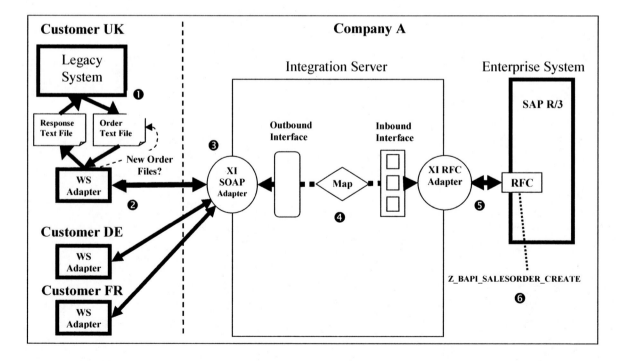

*Figure 4. XI messages*

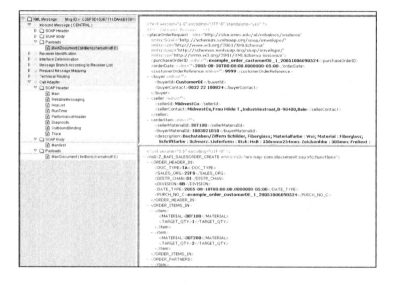

The response from the SAP R/3 system is passed back into XI via the RFC adapter, again transformed, and returned via XI's SOAP adapter to the calling customer's Web service adapter. The Web service adapter writes the response into a file readable by the customer's legacy system.

The SAP NetWeaver environment provides users with a number of tools to assist in the development and administration of the integration solution, including the tracking of messages that flow through the system. Figure 4 depicts an example of an actual message from the customer as it is processed in XI. The screenshots were taken from the message monitoring tool and show the SOAP message from the customer's Web service adapter as well as the message passed on to the R/3 system.

**Approach 2: Integration Broker and File Transfer**

The second approach tries to leverage existing FTP capabilities available in the customers' systems. Instead of using Web services to communicate with MMC's integration broker, the customer system sends files in a customer-specific file format to an FTP server located at MMC. The MMC's integration broker then processes these files through a file adapter. The following provides a step-by-step description of the message exchange between the customer and MMC using this approach.

1.  The customer's legacy system creates a text file containing the order information.

2.  An FTP client at the customer location sends this text file to the FTP server at MMC. Since the files reflect the data structures used in the customer's specific legacy system, they have a different format for each customer. To ensure confidentiality, each customer posts files to a private subdirectory on MMC's FTP server.

3.  MMC's integration broker leverages a file adapter to pull new order files from the FTP server and parses the file into an XML structure corresponding to the outbound interface for the customer. As the file format differs for each customer, a different file adapter and outbound interface must be configured in XI for each customer. The file adapter

*Figure 5. Approach 2: Integration broker and file transfer*

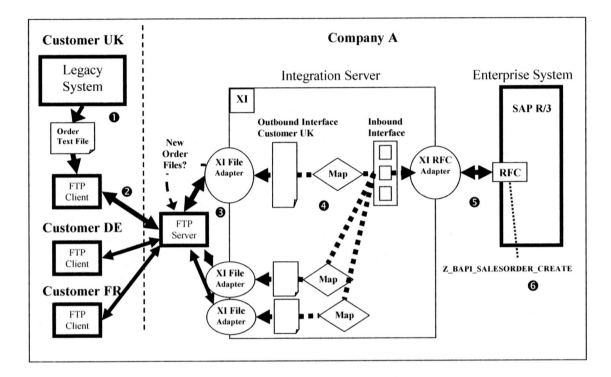

detects new files by polling the FTP server in specified time intervals. Each file adapter has its specific conversion instructions to transform the customer's file format into an XML file representing the customer data.

4. The XML message matching the outbound interface is then mapped to the inbound interface structure using a customer-specific mapping.

5. The message matching the inbound interface structure is then passed to the back-end SAP R/3 system using the RFC adapter.

6. The SAP R/3 system processes the order using a custom function module developed for creating and committing a sales order.

This approach was initially implemented as a one-way exchange. It is possible, however, to process the response from the R/3 system and use XI's asynchronous message processing capabilities to write the message back to the FTP server, where the customer's system, if it possesses the ability, could retrieve the response file and process it further. Figure 5 depicts the details of this solution.

## Approach 3: Using R/3 Web Services

The third approach eliminates the need for an integration broker by directly communicating with the SAP R/3 system using XML Web services. This approach takes advantage of the ability to expose function modules as XML Web services directly from SAP R/3. This is only possible, however, in the more recent version of SAP R/3 running on the SAP WAS.

*Figure 6. Approach 3: Using R/3 Web services*

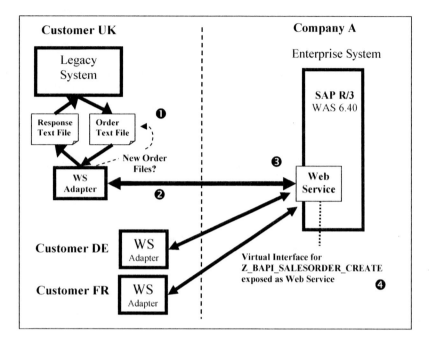

For this approach, it is again necessary to provide the customer with a Web service adapter that can communicate with the customer's legacy system. Instead of sending a SOAP message to the integration broker, the message is directly targeted at Web service interfaces on the SAP R/3 system. The following is a step-by-step description of each stage in this solution (see Figure 6).

1. The customer-specific Web service adapter detects and parses the flat files written by the legacy system and produces the data structures necessary for calling MMC's order Web service.

2. The customer's Web service adapter then sends a SOAP message to the appropriate Web service endpoint as defined in the WSDL for MMC's order Web service, which is now located directly on the Web applica-

tion server hosting the R/3 system. The Web service interface is different from the one used in Approach 1 since this interface now directly reflects the data structure used in the SAP R/3 function module.

3. The SOAP message is received by the Web service on the SAP WAS at MMC that hosts the SAP R/3 system.

4. Through a virtual interface defined in the SAP R/3 system, the corresponding remote-enabled function module is invoked and the order is processed in SAP R/3.

The result returned from the function module is passed back to the Web service via the virtual interface. The WAS handles the creation and sending of the SOAP response message back to the Web service adapter at the customer site. The Web service adapter writes the response into a file readable by the customer's legacy system.

To create a Web service on the WAS that hosts the R/3 system, it is necessary to first define a so-called virtual interface for an existing function module in R/3. The second step is to provide the appropriate configuration that exposes this virtual interface as a Web service on the WAS. For SAP WAS Version 6.40 and up, the development tools provide wizard support to create the virtual interface and the Web service. In previous versions of the WAS, more extensive manual configuration and programming are required to realize the Web service interface.

## EVALUATION

After completing a proof-of-concept implementation of the three approaches, we evaluated the relative advantages and disadvantages for each of the approaches. For both approaches involving the integration broker, it is necessary to install and correctly configure XI. While this is a substantial one-time effort, we focus our evaluation of the relative advantages of each approach on the implementation of a solution given a working XI system as this aspect has larger long-term business implications. Table 1 summarizes the evaluation of the three approaches.

### Approach 1: Integration Broker and Web Services

The current customer systems do not have Web services capabilities. Consequently, for this approach MMC would be required to develop and maintain a Web service adapter for each of its customers. While the adapter would be running on the customer's site, it is MMC's IT development and maintenance burden to support a greater

*Table 1. Evaluation of the three approaches*

|  | **Approach 1: Integration Broker and Web Services** | **Approach 2: Integration Broker and File Transfer** | **Approach 3: Using R/3 Web Services** |
|---|---|---|---|
| **MMC System Effort** | Low: One common interface and mapping | High: A separate file adapter and outbound interface and mapping for each customer | Low: Wizard tool creates Web service interface (WAS 6.40) |
| **Customer System Effort** | High: Different Web service adapter for each customer location; Needs to be maintained and managed by MMC | Low: Leverages existing FTP capabilities; No additional software component needed | High: Different Web service adapter for each customer location; Needs to be maintained and managed by MMC |
| **Flexibility** | High: Leverages XI capabilities such as flexible interface design, routing, and mapping | High: Leverages XI capabilities such as flexible interface design, routing, and mapping; Leverages built-in XI file parsing capability | Low: Few interface design, routing, or mapping capabilities; Only direct point-to-point connections |
| **Locale of Integration Processing** | Both on customer and MMC side | On MMC side only | Both on customer and MMC side |

number of different IT platforms for larger numbers of MMC customers. In our case, we only developed a Web service adapter for the .NET platform. There is the possibility in the future that customer systems will directly support Web services, thus eliminating the need for MMC to develop a Web service adapter.

The implementation effort on the integration broker is relatively low. With only one customer-facing interface, only one mapping needs to be developed as opposed to one for each customer in Approach 2. Furthermore, the integration broker provides flexibility. This includes flexible routing and message mapping. The integration broker allows MMC to decouple the customer-facing outbound interface from the data structures required in the back-end SAP R/3 system. This translates into a less complex interface for the customer, focusing only on the necessary data items. It also provides the opportunity to make changes in the back-end system without requiring changes for the customer. Changing the function module in SAP R/3 to process the orders, or moving to a different SAP system and using a different interface (i.e., using the XI protocol rather than the RFC adapter) will only require the definition of a new inbound interface and adjustments to one mapping. The integration broker also provides multiple options for securing and monitoring the message exchange. This provides MMC with technical flexibility while maintaining business continuity with the customer.

On the other hand, the Web service interface can be extended with relative ease to provide additional business functionality without disrupting the existing one. In our case, this could include added functionality that allows a customer to inquire about order status or to obtain a list of recent orders that were placed by the customer. Thus, the flexibility of the integration broker in combination with the use of Web services also contributes to increased business agility (Sambamurthy, Bharadwaj, & Grover, 2003).

In summary, although this approach provides high flexibility and scalability by leveraging the capabilities of XI and Web services, and requires a low implementation effort on MMC's integration broker, it also constitutes a substantial development effort to enable customers to use the Web service. Furthermore, the responsibility of MMC for integration processing at its own site as well as its customer's site poses additional staffing and management challenges.

**Approach 2: Integration Broker and File Transfer**

The key differences between the second approach and the first approach are a concern for the need to develop and support integration processing (e.g., Web service adapter) on the customer's site and the number of interfaces and mappings on MMC's integration broker. In this approach, the customer leverages its existing capability to transfer files to MMC using FTP. This eliminates the need for MMC to develop and maintain any integration processing (i.e., Web service adapter) on the customer's site. To facilitate communication with the customer, MMC only needs to set up a secure FTP server, which poses only a marginal effort.

More processing needs to be done in Approach 2 within XI since the order information arrives at the integration broker in a customer-specific format. The consequence is the need to implement and maintain a file adapter, an outbound interface, and a specific mapping to the back-end interface for each customer. The XI file adapter provides tools that facilitate the parsing of comma-separated value format files, and mappings are supported with graphical mapping tools. A large number of customers, however, may constitute a noticeable IT development and maintenance burden. This solution is particularly problematic in cases where customer interactions are ad hoc and short lived. The advantage of this approach is that all development integration processing occurs in XI at MMC, thus avoiding some of the potential management and staffing issues as noted in Approach 1. In comparison to writing a Web service adapter, configuring a file adapter requires lesser effort.

In summary, this approach also provides high flexibility and scalability by leveraging capabilities of XI. From a customer's perspective, Approach 2 is less invasive since no additional software components are needed and it leverages existing IT capabilities. On the other hand, more work has to be done in MMC's integration broker, but it eliminates the need for developing a Web service adapter and it also allows MMC more direct and central control in comparison to Approach 1.

**Approach 3: Using R/3 Web Services**

Approach 3 eliminates the need for an integration broker and it constitutes a point-to-point integration solution between the customer and MMC. Similar to Approach 1, a Web service adapter is necessary, with similar implications for IT development, maintenance, management, and staffing for MMC. This solution has the least amount of flexibility in terms of routing, mapping, and interface abstraction. As a consequence, the Web service interface is more complex as it reflects the interface defined in the SAP R/3 system. If an appropriate function module exists in R/3, a Web service interface can be quickly established using the wizard tool provided by SAP. In our particular example, however, we found that the WSDL created by the SAP tool for this interface was not entirely compliant with the standard and had to be manually adjusted to work with the .NET-based Web service adapter.

In summary, this approach is the least flexible and scalable solution but perhaps the quickest to implement, particularly if an XI integration broker has not yet been implemented and configured. This approach is more invasive from a customer perspective than Approach 2 and faces the same potential management and staffing issues as encountered in Approach 1 due to MMC's responsi-

bility for integration processing on the customer's site. This approach becomes substantially more difficult for older versions of the SAP WAS (< 6.40) as they lack the wizard support for creating the Web service interface.

## DISCUSSION

The three approaches that were examined in the proof-of-concept implementation provide MMC with a number of options to go forward and implement a solution to improve the order process. The advantages and disadvantages of each approach were outlined above. This section details the integration issues encountered during the development of the three approaches as well as the key lessons learned that apply to integration scenarios involving Web services and integration brokers in general.

First of all, it is important that MMC assesses its specific situation and considers its strategic goals before making a decision (Swanson & Ramiller, 2004). Since MMC's SAP R/3 system used for production is not a current version that supports exposing Web services directly, Approach 3 can only be considered after an upgrade of the SAP R/3 back-end system. This and some of the other disadvantages associated with using direct Web services make it likely that MMC will choose a solution involving an integration broker. The choice between Approach 1 and Approach 2 hinges largely on the ability of its customers to use Web services. This is an issue that any organization using Web services for B2B interactions needs to consider. In MMC's case, customer systems cannot currently leverage Web services, but this may change in the future. Thus, MMC needs to look ahead and make a prediction about the development of its customers in terms of their integration capabilities, especially related to Web services. This also encompasses tracking and understanding the development of standards related to Web services and their adoption. Independent of the integration solution vendor, it is important that MMC, or any other organization, understands which standards are supported in a particular integration product.

While MMC may go ahead with a single approach, it is quite reasonable to consider a combination of a Web-services- and a file-transfer-based solution. The flexibility and the tool support of the SAP XI integration broker make this a feasible solution as it is a relatively small effort to provide the additional Web service interface along with the interfaces needed for file processing. Multichannel access to information resources is a valuable feature of integration brokers in general. If Web services become a ubiquitous integration mechanism in future business applications (Hagel & Brown, 2001), MMC and its customers could benefit from the file transfer capabilities in the short term, and in the long term transition to Web services interactions once its customers' systems are Web service enabled.

Furthermore, the decision about whether XI should be obtained as an integration broker may not be influenced solely by considerations regarding the integration of external systems with the SAP enterprise system. This decision will also be driven by requirements for having XI available for internal SAP-to-SAP module integration. Already in the current release of SAP NetWeaver, XI is required for communication between some SAP products, and statements by SAP indicate that this requirement is likely to affect a larger set of components in future releases.

The experience of implementing the proof-of-concept solution for MMC showed that XI provides a manageable, flexible, and scalable infrastructure for integration that supports a number of integration mechanisms, including Web services and file transfer. The XI integration builder tool facilitated the definition of Web services interfaces and development of file conversions. Mappings can be visually developed with a graphical mapping tool or can be provided as XSLT or Java code.

The monitoring tools allow end-to-end tracking and examination of messages throughout the processing in the adapters and the integration engine. Setting up XI, however, is a substantial effort, and our experience emphasizes the need for experienced staff to configure XI correctly. Furthermore, our experience revealed that support for Web services is immature. For example, we needed to manually change the Web services endpoint in the WSDL documents generated by the Web service definition wizard. In addition, the WSDL generated for our scenario in Approach 3 was not WS-I compliant and needed further manual adjustments to work with the .NET Web service adapter.[5]

## CONCLUSION

As of the time this chapter was finalized, the approaches and implementations examined in our research remained proof-of-concepts while MMC was weighing its options. Our prototype implementation, however, demonstrates the opportunities for improving B2B transactions using a standards-based integration platform and clarifies the technical options that can be leveraged in MMC or other organizations facing similar B2B exchange problems and a similar IT environment. While no ultimate conclusion can be reported for this case, we believe that it provides practitioners with relevant insights in the ability of modern integration technology to improve B2B interactions and enhance competitive advantages by enabling a broader group of businesses to participate. In addition, the technical details and integration methodology principles presented in this case may be used as points of reference for other scenarios in which the same or a similar set of technologies is applied to advance the efficiency and effectiveness of B2B transactions.

# REFERENCES

Brynjolfsson, E., Yu, H., & Smith, M. D. (2003). Consumer surplus in the digital economy: Estimating the value of increased product variety at online booksellers. *Management Science, 49*(11), 1580-1596.

Clemons, E. K., & Row, M. C. (1988). McKesson Drug Company. A case study of Economist: A strategic information system. *Journal of Management Information Systems, 5*(1), 36-50.

Hagel, J., & Brown, J. S. (2001). Your next IT strategy. *Harvard Business Review, 79*(9), 105-113.

Havenstein, H. (2005). Sabre replacing EDI with Web services. *Computerworld, 39*(34), 12.

Iacovou, C. L., Benbasat, I., & Dexter, A. S. (1995). Electronic data interchange and small organizations: Adoption and impact of technology. *MIS Quarterly, 19*(4), 465-485.

McAfee, A. (2005). Will Web services really transform collaboration? *MIT Sloan Management Review, 46*(2), 78-84.

Meadows, B., & Seaburg, L. (2004, September 15). *Universal business language 1.0.* OASIS. Retrieved February 14, 2006, from http://docs.oasis-open.org/ubl/cd-UBL-1.0/

Mitra, N. (2003, June 24). *SOAP version 1.2 part 0: Primer.* W3C. Retrieved June 10, 2004, from http://www.w3.org/TR/2003/REC-soap12-part0-20030624/

Sambamurthy, V., Bharadwaj, A., & Grover, V. (2003). Shaping agility through digital options: Reconceptualizing the role of information technology in contemporary firms. *MIS Quarterly, 27*(2), 237-263.

Swanson, E. B., & Ramiller, N. C. (2004). Innovating mindfully with information technology. *MIS Quarterly, 28*(4), 553-583.

Teo, H. H., Wei, K. K., & Benbasat, I. (2003). Predicting intention to adopt interorganizational linkages: An institutional perspective. *MIS Quarterly, 27*(1), 19-49.

Wigand, R. T., Steinfield, C. W., & Markus, M. L. (2005). Information technology standards choices and industry structure outcomes: The case of the U.S. home mortgage industry. *Journal of Management Information Systems, 22*(2), 165-191.

World Wide Web Consortium (W3C). (2005). *Extensible markup language (XML) activity statement.* Retrieved March 10, 2006, from http://www.w3.org/XML/Activity.html

# ENDNOTES

[1] This research was supported in part by a grant by SAP Americas.

[2] Midwest Manufacturing Company is an alias name used to disguise the identity of the company.

[3] SOAP used to stand for simple object access protocol, but according to the current SOAP 1.2 specification, the acronym SOAP is no longer spelled out (Mitra, 2003).

[4] There are also efforts to develop a universal business language (UBL; Meadows & Seaburg, 2004) for core business documents, such as purchase orders.

[5] This assessment was made using the WS-I test in the Mindreef SOAPscope tool and was also corroborated by SAP OSS notes and articles found on the SAP Software Developer Network Web pages.

# Chapter VII
# The Changing Nature of Business Process Modeling:
## Implications for Enterprise Systems Integration

**Brian H. Cameron**
*The Pennsylvania State University, USA*

## ABSTRACT

*Business process modeling (BPM) is a topic that is generating much interest in the information technology industry today. Business analysts, process designers, system architects, software engineers, and systems consultants must understand the foundational concepts behind BPM and evolving modeling standards and technologies that have the potential to dramatically change the nature of phases of the systems development life cycle (SDLC). Pareto's 80/20 rule, as applied to the SDLC, is in the process of being drastically altered. In the past, approximately 20% of the SDLC was spent on analysis and design activities with the remaining 80% spent on systems development and implementation (Weske, Goesmann, Holten, & Striemer, 1999). Today, with the introduction of the Business Process Management Initiative (BPMI), Web services, and the services-oriented architecture (SOA), the enterprise SDLC paradigm is poised for a dramatic shift. In this new paradigm, approximately 80% of the SDLC is spent on analysis and design activities with the remaining 20% spent on systems development and implementation. Once referred to as process or workflow automation, BPM has evolved into a suite of interrelated components for systems analysis, design, and development. Emerging BPM standards and technologies will be the primary vehicles by which current systems portfolios transition to Web services and service-oriented architectures (Aversano, & Canfora, 2002). The Business Process Management Initiative's business process modeling notation (BPMN) subgroup is currently finalizing a standardized notation for business process modeling. Although the notation is still in working-draft format, system architects and designers should consider incorporating the concepts of BPM into their current and future systems analysis and design procedures.*

## Background

Adaptive organizations want to be able to rapidly modify their business processes to changes in their business climate including competitive, market, economic, industry, regulatory and compliance, or other factors. Meanwhile, enterprise architects within IT organizations have long dreamed of a repository for models that are interconnected and extend to support application delivery. No single tool exists that enables enterprise architects to connect the dots between high-level models geared toward a business audience and executable code to instantiate the vision (Carlis & Maguire, 2000).

Business process modeling (BPM) is both a business concept and an emerging technology. The concept is to establish goals, define a strategy, and set objectives for improving particular operational processes that have significant impact on corporate performance. It does not imply reengineering all business processes; rather, the focus is on business processes that directly affect some metric of corporate success. Business performance management and measurement emphasize using metrics beyond financial ones to guide business process management strategies (Delphi, 2001). Metrics related to customer value or loyalty are examples. Business process modeling is becoming the central point of organization for many systems. BPM as a concept is not new; multiple process management methodologies such as six sigma and lean manufacturing have existed for years. However,

new BPM technologies are fueling a renewed interest in process thinking (Ettlinger, 2002). New BPM technologies promise business modelers and managers a visual dashboard to manage and adjust, in real time, human and machine resources, as well as information being consumed as work progresses.

The business and IT worlds are taking more strategic and holistic views of IT and how it supports the business. IT strategy, business process improvement, and IT architecture are experiencing a renaissance. Enterprise architects have tackled the technical architecture effectively. Now, enterprise architects are looking to expand their efforts into the business architecture space. Enterprise business architecture (EBA) is the expression of the enterprise's key business strategies and their impact on business functions and processes (Adhikari, 2002b). Business architecture efforts in most organizations are limited to thematic project-level initiatives. Thematic business architecture artifacts generally fail to evolve once the projects are complete because little perceived value exists for keeping business architecture content alive. However, emerging standards show promise in keeping business architecture and associated artifacts alive to serve as key business strategy enablers.

The IT world is moving more toward a model of integrating pieces or components vs. building from scratch (Adhikari, 2002a). Organizations are looking to strategically optimize, automate, and integrate key processes to provide seamless service to more demanding customers in a multichannel world. To do this effectively, systems must be integrated at the process level as well as the data level. Integrating systems at the process level has been a challenge that, when unmet, leads to data duplication and inconsistency, and functional overlap (i.e., inefficient processes and processing; Reingruber & Gregory, 1994). Many companies are embarking on process improvement initiatives in hopes of increasing the efficiency or effectiveness of the organization.

Process improvement initiatives go by many names, including ISO certification, enterprise business architecture, business process improvement, six sigma, business process reengineering, and lean thinking, to name a few. Most of these initiatives utilize visual modeling techniques to understand current-state and future-state design. Several accepted standards exist for visual modeling (e.g., integrated definition [IDEF], American National Standards Institute [ANSI], event-driven process chains [EPCs]). None of these visual process modeling standards offers a seamless extension of visual models into executable script or source code. In response to this shortcoming, a new modeling standard (business process modeling notation, BPMN) is emerging that promises this seamless model to code integration (*BPMI.org*, 2005). This model to execute the goal is reminiscent of the failed promises of computer-aided software engineering (CASE) tools. Advances in the fields of computer science and mathematics are now making this goal attainable. BPMN offers a mechanism for true business- and IT-process modeling convergence. Notations and methods for process improvement and modeling within the business are mature, as are the notations and methods for application development and delivery within IT. However, the current business- and IT-process models and methods fail to align well with one another (Adhikari, 2002b).

BPMN is mapped to the business process modeling language (BPML) and business process execution language for Web services (BPEL4WS),

thus process models created with BPMN will automate much more rapidly than is currently possible with any other modeling notation (Ewalt, 2002). The business process query language (BPQL) has been introduced by the Business Process Management Initiative (BPMI) to query process repositories, similar to the manner in which SQL is used to query data repositories. Several concepts presented in BPMN should be considered for immediate adoption by system designers and integrators. First is the agreement on a standard notation and principle set for process modeling. Many modeling tools are incorporating support for BPMN, BPEL4WS, and BPML.

Technical modeling standards (e.g., entity relationship diagrams, unified modeling language) are being coupled with more business-oriented modeling approaches (e.g., BPMN, swimlane diagrams) to converge business and IT modeling. The goal of these efforts is to incorporate a set of standardized guidelines into all internal and external modeling initiatives. With increased attention to process integration with supply chain partners, organizations are quickly moving to a standardized approach for process modeling to support intra-organizational and interorganizational development and integration initiatives.

Software tools have long been used to automate tasks and reduce manual interactions. Thus, it is important to distinguish business processes from IT applications. A business process is a sequence of activities performed by people and machines necessary to produce a desired result (Adhikari, 2002b). It is initiated by the arrival of work (such as a phone call, faxed order, or timing event) that triggers the sequence of operational activities. An application is a logical grouping of tasks automated by a computer with the objective of reducing or augmenting human interactions. Thus, an application is a subset of the tasks of a business process. Many human-centric activities of a business process have largely been unautomated (though some organizations use workflow automation to manage electronic work queues).

A business process management suite (BPMS) is a new development environment that enables business users to collaborate with IT professionals in the design and development of optimized business processes (not applications), thus reducing the communication gap between business and IT. Ideally, a business process management suite supports a business process modeling environment that is shared by business analysts, process engineers, IT architects, and programmers. The modeling surface or palette exposes different capabilities to support each of these roles. Unlike earlier code-generating tools, the modeling environment creates XML (extensible markup language) metadata that describes how application functionality and information, and human activities should interact and be instantiated at run time.

The new run-time engines interpret the metadata at run time (Mangan & Sadiq, 2002). Increasingly, vendors are endorsing the business process execution language (BPEL) as the standard process description language. Thus, changing a business process requires changing the graphical model, regenerating the metadata, and redeploying process instances. This is a much simpler and faster approach to changing how work gets done. This faster rate of change to operational best practices and the ability to change activities dynamically are what make an organization adaptive. An adaptive organization can adjust its operational business processes in near real time to capitalize on opportunities, avoid threats, and maximize corporate performance.

## MODELING MADNESS: A BRIEF HISTORY

System models have long been a part of the systems analysis and design process, traditionally leveraging ANSI standard flowcharts to communicate logic flow. Modeling tools have evolved in response to the changing needs of business and advances in technology. Spreadsheets were (and still are) the primary modeling tool for many organizations. As a result, in many organizations, modeling has become a series of spreadsheets with what-if scenarios and is synonymous with financial modeling. In the IT world, modeling has become synonymous with a variety of visual system modeling techniques, which include the following.

- Entity relationship diagrams (ERDs)
- ANSI standard flowcharts
- Process models
- Data flow diagrams
- Unified modeling language (UML) diagrams
- Network diagrams
- CRUD (create, read, update, and delete) matrices
- IDEF charts
- Engineering process control (EPC) charts

In both the business and IT worlds, models have been used as mechanisms to describe complex ecosystems. IT-oriented modeling has historically focused on logic flow and data elements while business-oriented modeling has focused on quality and productivity metrics (e.g., cycle time, setup time, processing time, cost, and defect rates). Differences in modeling standards adopted by organizations have suboptimized the benefits of visual modeling, especially in interorganizational modeling efforts. A group of organizations, with strong interests in modeling, have joined forces to leverage the collective best practices of the member organizations to develop a common notation for business process modeling (BPMN) in the hope of unifying IT and business process modeling (*BPMI.org*, 2005).

Many of these best practices can and should be incorporated into the organizational modeling practice immediately. There should be an expectation that visual business models help expose characteristics of the underlying business process. For example, a business process related to entering purchase orders should be able to expose the underlying user interface for all data entry points of the process, as well as expose the underlying data tables related to the transaction. At a higher level of abstraction, the business models should be linked to entity relationship diagrams as well, which highlight the overall table structure.

Swim-lane diagrams are a process modeling technique that was popularized in the 1990s. This modeling technique graphically denotes responsibility for work with vertical or horizontal lanes (lines of demarcation) and is almost universally understood and accepted among process modelers. BPMN utilizes the concept of pools, which can contain lanes, to relate external entities. This functionality is extremely important in a world where the linking and automating of processes between customers and suppliers is essential (Aversano & Canfora, 2002). Incorporating pools and lanes enables responsibility to be assigned and recognized throughout the supply chain. The concepts of lanes and pools are important components in visual modeling that clearly bound business areas and define ownership and responsibility.

UML is maintaining its popularity as the modeling language of choice for software development. The Object Management Group's (OMG) model-driven architecture (MDA) is gaining traction, with the release of UML 2.0. However, UML was originally designed by technical people for techni-

cal people. OMG recognizes that UML does not provide a facility that actively addresses the needs of IT business modelers (Sadiq, Orlowska, Sadiq, & Foulger, 2004). OMG is currently considering BPMI's BPMN to fill this void. A merger, alliance, or convergence of the modeling standards efforts of OMG with those of BPMI would be a major advance for enterprise architecture. Joining BPMN with UML would do much to minimize redundant questioning by IT organizations during the system development life cycle (SDLC).

One of the strengths of the BPMN standard is its capability to support more ambiguity in the modeling environment. Events have traditionally been undermodeled by business process modeling teams, typically focusing on process triggers and end states. BPMN distinguishes between beginning, intermediate, and ending events, as well as event start and interruption. With prior modeling techniques, intermediate event information is typically buried in text annotations or symbol properties, essentially useless to the process modeler (*BPMI.org*, 2005). The BPMN standard contains several event types that are more comprehensive than most, if not all, other modeling standards. For example, exceptions are an event type that exists in the real world and has posed a great challenge to modelers. Earlier modeling notations are not able to adequately represent and handle exceptions. In order to address this shortcoming, BPMN was developed with a variety of event symbols related to exception handling.

Process modeling is much more than simply drawing business processes. Modeling requires adherence to standards when graphically depicting a process ecosystem, thereby allowing process models to be used for automation, simulation, integration, and refinement of existing processes (Smith & Fingar, 2003a). Process modelers typically have much discretion in modeling process flows, which can lead to inconsistent modeling practices with prior modeling standards. The

BPMN standard is simple and flexible, and enables consistent modeling of joins, forks, and decisions by formalizing process flow gates. For these reasons, BPMN is gaining wide acceptance in the business and IT worlds.

## BPMN AND UML

UML, conceived in 1996, was designed to provide a common modeling language to support software development. It was adopted and refined by OMG and 21 member companies. Since its inception, it has become the de facto standard for visual software development, spawning a host of new products to increase developer productivity and augmenting the functionality of existing modeling tools. As UML and XMI (i.e., XML metadata interchange, a standard enabling UML model interchange) advanced, interoperability was made practical among tools traditionally used for business process modeling and enterprise architecture, and those used for visual development environments (Carlson, 2001). Further advances in UML enabled modeling of more complex software systems.

UML, however, failed to provide the simplicity needed by non-IT business modelers. At best, a UML activity diagram could be created with swim lanes to provide the functional equivalence of swim-lane diagrams; however, adhering to UML standards might be overly burdensome to business modelers. OMG recognizes this and is currently moving up the stack to provide the appropriate level of standardized modeling language to enable business modelers to perform their needed tasks without undue effort. BPMN has been mapped to the UML standard (Torchiano & Bruno, 2003).

BPMN was developed to provide a common modeling notation to support a common process modeling language. At the outset, BPMN was designed to appeal to both IT and non-IT modelers of business processes. BPMN is gaining rapid acceptance; leading modeling-tool vendors support or plan to support BPMN (Smith & Fingar, 2004). This feat is accomplished by providing simple swim-lane diagramming ability to non-IT modelers and enabling the subsequent addition of technical detail to the same models as notational properties (i.e., the gory details are hidden from the nontechnical audiences within the tool).

If BPMN is accepted into UML, even in part, BPMI's efforts will be further legitimized. At present, much flux exists in the process modeling language space. Two languages, BPML and BPEL4WS, were viewed as competitors running a head-and-neck race for market acceptance. However, the broader vision of BPMI is to have a common modeling notation that generates a process modeling language that can be stored in a process management system, used to manage automated process change within and between organizations. If OMG adopts BPMN, even in part, the likelihood of BPMN as the notational support for BPEL4WS dramatically increases.

Long has been the vision of a modeled enterprise that maintains linkages from the conceptual business leaders to the operations. However, due to lack of standards and insufficient tool support, this vision has remained out of sight until now. As developers adopt service-oriented architectures (SOAs), they will continue to rely on UML. As enterprise architecture continues to evolve to include enterprise business architecture, enterprise architecture modeling will expand to include business modeling. If BPMN is incorporated into UML, simple business process models can rapidly be extended to generate BPML, BPEL4WS, or additional UML diagrams to extend down the MDA stack (Frankel, 2003). The vision of an enterprise repository of accurate models becomes reality as

UML and MDA enable automation with insurance against technology obsolescence.

## A NEW PARADIGM FOR PROCESS MODELING

With the introduction of BPMI, Web services, and SOA, the enterprise SDLC paradigm is poised for a dramatic shift. An SOA is a dynamic, general-purpose, extensible, federated interoperability architecture. Designing for SOA involves thinking of the parts of a given system as a set of relatively autonomous services, each of which is (potentially) independently managed and implemented, that are linked together with a set of agreements and protocols into a federated structure (Clark, Fletcher, Hanson, Irani, & Thelin, 2002). For enterprise application architects and developers, composite applications will be based on the SOA principle of dynamic, extensible, and federated interoperability and will be enabled by XML-based technologies such as Web services (Shegalov, Gillmann, & Weikum, 2001).

Mobile calculi, process algebras, and pi calculus are forms of mathematics that are of great importance to lines of business, IT professionals, and business architects. Process algebra and pi calculus are ushering in a new wave of process modeling and management systems. Pi-calculus-based systems are maturing and becoming the new wave of software infrastructure to manage business processes (Smith & Fingar, 2003b). These systems are being front-ended by BPMN. Enterprise business architects are embracing BPMN because it brings life and longevity to their modeling exercises.

Pi calculus systems are becoming prevalent, supported by BPMN. With the advent of third-generation languages (3GLs), application developers adopted a paradigm that was procedural and deterministic. This paradigm can be expressed with a form of mathematics known as lambda calculus.

3GLs tended to be sequential in nature, lending themselves to rigid codification of business activity. Pi calculus enables systems to more closely resemble real life by providing an environment that is parallel, nondeterministic, and tolerant of change and ambiguity. Pi calculus enables systems to treat things as processes and relationships. It changes the programming paradigm from procedural and deterministic to one of interrelated processes (Smith & Fingar, 2003b).

This shift in thinking enables relationships and processes to be automated in a manner that more closely resembles natural workflow (and in fact, how computers work at the machine level). Ironically, computers, at the machine level, are well-suited for a pi calculus paradigm; machine language, which manipulates operands within a computer's memory location based on the operator selected, atomically resembles a pi calculus paradigm (Smith & Fingar, 2004). BPML is a pi-calculus-based standard text language to define and automate processes developed by the Business Process Management Initiative. BPEL4WS is a pi-calculus-based standard language for automating business processes as Web services. This standard was jointly developed by Microsoft and IBM.

EBA is a creative process of future-state design of business processes, functions, and organizations. Business architects rely heavily on visual modeling to refine future-state design. Selected future-state business design must be implemented, which usually includes a certain degree of automation. One of the challenges of EBA has always been keeping models alive through the system development life cycle (including maintenance and enhancement; Smith, 2003). However, with BPMN, keeping the models current is critical to keeping the underlying BPMS functioning. Also, with BPMN, the underlying pi supports simulation, an art returning to popularity within businesses.

The BPMI formed a subgroup to develop a notation to front-end its pi-calculus-based language, BPML. The BPMN subgroup recognized three important things. First, modeling notations needed to exist that were interpretable to business audiences and could subsequently be used by solution delivery specialists; the BPMN subgroup chose a tiered approach, with the first tier resembling swim-lane diagrams. Second, a more detailed visual modeling notation needed to exist for system delivery; BPMN's tiered approach maps the swim-lane-like business process modeling notation to a more precise technical notation. Third, the BPMN subgroup recognized that BPML and BPEL4WS could merge, converge, or coexist, or BPML could be replaced by BPEL4WS; the BPMI proactively chose to map its visual notation to both BPML and BPEL4WS. Business architects and technologists can design models with the top-tier notation with business leaders, easily passing the baton to delivery specialists after a decision to proceed is made with little or no rework (Delphi, 2001). BPMN enables changing of the visual notation per audience without losing the requisite values, properties, or attributes necessary to extend to executable process language. This critical functionality was lacking or less robust in previous attempts to standardize on a notation-to-execute modeling approach (Smith & Fingar, 2003a).

The shift from the use of business process automation technology as a departmental workflow and process tool to that of a facilitator of enterprise change requires a planned approach for implementing processes that are aligned with enterprise business strategy. Utilizing this new paradigm and associated technologies, organizations will be able to effectively orchestrate and execute these processes and move beyond simple automation to business process optimization. Process automation, workflow, and business process management technologies will evolve into fully integrated BPMSs in the near future. BPM

technology suites will be the primary vehicles by which current systems portfolios are transitioned to service-oriented architectures (Smith & Fingar, 2004).

## PROCESS MODELING CHALLENGES

Faced with a rapidly changing business climate, many organizations are seeking new methods to enable rapid systems development and modification. The promise of universal connectivity offered by Web services coupled with the promise of technology-neutral systems offered by model-driven development offers a compelling vision of the future of systems design, development, and integration. This not-so-distant future will be process oriented, with systems designed and created by using a process model to direct the interaction of various systems and human actors. These systems will have their functions exposed as services to achieve maximum accessibility. The process engine technology will capture the semantics of the business process at various levels (O'riordan, 2002).

Currently, most organizations use business process automation in discrete areas, primarily as part of their integration environments. Leading organizations will enable process development across organizational boundaries, but most will struggle with the business (rather than technological) issues associated with these activities. The process model is becoming a standard part of the developer tool kit, and standards-based engines will shortly replace proprietary ones. Many organizations are beginning to make advances toward this future vision, particularly those organizations that are adopting service-oriented architectures. The issue of how to deal with the many challenges of the interim states and advancing this vision incrementally presents a great challenge to the IT organization.

The first roadblock organizations encounter on the path toward business process management is its basic value proposition. Although many projects using BPM tools have shown significant value (in terms of return on investment [ROI]), much of that value comes from the automation of manual processes. These processes, typically those that span organizational boundaries, involve manual touches to information and manual decision making that are not tracked through the formal automation systems embodied in the enterprise applications. The simplest means of creating value in this circumstance is to use process automation to automate these manual processes and their interfaces to enterprise applications and systems (Sharp & Mcdermott, 2001). This is useful since it creates automation for manual tasks and often will significantly reduce the costs and errors inherent in those tasks. However, the scope of this automation is clearly not the complete process, but only that portion of it that is reflected in these manual steps.

Even in this limited-scope case, there can be significant challenges. The manual processes often are developed to address exceptions and inconsistencies in the formalized processes embodied in the enterprise application portfolio. These processes often are either incomplete as implemented, or the business activities have changed yet the process assumptions on which the enterprise systems are based have not. In either case, one finds a situation where the manual parts of the process often are undefined or do not have clear rules, roles, or responsibilities. They may even contradict logic that is embodied in the formal systems. The BPM exercise will highlight the weaknesses for interaction in the existing model (Ettlinger, 2002). Furthermore, many of the manual parts of the process will retain manual components since they primarily focus on exception handling.

Compounding this issue is the challenge of the ownership and stewardship of business processes. Unless an organization has taken an aggressive, process-oriented view toward its business, there often is a tremendous amount of confusion about process definition and ownership (Adhikari, 2002b). Because the processes whose automation can provide the most value often span functional areas, there are usually no individuals with responsibility for the overall process. Instead, we have functional managers, each with individual responsibility for the subprocess performed by their areas. Effectively automating these processes requires the creation of new channels of communication as well as new decision processes to enable the organizations involved to reach an agreement on how to handle the processes.

Another issue affecting the achievement of this vision is that Web services are not complete. Although substantial progress is being made regarding the standards for and interoperability of Web services, the practical use of Web services within corporations is in its early stages (Charfi & Mezini, 2004). The universal connectivity that is required to link a process execution engine with the various actors and systems in the environment requires a substantial investment in integration technologies, and the minimal integration among various components in the infrastructure demands a substantial investment in the software platform. There is little doubt that creating business process models, deriving application code from those business models, and monitoring such models can create the right level of organizational linkage between IT and business executives. The problem is that much of this is an afterthought in a world that contains a collection of ERR best in class, and legacy applications that are either not modeled or modeled with multiple tools.

## CONCLUSION

Business process modeling is quickly becoming the central point of organization for many systems. Model-driven development frameworks are quickly becoming the preferred platform for service-oriented architecture, and Web services composite application design, development, and deployment (Smith & Fingar, 2004). Business-user-oriented modeling environments enable businesspeople to model the process as if it were an assembly line of swappable components (tasks) that can be reordered to achieve various results. A business process management suite exposes these components as XML Web services. BPMN will offer a common standard for enterprise process modeling that has the potential to make the model-to-execute vision more viable. Business analysts, process designers, system architects, software engineers, and systems consultants that utilize process modeling should begin their evaluation and adoption of emerging business process management suites that utilize BPMN.

These IT professionals must understand foundational BPM concepts as well as the emerging technologies and standards that enable these concepts. With this foundation, they will be able to better assess the importance of business agility, to better perform business process performance analysis, and to model business processes and systems in a manner that expedites development. More than technology, becoming an adaptive organization requires leadership and change management. The cross-boundary characteristic of BPM initiatives creates a unique opportunity for the leaders of the initiatives to become the enterprise change agents.

Corporations are beginning the transformation to a process culture and the implementation of business process management methodologies and technologies (Smith, 2003). The first steps can be problematic and political, even within organiza-

tions that have already committed to transforming to a process culture. Organizations are looking to strategically optimize, automate, and integrate key processes to provide seamless service to more demanding customers in a multichannel world. To do this effectively, systems must be integrated at the process level as well as the data level (Simsion, 2000). This environment requires system architects, consultants, analysts, integrators, and developers that have an organizational perspective and a diversified set of skills.

## REFERENCES

Adhikari, R. (2002a). 10 rules for modeling business processes. *DMReview*. Retrieved from http://adtmag.com/article.asp?id=6300

Adhikari, R. (2002b). Putting the business in business process modeling. *DMReview*. Retrieved from http://adtmag.com/article.asp?id=6323

Aversano, L., & Canfora, G. (2002). Process and workflow management: Introducing eservices in business process models. *Proceedings of the 14th International Conference on Software Engineering and Knowledge Engineering.*

*BPMI.org releases business process modeling notation (BPMN) version 1.0.* (n.d.). Retrieved April 5, 2005, from http://xml.coverpages.org/ni2003-08-29-a.html

Carlis, J., & Maguire, J. (2000). *Mastering data modeling: A user driven approach* (1st ed.). Addison-Wesley.

Carlson, D. (2001). *Modeling XML applications with UML*. Addison-Wesley.

Charfi, A., & Mezini, M. (2004). *Service composition. Hybrid Web service composition: Business processes meet business rules.* Proceedings of the Second International Conference on Service Oriented Computing.

Clark, M., Fletcher, P., Hanson, J. J., Irani, R., & Thelin, J. (2002). *Web services business strategies and architectures.* Wrox Press.

Delphi. (2001). *In process: The changing role of business process management in today's economy.* Retrieved from http://www.ie.psu.edu/advisory-boards/sse/articles/a4bd42eb1.delphi-ip-oct2001.pdf

Ettlinger, B. (2002, March 5). The future of data modeling. *DMReview.* Retrieved from http://www.dmreview.com/article_sub.cfm?articleid=4840

Ewalt, D. W. (2002, December 12). *BPML promises business revolution.* Retrieved from http://www.computing.co.uk/analysis/1137556

Frankel, D. S. (2003). *Model driven architecture: Applying MDA to enterprise computing.* Wiley.

Mangan, P., & Sadiq, S. (2002). On building workflow models for flexible processes. *Australian Computer Science Communications, Proceedings of the Thirteenth Australasian Conference on Database Technologies, 24*(2).

O'riordan, D. (2002, April 10). Business process standards for Web services: The candidates. *Web Services Architect.* Retrieved from http://www.webservicesarchitect.com/content/articles/oriordan01.asp

Reingruber, M. C., & Gregory, W. W. (1994). *The data modeling handbook: A best practice approach to building quality data models.* Wiley & Sons.

Sadiq, S., Orlowska, M., Sadiq, W., & Foulger, C. (2004). Data flow and validation in workflow modeling. *Proceedings of the Fifteenth Conference on Australasian Database, 27.*

Sharp, A., & Mcdermott, P. (2001). *Workflow modeling: Tools for process improvement and application development.* Norwood, MA: Artech House.

Shegalov, G., Gillmann, M., & Weikum, M. (2001). XML-enabled workflow management for e-services across heterogeneous platforms. *The VLDB Journal: The International Journal on Very Large Data Bases, 10*(1).

Simsion, G. (2000). *Data modeling essentials: A comprehensive guide to data analysis, design, and innovation* (2nd ed.). Coriolis Group Books.

Smith, H. (2003, September 22). *Business process management 101.* Retrieved from http://www.ebizq.net/topics/bpm/features/2830.html

Smith, H., & Fingar, P. (2003a). *Business process management (BPM): The third wave* (1st ed.). Meghan-Kiffer Press.

Smith, H., & Fingar, P. (2003b). *Workflow is just a pi process.* Retrieved from http://www.fairdene.com/picalculus/workflow-is-just-a-pi-process.pdf

Smith, H., & Fingar, P. (2004, February 1). *BPM is not about people, culture and change: It's about technology.* Retrieved from http://www.avoka.com/bpm/bpm_articles_dynamic.shtml

Torchiano, M., & Bruno, G. (2003). Article abstracts with full text online: Enterprise modeling by means of UML instance models. *ACM Sigsoft Software Engineering Notes, 28*(2).

Weske, M., Goesmann, T., Holten, R., & Striemer, R. (1999). A reference model for workflow application development processes. *ACM Sigsoft Software Engineering Notes, Proceedings of the International Joint Conference on Work Activities Coordination and Collaboration, 24*(2).

# Chapter VIII
# Business Process Integration in a Knowledge-Intensive Service Industry

**Chris Lawrence**
*Business Architecture Consultant, Old Mutual South Africa*

## ABSTRACT

*Knowledge-intensive administration and service activity has features favouring a particular architectural approach to business process integration. The approach is based on a process metamodel that extends the familiar input-process-output schema, and embodies the principle that the essential WHAT of a process is prior to any empirical and/or physical HOW. A structure of interrelated concepts can be derived from the metamodel. These can be used at logical level to define and analyze processes. They can also be implemented at a physical level—as an achievable and ideal integrated process architecture, or as a continuum of incremental control and integration improvements. Overall, the approach is to process what double entry is to accounting and the relational model is to data.*

## INTRODUCTION

The subject of this chapter is not business process integration in general, across all industries, but only in the context of a particular but increasingly important sector. The category of knowledge-intensive service industry covers most kinds of administration: all financial services, and many areas and levels of government and civil administration, education, law, tourism, and so forth. Although it excludes manufacturing and distribution, even these have significant administrative components (order processing and accounting,

for example), sharing important features of the intended focus area.

The category is not arbitrary. There are things fundamental to knowledge-intensive service industries that make a particular architectural approach to business process integration likely to succeed where others might fail.

A common obstacle to business process integration in service industries is that application systems are often incompletely process designed, or not process designed at all. Related to this, the systems often encourage the view of "business process" as primarily a string of activities performed on or by those systems and therefore defined in terms of those systems. A more "logical" alternative view of business process as an ordered satisfaction of a customer's need is often dismissed as unnecessarily pedantic, analytical, or purist—not "real world". In very "real-world" terms, radical critique of the business design implemented in an organization's IT portfolio can be unpopular if it looks an expensive direction.

But, sometimes you have to swim upstream. Today's process-centric enterprise cries out for a logical metamodel to do for process what double entry did for financial management and the relational model did for data.

This chapter outlines a metamodel that can revolutionize knowledge-intensive service industries in particular because so much of their production-line content is translatable into electronic form. The key is to define the process model at a logical level ("WHAT") free from any technical implementation ("HOW"). For a process-centric enterprise, that logical process model is the core of its business architecture. Business process integration is then a matter of achieving the best possible overall physical engine to implement that process model from available legacy applications, applied investment opportunity, and expert development resources.

## BACKGROUND

The process metamodel outlined here is essentially the same as that articulated in Lawrence (2005) and reproduced in part in Fischer (2005). In 2004, Old Mutual South Africa adopted it as the Old Mutual Business Process Methodology (OMBPM).

An important source of the metamodel is experience in designing and implementing the sort of process-designed systems described in Jackson and Twaddle (1997). It is in implementation in particular where the full business-transforma-

tional potential of integrated process architecture starts to show.

Where appropriate, the diagrams shown use the emerging BPMN (Business Process Modeling Notation) standard (Business Process Management Initiative [BPMI], 2004), as does Lawrence (2005). Although BPMN is more customarily used for depicting physical processes, no constraint has yet been expressed against extending its application to processes at a logical level. Indeed, part of the orthodoxy that this chapter challenges is that process analysis and design begin and end at the physical, empirical level.

## PROBLEM SPACE

### Administration

We start by inquiring into the nature of administration work: processing applications, granting approval, carrying out instructions, and so forth in sales, financial services, central and local government, education, tourism, and so on.

Administration work typically involves carrying out an implicit or explicit "request" from, or on behalf of, a "customer." It is usually governed by rules. There are right ways and wrong ways of doing things: rules about standard cases and exceptions, about sequence and completeness criteria.

Administration work is also increasingly supported by computer systems. The people involved increasingly deal with exceptions and special cases, and make rules rather than (just) follow them.

More analytically, we see that administration work can be treated abstractly without losing its essence. It can be translated into different formats (e.g., digitized), and so can its rules.

Take, for example, a life insurance policy. We could define this as a *legal contract between a financial organization and another person or or-*

*ganization in relation to one or more human lives.* Almost everything about a life insurance policy and its creation can be treated abstractly—in a translatable way. You could not say the same about making an armchair.

A life insurance policy is an abstract entity, translatable between formats. It must only exist as hard copy if the rules say so. An armchair is a concrete object. It cannot be translated into another format and stay an armchair.

## ANALYTICAL APPROACH

### Administration Processes

The approach starts with a fundamental decision: to see administration processes as things derived from rules rather than just as things discovered to exist in an organization. It is a decision to see administration processes as "essential" rather than "empirical."

To take the paradigm case of financial services, we would see the business processes connected with, say, a life insurance policy as derived from rules arising from the contractual obligations contained within the product sold to the customer—obligations that are themselves set within an overarching regulatory and/or legislative framework. The processes have a fundamentally "essential" component and are therefore not just things empirically discovered to exist inside the financial services company. ("Process" and "business process" will be used interchangeably throughout this chapter.)

The approach can be summarized as WHAT is prior to HOW. In a context like financial services, process architecture starts by defining and analyzing essential business processes (WHAT) so they can be optimally implemented in available or achievable application architecture (HOW). (A current HOW may of course be empirically discovered, for example, the way a contractually

essential insurance claim process may currently operate and be supported.)

A meaningful analogy can be drawn with data analysis and design. We can decide to consider only the actual physical databases an organization has, and what data entities and structures exist on those physical databases (empirical:HOW). Or we can understand what logical entities are important to the business, what the attributes are of those entities, and what numerical and modal relationships exist between those entities (essential:WHAT).

## PROCESS METAMODEL

### Request and Outcome

The decision to see processes in this way has fairly far-reaching consequences.

The "empirical" approach (found, for example, in many variants of business process reengineering) typically assumes an input-process-output metamodel.

There is nothing "false" or "wrong" about this model as there is nothing intrinsically false or wrong about empirically discovered information. The issue is more one of incompleteness and the architectural consequences of that incompleteness.

What I have called the "essential" approach leads to a small but significant refinement.

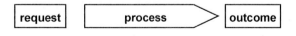

In this metamodel, the process begins with a requirement to generate an outcome, and therefore to carry out a piece of work. A business process is the work needed to carry out something for a recipient. A computer system may or may not be involved.

The requirement is typically, but not necessarily, a request or instruction from or on behalf of a "customer.' "Typically, but not necessarily" here means that a process instigated by a customer request is a paradigm case: it provides the model. The process ends with an appropriate outcome.

The appropriate outcome is normally the successful fulfillment of the request—in accordance with the relevant rules. It could also be the formal annulment of the original requirement, for example, the cancellation of the original request, again in accordance with relevant rules.

Example processes conforming to the model include the following:

Processing a customer's order to buy something

Handling a customer complaint

Processing an insurance claim

Processing an application to do the following

    Join an organization

    Open a bank account

    Invest money

    Borrow money

...etc

Examples like these show why a process may not have only one possible outcome. The goods the customer ordered may or may not exist, the insurance claim may be honoured or rejected, the loan company may accept or reject the loan application, and so on.

## Stable State and Unstable State

We shall now look at the model in a different light.

Imagine an organization (or department or system) whose only role is to carry out one type of request. Imagine the organization in a state where it has no request to carry out. It has nothing to do. It is doing nothing. We shall call this a "stable state."

We shall now imagine the organization hit with a request (from a "customer") of the type it can carry out. It starts doing what it knows how to do in conformance to its rules. Person A does *x*. This leads to Person B doing *y*, which needs to be approved by Person C. This then means that Person A can contact the customer and tell him or her such and such until the request is either met or cancelled in some well-formed way.

When that has happened, stability is restored.

The organization is now in the same state of stability as before. The period in between (Person A doing *x*, Person B doing *y*, etc.) was a period of "instability."

So this is another way of looking at the business process.

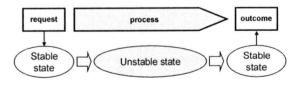

The process is all the events and activities (*x*, *y*, etc.) that take the organization (department, system, etc.) from one state of stability to another: from the initial destabilising event to the restoration of stability.

## Instance Level

We are talking about one request—one instance. The process is all the activities needed to restore stability in respect to that one initial destabilising event.

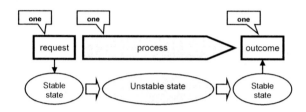

The complete description of the process may include things only applicable to some instances. The complete description of an insurance claim process might need to include sets of *ifs*.

If the claim amount is more than £1,000, then inform a claims assessor.

If the customer has a no-claims bonus, then calculate the bonus reduction.

However, the process itself is not a "batch" thing. It is not all the insurance company's claims, or all the claims of a particular type, or all the claims coming in on a particular day.

This definition of "process" is not arbitrary. A process is not an arbitrary sequence of actions that happen to produce some kind of output from a particular set of inputs. It is everything that has to be done to restore the original stability that the event or request disturbed. If stability has not been restored (if the request has not been met or annulled), then the process has not finished.

## Intentionality

To start with a "request" rather than just an "input," and to end with a "well-formed outcome" rather than just an "output" is to see a thread of intentionality running through the process. An input would not necessarily be described in the same words as the resultant output. But, "request" and "outcome' share an identity: The (principal) outcome is the thing requested, and the request is for the intended outcome.

This thread of intentionality is part of what it means to say the process is not arbitrary. The customer has an intention, which is expressed in the request. The supplier intends to carry out the request. At the logical level, this intentionality drives the process to completion.

## Multiple Process Instances

Now consider an order processing department. The example is deliberately chosen for simplicity and familiarity, in preference to a more involved scenario from, for example, financial services.

There will be many orders from many different customers. The department as a whole will be in a constant state of instability with respect to at least some orders. Individual process instances will be at different points. Some will have just come in. Some will be awaiting credit check. Some will be matched against stock. Others will be waiting to be matched, or waiting to be fulfilled or dispatched.

I now want to make two fairly obvious assumptions explicit.

Assumption 1 is that a business area should be well-managed. It is good to know where things are rather than not know. Order is better than chaos.

Assumption 2 is that it is good to be customer centric. It is good to look at things from the customer's point of view, and to care about the service customers are getting.

In a competitive, commercial environment, these two assumptions are both obvious and obviously linked.

We can now list some desirable features of a business area that lives by these assumptions. It follows rules. It knows how to process different kinds of orders. It follows set procedures. It knows where bottlenecks may occur, and manages its work to minimize bottlenecks.

It will want to complete each process instance as fast as possible. This benefits the customer, who will get what he or she wants as soon as possible. It also benefits the business. It improves cash flow, and also the less work in progress, the less queries, follow-ups, mistakes, and rework. So, it will want to get through each process instance as quickly as possible, and by as standard a route as possible.

This brings us to the concept of "subprocess," which is an important consequence of the concept of "process" articulated so far.

## Subprocess

A business process can normally be broken down into a finite series of steps or subprocesses:

The following generic statements can be made about this breakdown:

*The subprocesses will be arranged in a fixed pattern, typically sequential, although subprocesses can also occur in parallel.*

*Subprocess 1, Subprocess 2, and so forth can be described in purely business terms in relation to what the process is about, for example, authorise order, match against stock, and so on. Subprocesses have to happen regardless of whether a computer system is used or what system is used.*

*Boundaries between subprocesses often correspond to handoffs and/or break points where external interaction is needed, for example, missing input or authorisation. (Since subprocess descriptions will be in business rather than system terms, the break points should also be dictated by business rather than system constraints.)*

*The boundaries between subprocesses often correspond to bottlenecks, for example, x instances are awaiting authorisation, y instances are awaiting credit check, and so forth.*

Subprocesses are not arbitrary collections of actions, nor are they events only meaningful in terms of a particular computer system. A subprocess is something like "check customer's credit rating," not "run job C123." So, "subprocess" is not an arbitrary definition.

A subprocess is not primarily a piece of functionality, either. It is the work needed to get from one recognized status to the next, where status can be described in purely business terms, and where the transition from one status to the next is significant in business terms. Just as a process is seen in terms of a (single) initiating request leading to a (single) well-formed outcome, then a subprocess will also be seen in those terms.

Take the order process in Figure 1. The notation is the emerging BPMN standard (BPMI, 2004). The symbols themselves need not concern us.

The subprocess "check credit rating" is the work needed to get an order that has been "matched against stock." (See Figure 2.)

In order to break the process down to a more granular level, we need a few more analytical concepts.

## Subject Entity

This is the object or request moving through the process. In the order process, it is the order itself. See Figure 3.

It is typically a request (implicit or explicit) for something to be done. The "something to be done" is the work of the process.

There will only be one subject entity per process instance, for example, Order Number 123.

It may be linked to a physical object (e.g., an order form or an insurance claim form), but it does not have to be.

*Figure 1.*

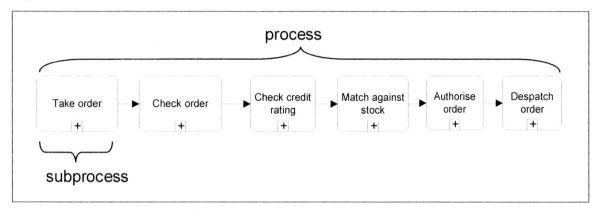

*Figure 2.*

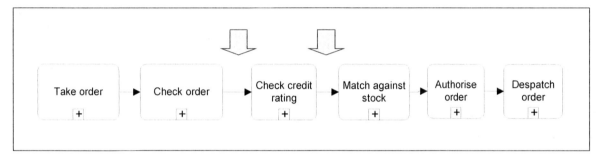

*Figure 3.*

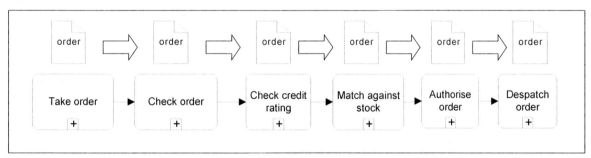

It may be linked to a data record or data set (e.g., a policy application, or the details of a switch request), but it is not primarily a collection of data. It is primarily an instance of work: a request to carry something out.

However, the subject entity (request) needs to be carefully defined to determine the relevant linked data set. To determine what level the process is operating at (order, customer, portfolio, policy, benefit, etc.), it is important to define what level the request is at. (This may seem obvious in the abstract, but it is remarkable how often processes are implemented at levels different from the ones the defining requests are formulated at. This can happen when standard indexing is applied in a generic workflow implementation.)

## Workflow Status

In the example order process, the only way to move an order to the state of "awaiting Check order" is by the subprocess "Take Order." "Awaiting check order" just means that point in the process where the "Take Order" subprocess has happened, but the check-order subprocess has not yet happened. We shall call "awaiting Check order" a workflow status (or process status). See Figure 4.

Similarly, the subprocess "Check order" moves the order from workflow status "Awaiting Check order" to workflow status "Awaiting Check credit rating."

Workflow status is expressed in business terms: "awaiting check order," not "awaiting Job C123."

The thing that has the workflow status is the order: the subject entity. When we get to task level (below), we will see that workflow status can also be expressed as awaiting a particular task. Since parallel processing is possible at a logical level (even if unavailable in a particular physical architecture), a subject entity can have more than one workflow status at the same time.

## Business Rule

Much of the discussion so far could be expressed in terms of business rules. The diagrams of pro-

cesses and subprocesses are effectively rules in graphic notation.

There are rules about what processes need to occur, rules about what subprocesses a process consists of, rules about the sequence of subprocesses, and rules about what happens inside a subprocess. The "Check order" subprocess could include the following rules.

*All orders must be for a known customer.*

*Items must be identifiable as goods the business trades in.*

*Quantities must be specified.*

It is not that processes and subprocesses come first, and then we decide what the rules are. Rules come first. The definition of a process is a rule. To say where the process starts and stops, and how the process divides into subprocesses, is to state rules.

However, some rules do fit inside other rules. There is the rule that you have to do $x$, and then rules about how to do $x$.

## Task

"Process" and "subprocess" represent the first two levels of process analysis. Following the

*Figure 4.*

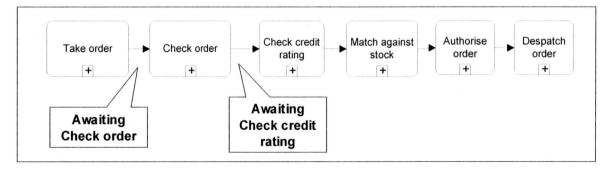

Figure 5.

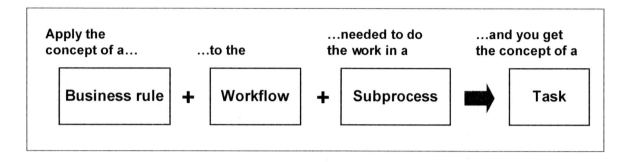

Figure 6.

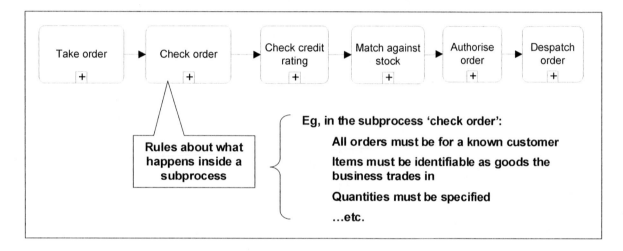

discussion of subject entity, workflow status, and business rule, all the foundations are now in place for articulating the third level, the task.

Task is a component of subprocess, just as subprocess is a component of process. Its place in the overall process architecture can be formulated as in Figure 5.

To explain this, consider the subprocess "Check order" and example business rules. See Figure 6.

Imagine these rules being run against the order "automatically"—by a computer program or by an employee whose only job is to validate the order against the rules.

Every order would end up in one of two categories: those that obeyed the rules and those that did not. The ones that obeyed the rules would go to the next subprocess: "check credit rating." See Figure 7.

But what about the ones that did not pass? Do they get rejected and discarded? That would be irrational. An order could fail for many reasons: it may have been taken wrongly; it may have gone to the wrong company; it may have one letter missing in the customer name or product description; and so forth.

In general, errors and anomalies would need to be sorted out "manually"—by using discretion to see if there is any chance of correcting them and

*Figure 7.*

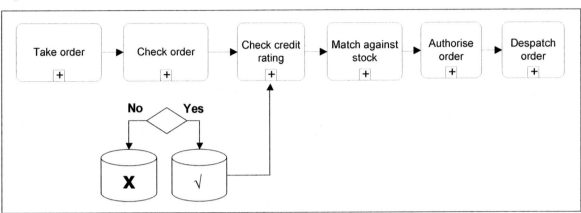

*Figure 8.*

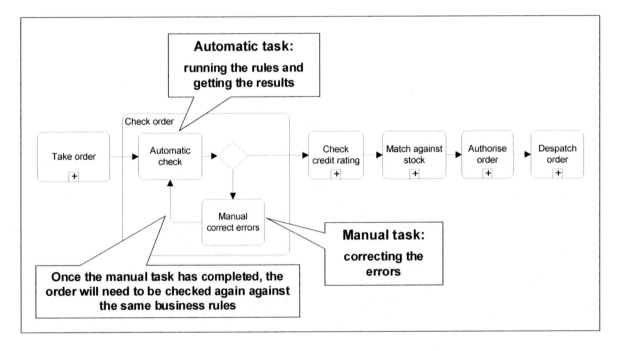

saving the order. In contrast, applying the rules themselves can be seen as "automatic" work.

We call the two activities "tasks." See Figure 8.

Adding these two tasks to the "Check order" subprocess (or, perhaps, more accurately, analyzing the "Check order" subprocess into these two tasks) shows that all the work of the subprocess is actually contained within the tasks. The subprocess is only a container for the tasks, or alternatively, an analytical concept required on the way to defining the tasks. (We shall see later that "subprocess" is not essential at *implementation* level.)

Routing will depend on the attributes of the order:

*All orders will go through the automatic task.*

*Perfectly correct orders will go on to the subprocess of "check credit rating."*

*Imperfect orders will be routed by the automatic task to the manual task. The manual task will then route them back to the automatic task for rechecking.*

The simple example above demonstrates an important feature of process-architected design in support of business process integration.

Consider an extreme case of bad data: an order failing every rule in the "Automatic check" task. It would not be able to move past the subprocess without every error being corrected. However, it did get captured. The business does know about it. It is not sitting around on someone's desk—outside the system. The order has been rejected as *data*, but not rejected as *work*.

This is important for control, measurement, and completeness. A process designed in this way may let "garbage in," but it will not let "garbage through." This is because it is based on a view of business process as primarily an ordered satisfaction of a customer's need, not a string of activities performed by system components. The "garbage in, garbage out" design principle derives from system integrity priorities. There is nothing wrong with system integrity, but system or component design decisions should not dictate how business processes are conceived and therefore how process instances are recorded.

The concept of task is no more arbitrary than that of process or subprocess. Whether a subprocess is divided into one, two, or more tasks is decided by the possible attribute values of the subject entities and related data sets (cases), and what needs to be done to move those possible objects to the next subprocess. It is not that each of the two tasks represents roughly half the work of the "Check order" subprocess. That would be arbitrary.

To reinforce this important development of the metamodel, we shall repeat the analysis with the next subprocess: "check credit rating." Again, the analysis into tasks will depend on business rules. But here the rules go beyond a simple pass or fail.

We shall assume the rules are as shown in Box 1.

The rules tell us what events and conditions the subprocess needs to allow for.

*If the criteria of Rule 1 or Rule 2 are met, the subprocess should be passed automatically.*

*Rule 3 needs the subprocess to include manual approval.*

*Rules 4 and 5 need the subprocess to include writing to the customer, and therefore recording and acting on the reply, and following up if no reply is received.*

We could therefore model the subprocess as in Figure 9.

An initial automatic task would apply Rules 1 to 5. Any order passing Rule 1 or Rule 2 would go to the next subprocess with workflow status "Awaiting Match against Stock." An order failing Rules 1 and 2, and therefore meeting the criteria of Rule 3, 4, or 5 would be routed to a manual task.

For simplicity's sake, we shall assume the same manual task allows for both approval (Rule 3) and writing to the customer (Rules 4 and 5).

An order that is now manually approved (Rule 3) is routed back to the automatic task, which then allows it to pass to the next subprocess.

Rules 4 and 5 need a manual task for acting on the reply ("manually record documents") and an "Automatic follow-up" task to loop back if no reply is received after a fixed period.

It should now be clear what was meant by the formula:

*Box 1.*

---

**Rule 1**

    IF
        Customer has sent cash with order
    THEN
        Pass credit check

**Rule 2**

    IF
        Customer has not sent cash with order
    AND
        Amount of order not > total current unused credit
    AND
        Customer is not in arrears
    THEN
        Pass credit check

**Rule 3**

    IF
        Customer has not sent cash with order
    AND
        Customer is not in arrears
    AND
        Amount of order > total current unused credit
    AND
        Customer has enclosed a bank reference justifying the credit increase
    THEN
        Credit increase needs to be manually approved

**Rule 4**

    IF
        Customer has not sent cash with order
    AND
        Amount of order is not > total current unused credit
    AND
        Customer is in arrears
    THEN
        Write to customer requesting payment before accepting order

**Rule 5**

    IF
        Customer has not sent cash with order
    AND
        Customer is not in arrears
    AND
        Amount of order > total current unused credit
    AND
        Customer has not enclosed a bank reference justifying credit increase
    THEN
        Write to customer requesting bank reference before accepting order

---

*Figure 9.*

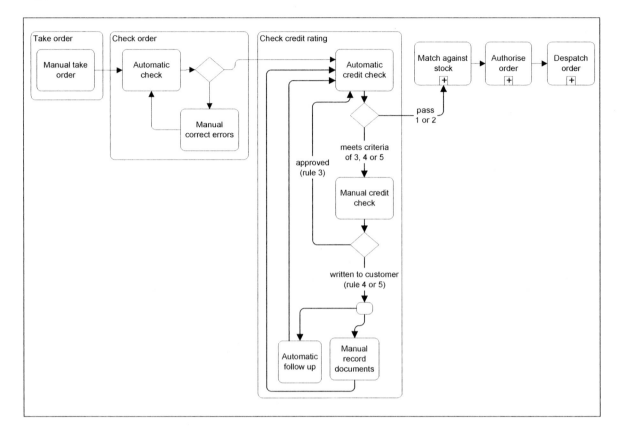

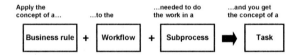

The logic and functionality within a task can be analyzed further. However, this will not affect workflow. Workflow is established at the task level. (The concept of task employed here makes this true by definition.)

Workflow at the *subprocess* level comes from logical sequencing of process rules, ignoring the variety of individual subject-entity instances (cases). Workflow at the *task* level comes from the logistics of applying process rules to the variety of possible cases.

## Process Model

Having analyzed an individual process down as far as we need to, we shall now complete the picture by going up in the opposite direction.

A trading organization is unlikely to have just an order process. There will be a number of process types with relationships (interactions, dependencies) between them: recording new customers, handling customer data changes (bank details, address), ordering stock, billing and settling, recovering bad debts, and so on.

Many people in the organization will be involved in more than one process, and many data entities will be relevant to more than one process: customers; stock item; sales representative; and so forth.

Figure 10.

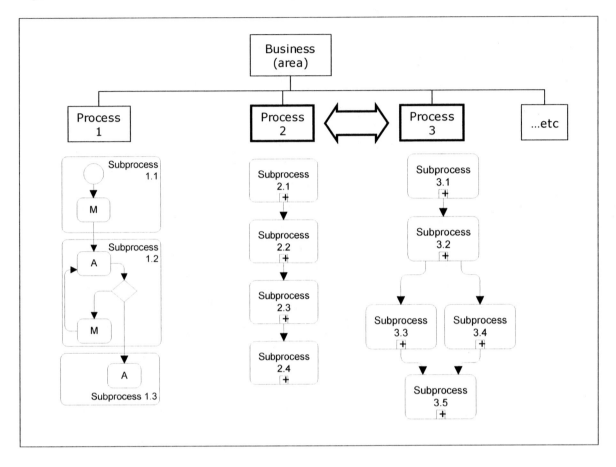

From a process-architectural perspective, a business or business area can be analyzed into a finite set of processes. Each process can be analyzed into a finite set of subprocesses, and each subprocess can be analyzed into a finite set of tasks—all following the rules of the business.

The number, nature, and routing of processes, subprocesses, and tasks are determined by the work needed to apply the rules to the attribute values of possible cases, where the paradigm "case" is a customer request.

This analysis is the process model (see Figure 10).

The double-headed block arrow indicates that processes interact with each other. One may initiate another (by creating the subject entity of the initiated process); or one may determine or influence the outcome of another process instance already begun by terminating it, suspending it, releasing it from suspense; or creating, amending, or deleting data the other process will operate on.

The process model is ultimately a structure of rules. The rules identify a finite set of processes. The processes are typically the carrying out of requests, which represent what the organization is there to do. The rules define possible interactions between processes: how each process breaks down into subprocesses, how each subprocess breaks down into tasks, and the number, type, and routing of those tasks.

The process model can be depicted in many ways, but it would typically be a combination of diagrams and text. Depending on how detailed (granular) it was, it could be a complete description of the business or business area at the WHAT (rules) level. It can provide an overall single framework for business analysis, change implementation, change management, business management, risk assessment—and computer system and architecture design.

## Process Model and System Design

The process model as conceived above is a structure of rules wholly at the WHAT level. It could not be built using a concept of business process as a sequence of activities performed on or by components of given physical systems. This is because those physical components are necessarily at the HOW level.

This does not mean that systems cannot be designed based on this concept of process model—precisely the reverse. Systems can be (and have been) designed, built, and implemented on the basis of logical process architecture, with all the logical concepts physically implemented in solution architecture. Whether this option is available to an organization is a different question, depending on that organization's strategic options and imperatives at the time.

However, the logical model can represent an attainable ideal, and can therefore be a yardstick for evaluating current architecture and the degree to which (and the manner and success by which) "process" is implemented in that architecture.

Designing, building, and implementing process-architected systems awaken the realization that process components and functionality components are actually different things. Process components (process, subprocess, task) are transitions in the workflow status of a subject entity. Functionality components (object, program) are what effect those transitions.

As mentioned earlier, "subprocess" is not strictly necessary at implementation level, although it is crucial in process analysis and design. It can have

value, particularly in management information, simulation, and so forth, and in implementing traceability between a logical model and a physical model. But while physical process architecture cannot work without process and task constructs, it can work without a subprocess construct.

But to focus on the undeniably necessary concept of "task," what do we mean by saying that Task X and the component or components implementing Task X are different things?

Component functionality implementing a task ("get customer details," "get product details," etc.) is more concerned with the data set (e.g., the details of the order or of the life insurance application) linked to the subject entity than with the subject entity itself—the request. The subject entity, the request, is primarily concerned with intentionality (monitoring stages in carrying out the intention), not functionality (what has to happen in order carry out the intention).

Keeping these domains separate at the logical level is key to designing process-architected applications. The result is that, within the boundary of a particular system, the business process as implemented in that system can be the business process as business users and, where relevant, customers understand it. This is because the concepts of process, task, and (possibly) subprocess will be physically implemented as generic design constructs.

So, the physical data model will have a ProcessType entity and a ProcessInstance entity, a TaskType entity and a TaskInstance entity, and possibly a SubprocessType entity and a SubprocessInstance entity as well. ProcessType will identify the processes the organization recognizes (order, customer creation, payment, etc.), and ProcessInstance will record the current and historic process instances (of the various ProcessTypes) created in response to the received requests (subject entities, cases). TaskType will include, for example, Automatic check, Manual correct errors, Automatic credit check, and so on, and TaskInstance will hold the current and historic task instances (of the various TaskTypes) created in response to the rule-determined routing of the subject entity instances.

The logical data model requirements are detailed in Lawrence (2005), which also includes a full case study. However, from the snapshot given here, the reader should appreciate the generic and structured implementation of "process" which the approach entails. Because the implementation is generic, the process dynamics can be supplied by a generic process engine or workflow engine that understands the constructs and articulates each instance (of process, subprocess, and task) as necessary.

The required process or workflow engine is no different in principle to the generic functionality many proprietary workflow packages contain. The difference in practice is not so much the engine, but the architectural positioning of proprietary workflow functionality as necessarily external to an organization's "administration" system(s)—an often frustratingly inadequate response to the lack of process awareness in many legacy applications.

## PROCESS INTEGRATION

The section immediately above indicates how the metamodel can be used to guide the design of a process-architected system. Because systems can be built from preexisting components rather than purely from scratch, it can also guide the reengineering of existing applications into a process-architected (or more process-architected) result.

"Reengineering existing applications into a process-architected result" and "process integration" are not far apart. Whether one regards them as identical or as overlapping areas on a continuum will largely depend on one's view of process architecture and its role in operational and strategic business activity.

The present author would claim the metamodel can guide and support process integration precisely because it can define the target that process integration efforts should aim at. Regardless of what technology outcome is actually feasible (by considering legacy constraints, investment options, timescales, development resources, and development priorities), it is always possible to define the thing that process integration is about: the business process model.

The metamodel can be used to define and analyze the processes that interact together to create the organization's process model. What happens next will depend on business priorities and development opportunities. At one extreme, there could be a fully process-architected solution encompassing the whole of the organization's activities, driven and articulated by a single process engine. Somewhere near the other extreme will be a heterogeneous assortment of applications supporting the same breadth of activities, but in different ways and displaying different approaches to "process" (including ignoring "process" altogether). However, inroads can be made even in a scenario like this, as in the following examples:

*The subject entity concept is always useful to define the level at which to implement a process.*

*Where a process implemented in one system initiates a process implemented in another system, it may be beneficial to design the interface around the subject entity generated by the one and processed by the other.*

*Where a process implemented in one system employs subordinate functionality in the other, it may be possible and beneficial to define tasks in the one which are implemented and/or supported by functionality in the other.*

*Where one of the applications has generic process-engine functionality, it may be possible and advantageous to extend its operation to control processes implemented in other applications.*

*If an application with process-engine functionality initiates a process implemented in another system, also with process-engine functionality, then it would be advantageous if possible to define subject entity instances as the formal integration objects between them.*

These suggestions all have the same intention: implementing control on the basis of identified business processes. Administration systems very often work against this, for example, by providing rigid and automated processing of "standard" cases but inadequate control over exceptions. (An example is where the garbage in, garbage out design principle prevents exceptions from being captured at all.)

Another familiar example is where different solutions (A and B) contain different portions of the organization's data model, but in order to provide further automation to a process supported by Solution A, access is needed to data contained in Solution B. Without an overall business process paradigm, the result can be ad hoc local integration functionality or data replication. The approach gets repeated and then rationalized into sophisticated "interoperability" and "middleware" solutions, which assume more and more of a controlling role. The business process gets harder to discern, often only observable in "workflows" implemented in external workflow packages that focus primarily on routing links to digitized documents from one person to another.

Mention of external workflow packages brings us to what may be the best practical option many organizations will have for implementing or improving their business process integration. This is not because external workflow packages are optimally architected, but because historically they tend to enjoy an extensive footprint in both user and support communities. Business processes

are frequently seen in terms of implemented workflows. In theory, this is right. In practice, however, the external positioning of proprietary workflow can constrain and frustrate.

To base a process integration initiative on an external workflow package is not an undertaking to be considered lightly. Logically, it leads to an outcome where the workflow package controls everything. All (or most) of the organization's functionality would end up logically "inside" the workflow package, in the sense of being rearchitected into components called only by workflow components, directly or indirectly. The workflow package data model should be evaluated to ensure it is rich enough to reflect the constructs and relationships successful process architecture needs. The evaluation should presuppose a metamodel like the one presented here.

Process integration by way of a "business process management system" (BPMS) is in principle the same as the workflow package route. The difference is more in the extent and sophistication of the process management functionality. Administration functionality (the functionality components contrasted earlier with process components) has to be supplied from somewhere: built from scratch, supplied by preexisting administration systems and linked by more or less sophisticated integration, rearchitected from those preexisting administration systems into componentized services; or combinations of these.

## FUTURE TRENDS AND RESEARCH

Business process integration is an essentially practical exercise. Either you integrate processes or you do not, and you can do it more or less extensively and/or tightly. Much of the decision making about whether or not to use a particular metamodel to guide process integration will be economic and logistical, depending on how costly it is, how familiar or alien it is to established development and support teams, and so forth.

Considerations like these may be quite independent of how sound the metamodel is. As argued previously, process-architectural thinking is new, and established application architecture presents challenging obstacles.

On the other hand, a significant groundswell is building up in the BPMS and BPM industries. I have deliberately avoided these waters because I have been intent on mapping the logical, business-architectural terrain irrespective of technology or product sophistication.

However, there is scope here for further navigation, for example, whether a BPMS implementation needs an overarching process-architectural methodology, or whether it is better without, and if so, why. A line of particular interest is about the BPMN standard: why it is so rarely used for "logical" process modeling even though no obvious alternative candidate appears to exist.

There is also considerable apparent scope for investigating the relationship between rule-based process architecture as outlined here and the formidable business rules industry building up in the IT environment (Date, 2000; Ross, 2003).

The most intriguing question, though, concerns process architected systems themselves. From personal experience of designing and implementing several of them, I can honestly say I have not yet come across a better paradigm. Yet, why are they so rare? Is it immaturity of process thinking in the IT community that causes suboptimal implementations, and therefore the paradigm itself loses credibility? Is it the weight of established IT legacy, and the still-prevailing investment principle of reuse, then buy, then build? Is it that, despite its comprehensibility and comprehensiveness at the logical level, true integrated process architecture resists translation into marketable product?

## CONCLUSION

Knowledge-intensive administration and service activity has features favouring a view of "business process" as an ordered satisfaction of a customer's need. This differs from, or enhances, a view of "business process" as a string of activities performed on or by the particular set of systems an organization happens to have, and therefore defined in terms of those systems.

Administration work can be treated abstractly without losing its essence. It is architecturally advantageous to view administration processes as essential rather than empirical. This view can be summarized as: WHAT is prior to HOW.

An empirical approach (as in, e.g., business process reengineering) typically assumes an input-process-output metamodel. The more essential approach favours an enhanced metamodel where "customer request" replaces "input," and "appropriate outcome in terms of that request" replaces "output." This enhanced model introduces an element of intentionality into the input-process-output model.

This rule-based metamodel can then generate a sequence of interrelated analytical concepts: process, subprocess, task, subject entity, workflow status, process interaction, and process model.

Where there is development and transition opportunity, these analytical concepts can translate into physical design constructs to implement integrated process architecture, on the basis of a structured physical data model articulated by a workflow or process engine.

Where less opportunity exists, the approach still provides a continuum of possibilities including the evaluation of current application architecture from a process perspective, pragmatic recommendations and design patterns for the allocation of control and incremental integration, and an ideal logical framework to serve as a strategic target.

# REFERENCES

Business Process Management Initiative (BPMI). (2004). *Business process modeling notation (BPMN) specification* (Version 1.0).

Date, C. J. (2000). *What not how: The business rules approach to application development.* Boston: Addison-Wesley.

Fischer, L. (Ed.). (2005). *Workflow handbook 2005.* Lighthouse Point, FL: Future Strategies Inc.

Jackson, M., & Twaddle, G. (1997). *Business process implementation: Building workflow systems.* Harlow, England: Addison Wesley Longman Limited.

Lawrence, C. P. (2005). *Make work make sense: An introduction to business process architecture.* Cape Town, South Africa: Future Managers (Pty) Ltd.

Ross, R. G. (2003). *Principles of the business rule approach.* Boston: Addison-Wesley.

# Section III
# Integration Methods and Tools

# Chapter IX
# A Systematic Approach for the Development of Integrative Business Applications

**Michalis Anastasopoulos**
*Fraunhofer Institute for Experimental Software Engineering, Germany*

**Dirk Muthig**
*Fraunhofer Institute for Experimental Software Engineering, Germany*

## ABSTRACT

*Modern business applications consist of many subsystems (or components) potentially developed and maintained by diverse organizations. Generally, there are three different points of view. First, organizations using business applications are interested in the unified look and feel of composed applications, the maximum interoperability and synergetic features among subsystems, the high availability of all subsystems, and quick and seamless updates after new releases or bug fixes. Second, organizations providing single subsystems want, on the one hand, of course, to satisfy their customers and business partners, but on the other hand, to minimize their overall effort. Third, organizations integrating single subsystems aim at a uniform and cost-efficient integration architecture. This chapter takes the two latter viewpoints and describes a methodology for organizations integrating their subsystems with many business applications and all relevant types of subsystems, as well as with the whole family of subsystems from different vendors. The methodology is a product-line approach optimally tailored to the needs of such organizations. It views subsystems delivered by a single organization with all corresponding integration contexts and requirements as a family of similar systems, and engineers this family by taking systematical advantage of common characteristics and proactively considering differences in anticipated future scenarios. The methodology is based on Fraunhofer PuLSE™ (PuLSE™ is a trademark of the Fraunhofer Gesellschaft), a customizable product-line approach validated in practice by many industry organizations since 1997. The integration methodology has been developed in the German research project UNIVERSYS by tailoring Fraunhofer PuLSE™ together with industry partners to the integration context described.*

# INTRODUCTION

Modern business applications consist of many subsystems (or components) potentially developed and maintained by diverse organizations. Consequently, installing such a business application includes a significant effort in integrating all required pieces, typically subsumed by the term enterprise application integration (EAI).

There are generally three different points of view for looking at EAI solutions. First, organizations using business applications are interested in the unified look and feel of composed applications, the maximum interoperability and synergetic features among subsystems, the high availability of all subsystems, and quick and seamless updates after new releases or bug fixes. Second, organizations providing single subsystems, on the one hand, want to satisfy their customers and business partners; on the other hand, however, they also want to minimize their overall effort. Third, organizations integrating single subsystems aim at a uniform and cost-efficient integration architecture.

This article takes the two latter viewpoints and describes a methodology for organizations integrating their subsystems with many business applications and all relevant types of subsystems, as well as with the whole family of subsystems from different vendors. The section titled "EAI Viewpoints" introduces the two viewpoints in more detail.

EAI solutions are of special importance to small and medium-sized enterprises (SMEs), which rarely can sell their products as isolated single systems due to their special but limited focus. By their very nature, SME applications are favorably seen more as one element integrated into bigger application suites.

Technically, EAI solutions are to date being realized with the help of special EAI frameworks or tool infrastructures. There are numerous products in this area. Apart from the fact that most of these products are not affordable for SMEs, they mainly focus on the technical realization of integrative solutions and do not provide concrete methodological support. Additionally and maybe most importantly, existing EAI frameworks explicitly deal neither with the different EAI viewpoints nor with the flexibility expected from an EAI solution. Therefore, the section "Related Work" provides a survey of existing EAI methodologies and identifies their support for the two viewpoints addressed in this chapter.

The approach presented in this chapter addresses explicitly the two different viewpoints discussed above, namely, the composition and the integration viewpoint, and furthermore provides a concrete methodology for the creation of integrative software architectures. Finally, the approach addresses the flexibility issue by viewing subsystems delivered by a single organization with all corresponding integration contexts and requirements as a family of similar systems. The approach engineers this family by taking systematical advantage of common characteristics and proactively considering differences in anticipated future scenarios. The methodology is based on Fraunhofer PuLSE™ (Product Line Software Engineering), a customizable product-line approach validated in practice by many industry organizations since 1997. The section "Engineering Approach: Fraunhofer PuLSE™" introduces the original approach; the section "Developing Integrative Business Applications" then presents a version of it tailored to the addressed viewpoints and scenarios.

The section titled "Case Studies" then presents two real case studies covering both viewpoints; that is, each case study is described by one of the two special viewpoints. The section "Conclusion and Future Work" concludes the chapter with some final remarks and a brief outlook on future activities.

## EAI VIEWPOINTS

Each EAI project can be seen either as a composition or as an integration project. The composition viewpoint approaches a single integrated system and focuses on customer satisfaction and the minimization of the effort needed for its construction. The integration viewpoint starts from a single subsystem, which should be reused as often and as efficiently as possible for creating integrated applications. Hence, the integration viewpoint aims at a uniform and cost-efficient integration architecture.

### Composition Viewpoint

In this composition viewpoint, the composite application under development plays the role of an aggregate that integrates external applications. A typical example of a composite application is a process orchestration application that coordinates the interaction of various underlying applications along the lines of a well-defined process. In this case, the composite application is being developed with reuse. This means that the major interest lies in the discovery and efficient selection of various external applications. In other words, the composite application must specify the applications to be reused in a flexible way. An example of a composition is illustrated in Figure 1.

### Integration Viewpoint

On the other hand, an integration project aims at the realization of component applications. The latter are developed for reuse. In this case, the interests lie in the creation of qualitative components that are worth reusing as well as flexible components that can be reused in many different situations. Figure 2 illustrates the integration viewpoint.

*Figure 1. Composition viewpoint*

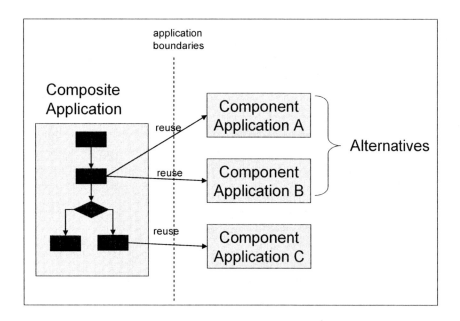

*Figure 2. Integration viewpoint*

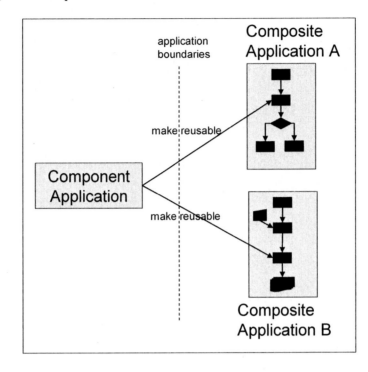

## RELATED WORK

In spite of the fact that EAI has gained much importance in the last years, the focus of the EAI community lies more in the technology and the technical realization and less in the methodology. There are numerous publications in the EAI domain that provide insights into the principles, patterns, and tools that can be employed without discussing engineering approaches to integration problems. Often, EAI solutions are seen as an art and less as software engineering science (Schmidt, 2002).

Although the differences between a conventional and an integrative software development approach haven't already been identified, none of the existing EAI approaches has reached wide acceptance. In the following subsections, two of the most promising approaches will be presented.

The subsequent approaches and the PuLSE™ DSSA approach, discussed in the sections "En-

gineering Approach: Fraunhofer PuLSE™" and "Developing Integrative Business Applications," are not mutually exclusive. The following EAI approaches rely on rich experiences with EAI solutions and therefore can be sensibly combined with PuLSE™ DSSA, a customizable process, which deals explicitly with variability and generally aims at a complete software engineering approach.

Both of the subsequent approaches reflect the typical way EAI problems are tackled. That is, the composite and component applications are known up front and the resulting EAI architecture is built up upon this knowledge. This is the major differentiation point from the PuLSE™ DSSA approach, which aims at engineering flexible, generic EAI solutions for varying composite or component applications (depending on the viewpoint).

## Enterprise Integration Methodology

The enterprise integration methodology (Lam & Shankararaman, 2004) provides a practical, step-by-step approach to enterprise integration problems. The approach comprises the following phases.

**Understand the Business Process**
In this phase, the envisioned business process integrating various applications is being analyzed. The respective stakeholders as well as the use cases are specified.

**Map onto Components**
After clearly specifying the collaborative business process, the mapping of the activities contained therein onto application components takes place. This phase can be seen as a feasibility study eventually leading to the adaptation of application components.

**Derive the Requirements**
In this phase, the requirements on the application components are being concretized. The method provides checklists for typical technical characteristics of an EAI solution (e.g., the selection of SOAP [simple open access protocol] for the communication) that can be applied for each of the involved applications.

**Produce the Architecture**
*In this phase, the overall integration architecture is being modeled.*

**Plan the Integration**
Based on the integration architecture, the implementation steps to follow are being defined. This involves project management, infrastructure acquisition, detailed architectural design, implementation, and eventually reengineering of existing applications, and finally quality assurance.

The approach provides valuable support with respect to the technical realization of an EAI solution. The lists of common EAI requirements as well as the various guidelines are especially helpful. However, some parts of the approach are not described very precisely. This applies in particular to the transition between phases as well as to Phases 4 and 5 of the approach.

## OpenEAI Methodology

The OpenEAI methodology (Jackson, Majumdar, & Wheat, 2005) has been derived from the OpenEAI project, which aims at the analysis and exploitation of common EAI practices and principles. The OpenEAI project defines among other things a series of templates for the documentation of EAI solutions as well as a protocol for the message exchange. All project results are freely available under the GNU Free Documentation License.

The OpenEAI methodology comprises the following top-level steps, which are broken down into more detailed activities.

**Analysis**
The systems to be integrated are identified, and the according analysis documentation templates are filled in, thereby specifying the necessary data and control flows.

**Message Definition**
The data flows from the previous step are refined in terms of XML (extensible markup language) documents.

**Creation of the Message Objects**
The XML messages from the previous step are automatically transformed into Java objects that are optimized for the communication via the Java message service (JMS; i.e., with the help of tools made available by the OpenEAI application programming interface [API]).

### Development

The JMS applications are implemented. At this point, the approach is flexible with respect to the chosen implementation approach (including the provided OpenEAI API) and provides some general guidelines.

### Update the Documentation

The implementation results are documented using a series of available documentation templates.

### Deploy

The integration solution is deployed and launched. To this end, the methodology defines a set of deployment patterns illustrating different alternatives for the physical structuring of the solution.

### Monitoring

Monitoring is a continuous process that controls the evolution of the integration solution. Analysis templates are used as the main instrument for this phase.

The main contribution of the OpenEAI methodology is the variety of available documentation templates, which explicitly address the concerns of EAI projects. Yet, the respective documentation schemata are missing that would describe how the different documents can be structured and packaged. The absence of the schemata raises the effort for understanding the templates, which in some cases are quite complex. Finally, it is difficult to estimate whether the proposed API brings advantages compared to other EAI programming frameworks since there are no published experiences about the exposure to the OpenEAI API.

The OpenEAI methodology can also be combined with PuLSE™ DSSA and in particular with the documentation subprocess of the DSSA methodology (see section titled "Documentation").

## ENGINEERING APPROACH: FRAUNHOFER PULSE™

### Software Product Lines

Software development today must meet various demands such as reducing cost, effort, and time to market; increasing quality; and handling complexity and product size. Moreover, organizations must address the need of tailoring software for individual customers.

Organizations trying to comply with the above demands are often led to the development and maintenance of a set of related software products that have common features and differ in context-specific characteristics. However, the different software products are often developed separately. This means that product variations are handled in an ad hoc way and product similarities are not fully exploited. In other words, each product in such a family of products is developed as if it was a single one-of-kind system. This leads to uncontrolled redundancy, unnecessary development overhead, and finally failure in responding quickly to new customer requirements and market needs.

Software product-line engineering (PLE), which was originally introduced by Parnas, promises to avoid these problems by handling a family of software products as a whole. So, a software product line is defined as a family of products designed to take advantage of their common aspects and predicted variability. A product-line infrastructure aims at developing generic software assets that are flexible enough to cover at the same time various customer requirements. Therefore, these assets contain variability, which is resolved (or instantiated) upon creating a customer-specific solution.

### PuLSE™ Overview

PuLSE™ is a method for enabling the conception and deployment of software product lines within a large variety of enterprise contexts. This is

*Figure 3. PuLSE™ overview*

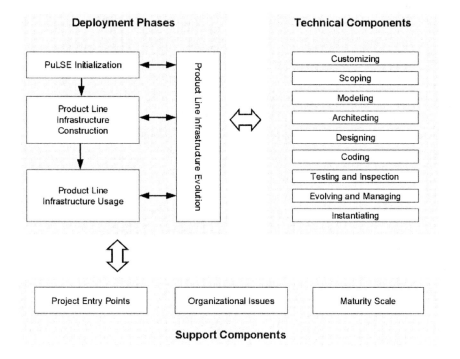

achieved via a product-centric focus throughout the phases of PuLSE™, the customizability of its components, incremental introduction capability, a maturity scale for structured evolution, and adaptations to a few main product development situations (Bayer et al., 1999).

PuLSE™ consists of technical and support components, which accompany its deployment phases as shown in Figure 3. Due to space limitations, we will concentrate only on the main issues here and especially on the pervasive computing perspective. Readers more interested in the PuLSE™ method are referred to *PuLSE™* (2005).

## PuLSE™ Customization

Methodologies up to date are often criticized for being too heavyweight by imposing, for example, a lot of documentation burden that slows down

development. For that reason, PuLSE™ was designed to be customizable. The PuLSE™ initialization phase shown in Figure 3 analyzes the given product-line situation in an enterprise and accordingly customizes the process. The main idea behind the initialization phase consists in the customization factors and decisions attached to the individual technical components. The customization factors are characteristics of great importance for the product line under development.

## PuLSE™ DSSA

The architecture component of PuLSE™ is realized by the PuLSE™ DSSA approach, which integrates architecture development with the analysis of existing systems in order to obtain a reference architecture that takes optimal advantage of existing systems.

The approach starts with the understanding of the business goals and requirements for the new reference architecture as this will, together with the scope of the product family, determine the functionality and qualities that have to be provided. The information from existing systems and experiences made while developing them support the design of the reference architecture.

This information is obtained by request-driven reverse architecture activities. To achieve our goals, we apply the following underlying concepts.

- **Top-down design:** The architecture is designed from the top level and detailed in several iterations.

- **Bottom-up recovery:** Information from existing artifacts is first extracted and then abstracted to higher levels.

- **View-based architectures as interface:** Architectural views are a means of communication between design and recovery, and among stakeholders.

- **Scenario-based reference architecture definition:** The reference architecture of a product family is designed and evaluated with the help of prioritized scenarios.

- **Request-driven reverse architecture:** Analysis of the systems and their artifacts, performed in a request-driven manner, means the right information is provided when it is needed.

Figure 4 illustrates the PuLSE™ DSSA approach.

*Figure 4. The PuLSE™ DSSA approach*

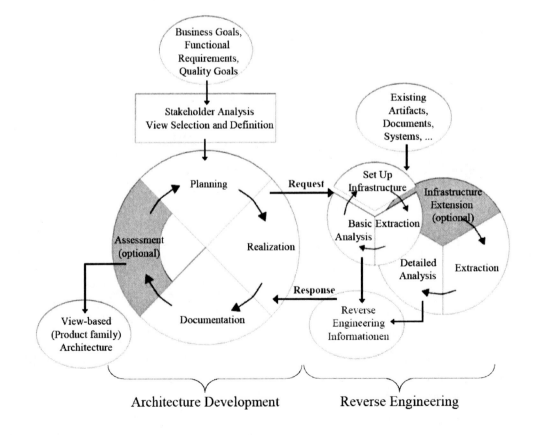

## PuLSE™ DSSA Customization Support

As part of PuLSE™, the DSSA approach is also customizable. There are cases in which standard or predefined architectural views in a predefined view set are not fitting to the respective context. Therefore, DSSA provides an approach for extending existing view sets that elicits and defines the views necessary in a given context based on an approach for view-based software documentation.

The goal of the view customization activity is to define an optimal set of architectural views that are used to document the architectures of the planned products. There are cases in which an existing view set can be used without adaptation. However, often the proposed views are not sufficient to describe all relevant architectural aspects. Then, a new view set must be defined.

The process for customizing architecture views is shown in Figure 5. To prepare the view set definition, business and quality goals are elicited and documented. These are, together with an existing view set that is to be extended, the input to the actual view elicitation and definition activities.

## DEVELOPING INTEGRATIVE BUSINESS APPLICATIONS

The proposed approach is based on the PuLSE™ DSSA architecture presented in "Engineering Approach: Fraunhofer PuLSE™." Therefore, the following subsections reflect the phases of PuLSE™ DSSA. The reverse engineering part of the approach has been left out as it was not in the focus of this work.

### Stakeholder Analysis

The customization support of PuLSE™ DSSA as described in the section "PuLSE™ DSSA Customization Support" is being employed for adapting the approach to the characteristics of the EAI domain. The following sections will therefore elaborate on the qualities required by the involved stakeholders and the respective views.

### Quality Model

The first step toward the customization of PuLSE™ DSSA lies in the definition of a quality model. To this end, we employ the first step of the GQM

*Figure 5. PuLSE DSSA customization process*

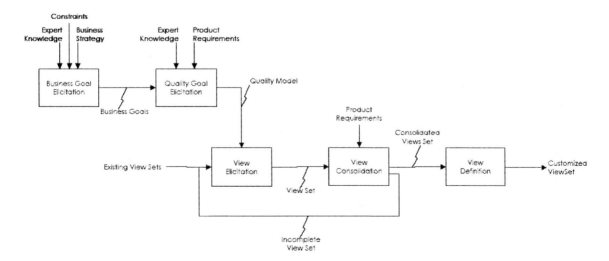

*Table 1.*

| Goal ID | G1 |
|---|---|
| Analyze | the architecture of a composite application |
| For the purpose of | understanding |
| With respect to | the reuse effect on the development effort |
| From the viewpoint of | the developer of a composite application |

*Table 2.*

| Goal ID | G2 |
|---|---|
| Analyze | the architecture of a composite application |
| For the purpose of | understanding |
| With respect to | the maintainability of the composition |
| From the viewpoint of | the developer of a composite application |

approach (Basili, Caldiera, & Rombach, 1994) for the composition and integration scenarios.

## Composition Viewpoint

For the composition viewpoint, we can define the following initial goal, which is related to the reuse of an external application and to the effect of this reuse to the effort required for implementing a composite application. (See Table 1.)

For this goal, the following questions can be formulated, which when answered indicate the degree of fulfillment of the above goal.

*Q1. Which external applications have been already integrated?*

*Q2. What was the role of the integrated applications in the composite application?*

*Q3. What integration technologies are supported?*

*Q4. What was the effort connected to each integration activity?*

*Q5. What was the experience or background of the involved developers?*

*Q6. How heterogeneous was the external application?*

*Q7. What was the visibility to the external application (e.g., black box)?*

*Q8. Which level of integration has been reached?*

Apart from the effort of a composition project, another important quality that must be considered is the maintainability of the solution. Therefore, the following goal can be defined. (See Table 2.)

This goal leads to these subsequent questions.

*Q9. What is the impact of changing the interface of the external application?*

*Q10. What is the impact of changing the interface of the composite application?*

*Q11. What is the impact of changing the business logic of the external application?*

*Q12. What is the impact of changing the business logic of the composite application?*

*Q13. What is the impact of omitting the external application?*

**Integration Viewpoint**
The goals and questions for the integration viewpoint can be obtained in a similar way. (See Table 3.)

*Q14. Which composite applications are supported?*

*Q15. Which component services are meant for integration with external applications?*

*Q16. Which integration technologies are supported?*

*Q17. What was the effort connected to each integration activity?*

*Q18. What was the experience or background of the involved developers?*

*Q19. How heterogeneous was the composite application?*

*Q20. What was the visibility to the composite application (e.g., black box)?*

*Q21. Which level of integration has been reached? (See Table 4.)*

The questions related to the latter goal are identical to questions Q9 to Q12.

*Table 3.*

| **Goal ID** | *G3* |
|---|---|
| ***Analyze*** | the architecture of a component application (i.e., a part) |
| ***For the purpose of*** | understanding |
| ***With respect to*** | the reusability of the component application |
| ***From the viewpoint of*** | the developer of a component application |

*Table 4.*

| **Goal ID** | *G4* |
|---|---|
| ***Analyze*** | the architecture of a component application (i.e., a part) |
| ***For the purpose of*** | understanding |
| ***With respect to*** | the maintainability of the component application |
| ***From the viewpoint of*** | the developer of a component application |

## Required Views

Following the definition of the goals and questions, we can derive the following minimum set of architectural views, which provides the required information for answering the above questions. The views proposed by the UML 2.0 (unified modeling language) specification are used as a reference at this point. However, the selection of the UML as the notation is not binding.

Table 5 contains a column on the tailoring required for explicitly addressing the goals defined in the previous sections. This tailoring can be accomplished by using the common UML extensibility mechanisms through special features of the employed modeling tool or through the definition of custom views. Furthermore, the column EAI Layers relates the derived views to common EAI realization layers as described in Themistocleous and Irani (2003). Finally, the rightmost column maps the views to each of the EAI viewpoints that we consider. For instance, for the composition viewpoint, the activity view is recommended, while this view is not required in the integration viewpoint.

## Planning

In the planning phase, the goals of the EAI project are being refined in terms of scenarios. Here the use-case templates proposed by Cockburn (2001) come into play. For the definition and refinement

*Table 5. Required views for integrative development*

| View | Related Questions | Required Tailoring | EAI Layer | EAI Viewpoint |
|---|---|---|---|---|
| Activity | Q2, Q4, Q5, Q7, Q10, Q11, Q12 | – Information on the required effort <br><br> – Information about the involved developers | Process automation | Composition |
| Deployment | Q5, Q19 | None | Transportation, Translation | Composition, Integration |
| Structure (class) <br><br> Interaction | Q1, Q3, Q4, Q5, Q6, Q7, Q8, Q9, Q12-Q21 | – Information on connectors <br><br> – Information on translation between heterogeneous data schemes <br><br> – Information on the required effort <br><br> – Information about the involved developers | Connectivity | Composition, Integration |

of scenarios, the following scheme can be taken into account.

- **Data integrationscenarios:** Scenarios that deal with integrating heterogeneous data sources
- **Application-logic integration scenarios:** Scenarios that deal with the invocation of external application logic
- **Process integration scenarios:** Scenarios that deal with the development of collaborative business processes (i.e., across the boundaries of a single enterprise)
- **Real-time enterprise scenarios:** Scenarios that deal with the synchronization of different business processes connected via a collaborative process
- **Complex event processing scenarios:** Scenarios that deal with the handling of multiple event streams arising from the integrated applications

After definition, the scenarios are prioritized and mapped to the iterations that will follow. For prioritizing scenarios, PuLSE™ DSSA provides some common heuristics like the economic value, the typicality, or the required effort. The latter can be supported by the above classification scheme. In general, it makes sense to start with data integration scenarios and to move on toward the more complex event handling scenarios.

Scenarios developed in this phase will eventually contain variability. In other words, the scenario descriptions can contain optional or parameterized sections that will be selected or set according to the variability decision that will be made for a specific integration solution. Such a variability decision could be, for example, the type of external system to be integrated. Furthermore, all unresolved variability decisions for a scenario are collected in the scenario's decision model. On the other hand, the set of resolved decisions, which reflect a specific integration solution, are collected in the scenario's resolution model.

## Realization

In the realization phase, the requirements posed by the integration scenarios are realized by making the necessary design decisions. Here the EAI approaches (as described in the introduction), common architectural patterns, and design principles can come into play. The EAI literature provides a great amount of help in this regard. For example Themistocleous and Irani (2003) define the following layers, which should be part of every integration solution.

- **Process automation:** Deals with the business processes that consist of activities provisioned by the external system
- **Transport:** Deals with the transport of messages between heterogeneous systems
- **Translation:** Deals with the translation between heterogeneous data models
- **Connectivity:** Deals with the access to external systems

Regarding EAI design decisions, Keller (2002) proposes the following schema.

- **Communication models:** Selection of synchronous or asynchronous communication
- **Integration method:** Integration over the data layer, the graphical user interface, the application interfaces, or over application components
- **Infrastructure:** Selection of a communication infrastructure (e.g., object brokers, message-oriented middleware, enterprise service bus; Chappell, 2004)

Finally, the literature (e.g., Chappell, 2004; Fowler, 2005; Hohpe & Woolf, 2004) proposes a series of design patterns as well as workflow patterns (Wohed et al., 2003) that are relevant for the realization phase.

*Figure 6. Realization phase*

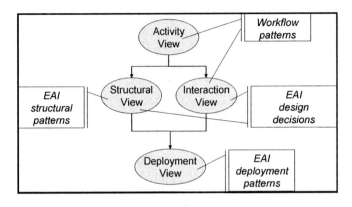

*Figure 7. Documentation process*

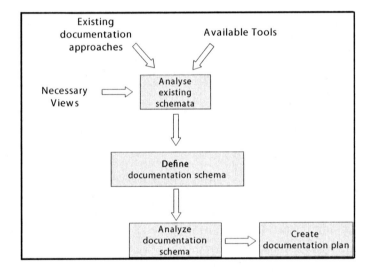

The realization phase is also responsible for fulfilling the variability, which is eventually contained in the scenarios. In this case, the variable sections in the scenarios are to be transformed to variable architectural models. In other words, the resulting models also contain optional or parameterized sections, which must be set according to the scenario resolution models.

Figure 6 depicts the ordering of activities in the realization process as well as the according patterns and models that can be used as input.

## Documentation

In the documentation phase, the design decisions made during the realization are documented according to a documentation scheme, which is derived as shown in Figure 7. The scheme defines the types of documents to be produced as well as their structure and interrelations (cf Bayer, 2004). In other words, the scheme can be seen as a synonym of the view set discussed in the section titled "PuLSE™ DSSA Customization Support." As shown in Figure 7, one of the inputs to the documentation process is the group of views mentioned in "Required Views."

Since PuLSE™ DSSA is an iterative approach, the documents are to be updated and refined iteratively. For illustrating the variability in the models, special notations can be used, for example, UML stereotypes or special coloring (cf Flege, 2000).

The consistency and correctness of the produced documents can be checked by inspection techniques as proposed, for example, in *Software Inspektion* (2006). Perspective-based inspections (Laitenberger, 2000) are especially of interest at this point as they can enable the inspection of integration architectures from the special viewpoint of an EAI stakeholder (i.e., composite application developer). In this case, the inspection process defines the different inspection scenarios based on perspectives and provides concrete guidelines. For example, the composite application developer's perspective could be described as shown in Box 1.

## Assessment

In the assessment phase, the integration architecture is being judged against the integration requirements. In other words, the goal is to check to which degree the delivered architecture meets the scenarios of the planning phase as well as additional scenarios that reflect nonfunctional requirements like performance or robustness. The latter can also comprise special interoperability qualities as described, for example, in IEC TC 65/290/DC (2002).

For the assessment, context-based architecture evaluation methods such as SAAM or ATAM are typically used (Clements, Kazman, & Klein, 2002). Furthermore, the GQM method discussed in the section "Quality Model" can come again into play by establishing a measurement program that iteratively measures the satisfaction of the composition or integration goals.

*Box 1.*

---

**Composite Application Expert**

*Introduction*

Imagine you are the expert of the composite application under development. Your interest lies in the efficient reuse of the integrated external applications.

*Instructions*

For each of the produced architectural documents, do the following.
- Determine the integrated external applications.
- Determine the design decisions made for the integration.

Identify defects in the architecture using the following questions.
- Are the available interfaces to the external applications fully exploited?
- Is there any customization interface of the external applications that has not been used?
- Is the translation between heterogeneous data types being considered?

---

153

## CASE STUDIES

### Introduction

This section will compactly demonstrate the application of the PuLSE™ DSSA approach for two of the cases, which arose in the context of this work. The first scenario reflects the composition viewpoint and deals with an ERP (enterprise resource planning) application that aggregates an external GIS (geographical information system) and a PCS (process control system). The other case takes the opposite view and deals with the integration of the PCS into the ERP system.

### Composition Viewpoint

The ERP system under consideration enables the management of human resources and objects within an organization. One central system feature is the observation and planning of tasks, which are carried out on objects. Examples of objects are devices, sensors, or actuators, and examples of tasks are repairs, replacements, or measurements.

Since objects are often scattered on the organization's premises, the necessity of integrating the geographical information of the controlled objects becomes apparent. This information can be integrated into the workflow subsystem, which drives the task management and in so doing can facilitate the localization of an object. For the same reason, the ERP system aims at the integration of process control for accessing devices remotely and for avoiding costly and error-prone measurements (e.g., meter reading) or calibrations on devices.

### Composition Planning

During the planning phase, the functionalities of the external applications have been captured in terms of scenarios. These scenarios practically represent activities in the overall ERP workflow,

which will be realized by the external applications. For simplicity, the following list contains only a part of the collected scenarios.

1. Display the position of a selected object in a map obtained from the GIS.
2. Obtain the current values of a selected object from the PCS.
3. Synchronize the ERP system data repository with the repositories of the partner applications.

Each of the scenarios contains variability, which reflects the different types of external systems that can be integrated. The first scenario, for instance, contains an optional step, which allows setting the resolution of the returned object bitmap. The third scenario is the most generic one as it provides for data synchronization with any connected system.

After the scenario definition, the scenario prioritization took place. The third scenario received the greatest priority since the content integration on the basis of common metadata is known as a foundational step in each integration project (Martin, 2005). The realization of the third scenario requires the definition of a common metadata scheme, which is used in later iterations for the other two scenarios. The latter have received equal priorities.

In the following, we will focus on the data synchronization scenario, which due to its priority has been chosen for the first iteration of the integration architecture. The variability of this scenario is captured in the following decision model. Only the variability that directly affects the integration solution is listed here. For instance, the scenario provides for manual or scheduled initiation of the synchronization procedure. Yet, since this decision will be assumed by the ERP system and does not affect the external GIS or PCS, it is not listed in Table 6.

*Table 6.*

| ID | Decision | Possible Resolutions |
|----|----------|----------------------|
| D1 | Which external systems are connected? | GIS or PCS |
| D2 | What types of modifications can be synchronized between data sources? | Changes or Deletions |
| D3 | Is the synchronization connection secured? | No or Yes |

*Table 7.*

| Process Execution | Initiate synchronization, show the synchronization data to the user, and update the ERP repository. |
|-------------------|----------------------------------------------------------------------------------------------------|
| Translation | Translate the extracted data to the ERP system environment (e.g., .Net framework). |
| Transportation | Transport the extracted data to the ERP system. |
| Connectivity | Send synchronization requests to the external systems. |

## Composition Realization

For facilitating the realization of scenarios, the layers discussed in the section "Planning" come into play. Here scenarios are being refined as shown in Table 7. The variability of the decision model comes into play again in the later phases of the realization and therefore is not reflected in the table, which for that reason is variance free.

The realization in the composition viewpoint concentrates on reusing external synchronization services. The first scenario relies on the creation of the activity views. The latter are directly associated with the process execution layer of the previous section. The top-level activity model for the synchronization scenario is depicted in the UML 2.0 activity diagram in Figure 8. As it can be seen, the diagram is generic as it considers the open decision D1.

In the event of integrating both GIS and PCS (i.e., resolving decision D1 in GIS and PCS),

the diagram in Figure 8 is instantiated as in Figure 9.

As shown in Figure 9, the activity of initiating synchronization is nested (symbol in the bottom-right corner). For the realization of this activity, the Web service pattern realization for .Net has been selected. Therein, the activity iterates over all external systems, against which the synchronization is necessary, and for each case it uses proxy objects for sending the according synchronization messages. The activity is shown in more detail in Figure 10. Again, the activity is generic and can be instantiated depending on the resolution of the decision D1.

Figure 10 contains a conditional node that relates to the decision model discussed in "Composition Planning." The activity in the *if* clause examines the current resolution model, which is an instantiation of the decision model where the according decisions have been resolved. As shown in the picture, the resolution model is given as an

*Figure 8. Generic synchronization activity*

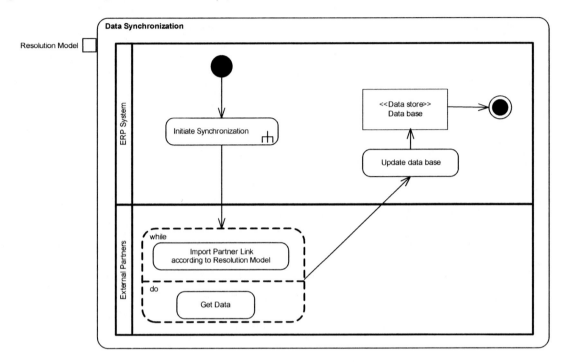

*Figure 9. Synchronization workflow*

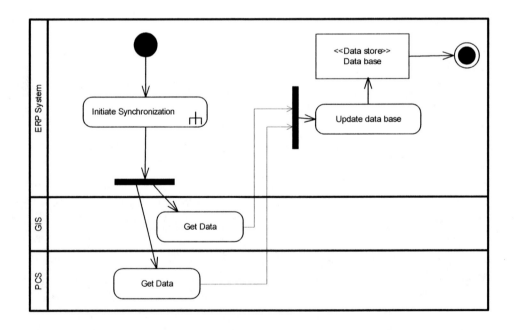

*Figure 10. ERP system's "initiate synchronization" activity*

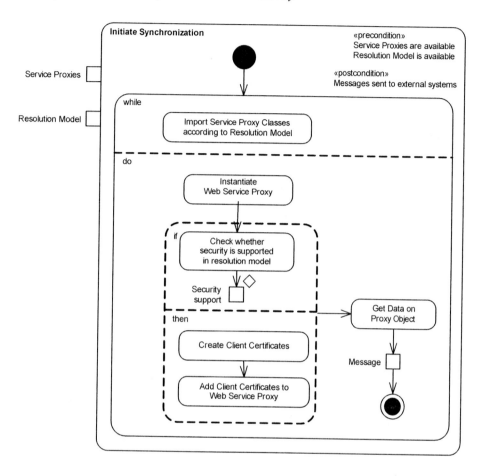

input to the synchronization activity and must be available for the activity to be instantiated properly. The examination of the resolution model returns a Boolean value denoting whether security is supported in the current setting. If this is the case, then the two activities in the *then* clause are being executed and certificates are being added to the service proxies, thereby securing the resulting synchronizations.

From the generic synchronization workflow, we can now derive the following excerpt of the ERP system architecture. The latter is also generic as it uses the generic ServiceProxy class that can be instantiated for the different types of partner systems. Figure 11 provides insights into the connectivity layer discussed in the sec-

tion "Composition Planning." The transport and translation layers are not modeled in this case since they are realized automatically by the underlying infrastructure (i.e., .Net framework).

In essence, Figure 11 specifies the remote synchronization service. The diagram can be used for deriving a more detailed service description (e.g., Web services description language, WSDL), which must then be fulfilled by the external applications. In the context of this work, it has not been necessary to provide any additional translation logic that maps the above specification to the external systems. Yet, in many cases this step will be necessary.

For the implementation of the above architecture in .Net, the developer must use an alternated

*Figure 11. Synchronization service specification*

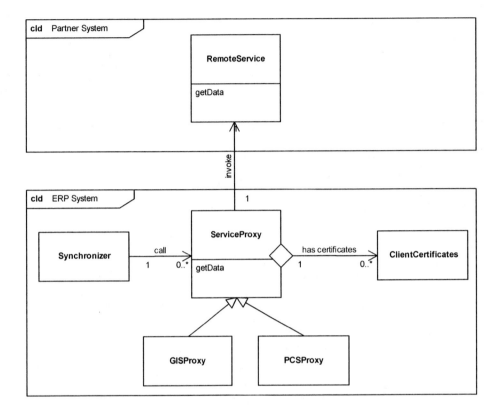

version of the typical Web service client development process with the .Net framework. The latter begins with the addition of a service reference to the client development project, followed by the automatic generation of the proxy classes. For the above genericness to be accomplished, the developer must first create the service descriptions and/or proxy classes, and then bound them to concrete implementations.

## Composition Assessment

For the assessment of the synchronization architecture, the design decisions that have been taken must be evaluated against scenarios that reflect the goals of the integration. As a starting point for the derivation of such scenarios, common interoperability standards have been employed. For example, the interoperability standard IEC TC 65/290/DC (2002) defines different compatibility levels. The architecture presented in the section "Composition Realization" would fail to reach the fourth level of compatibility, which foresees data transfer between heterogeneous applications. The architecture developed so far has support only for the exchange of object data and not for the exchange of the according data types. In other words, if a new object type is created in the PCS, it is not possible to synchronize the respective object data across the PCS boundaries. Therefore the synchronization architecture had to be extended with additional operations for the exchange of metadata (i.e., object types), and furthermore a common metadata scheme (i.e., common structure of an object type) had to be defined.

## Integration Viewpoint

The PCS under consideration enables the remote monitoring and diagnostics of facilities (e.g., electricity and water supply or long-distance heating facilities). The core of the system is based on so-called process variables. These are variables that represent the status of the monitored facility and the objects contained therein. An example of such a process variable is the fill level of a tank.

Since the PCS provides low-level technical information, the integration with high-level applications is sensible. The technical information can be encapsulated and delivered as value-added information to the end user. In particular, it definitely makes sense to support the integration with ERP systems, which thereby can reach a high degree of process automation and management. For that reason, the PCS aims at enabling the connectivity with various ERP systems including, of course, the ERP system presented in the section "Composition Viewpoint."

### Integration Planning

As prescribed by the methodology, the planning phase included the derivation of integration scenarios. Some of these scenarios are listed below.

1. Read and write access to PCS process variables from the ERP system
2. Subscription to PCS events from the ERP system
3. Read access to the PCS configuration from the ERP system (the configuration contains the set of objects to be monitored from the PCS)

For the same reasons as in the "Composition Planning" section, the third scenario, which practically also deals with data synchronization, received the highest priority. It is interesting to note here that although both synchronization scenarios have the same semantics, there is a difference in the terminology. Such an issue can be addressed by semantic integration, which is an evolving discipline in the EAI area (Castano et al., 2005).

In the following, we will focus on the second scenario, namely, the event handling across the PCS boundaries. The decision model for this scenario is given in Table 8.

### Integration Realization

The refinement of the scenario to the EAI layers is shown in Table 9.

*Table 8.*

| ID | Decision | Possible Resolutions |
|----|----------|---------------------|
| D1 | Which external systems are connected? | ERP system or<br>∅ |
| D2 | How does the subscription take place? | ERP system accesses the PCS for subscribing to PCS events or<br>PCS accesses the ERP system for connecting PCS events (e.g., threshold value overrun on Object X) with high-level ERP event definitions (e.g., Object X alert) |

*Table 9.*

| Process Execution | Notify subscribers about PCS events. |
|---|---|
| *Translation* | Translate the event data to a common event scheme. |
| *Transportation* | Transport the event data to the external system. |
| *Connectivity* | Register the ERP system as a subscriber to the PCS. |

*Figure 12. PCS event service realization*

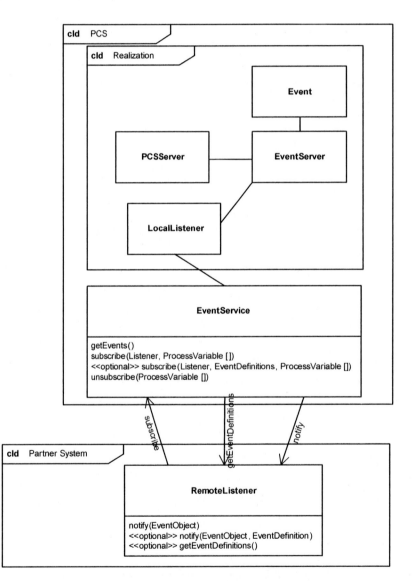

*Table 10.*

| ID | Decision | Possible Resolutions |
|----|----------|----------------------|
| D3 | For which events can a partner application subscribe? | All or Events marked as external |

In the integration viewpoint, the realization concentrates on the development of a reusable event service that can be used by many external applications. In this case, the activity view is not necessary and the developers can focus on the service usage view, which is depicted as a structural diagram in the Figure 12.

Figure 12 contains stereotyped elements that reflect the variability of the decision model. Moreover, the picture depicts the internal realization of the event service, which if necessary can be hidden from the service user. The diagram provides the connectivity view of the event service.

## Integration Assessment

During the assessment of the event service realization, it was noticed that in some cases it is not feasible to subscribe an external application to all PCS events, as the latter can be numerous. For that reason, it was decided to introduce an additional variability decision.

This in turn has led to the introduction of an additional optional class to the design that represents external events (i.e., events that are visible to external applications).

## CONCLUSION AND FUTURE WORK

This chapter has presented a methodology for the creation of EAI solutions based on the concepts of software product-line engineering. The overall idea was to view the external applications with which a given system wants to integrate as a family

of systems. In this way, the flexibility required by EAI applications can be assured. The proposed PuLSE™ DSSA approach and its customization to the EAI domain enables the creation of EAI applications that are flexible and contain variability decisions. The latter are resolved when a concrete EAI setting is at hand, thereby producing a concrete member of the product family.

The application of the approach has shown that the EAI domain is characterized by great variability, which makes the explicit management of the variations necessary. Moreover, the customization of the PuLSE™ methodology has shown that the EAI domain has special concerns with respect to architectural design, quality assurance, and implementation. While the implementation concerns enjoy adequate support in the EAI community, the architecture and quality assurance issues are not systematically considered. The proposed approach addresses these issues explicitly.

There are various research topics that are subject to our future investigations. The reverse engineering part of PuLSE™ DSSA must be definitely incorporated in the approach. To this end, the reflection model approach (Murphy & Notkin, 1995) for obtaining the architectures of existing systems as well as for comparing the realization of an EAI solution against the originally envisioned architecture can come into play. For this, current reflection tools must be extended toward special EAI characteristics like asynchronous messaging or application-server-based solutions. Furthermore, extending current business process technologies like BPEL (business process execution language) with variation management support is also a challenge (see also *Process Family Engineering in Service-Oriented Applications [PESOA]*, 2005). Finally, since EAI applications are often seen as complex adaptive systems, it will be of great interest to investigate whether the flexibility discussed in this chapter can be completely shifted to the run time, thereby enabling the dynamic adaptation of EAI systems and composite applications in particular.

## REFERENCES

Basili, V. R., Caldiera, C., & Rombach, H. D. (1994). Goal question metric paradigm. *Encyclopaedia of Software Engineering, 1*, 528-532.

Bayer, J. (2004). View-based software documentation. In *PhD theses in experimental software engineering* (Vol. 15). Stuttgart, Germany: Fraunhofer IRB Verlag.

Bayer, J., Flege, O., et al. (1999). PuLSE: A methodology to develop software product lines. *Proceedings of the Fifth ACM SIGSOFT Symposium on Software Reusability (SSR'99)*, 122-131.

Bayer, J., et al. (2004). *Definition of reference architectures based on existing systems, WP 5.2, lifecycle and process for family integration.* Proceedings of the Eureka Σ! 2023 Programme, ITEA Project ip02009.

Castano, S., et al. (2005). Ontology-based interoperability services for semantic collaboration in open networked systems. In D. Konstantas, J.-P. Bourrières, M. Léonard, & N. Boudjlida (Eds.), *Interoperability of enterprise software and applications.* Springer-Verlag.

Chappell, D. A. (2004). *Enterprise service bus.* Sebastopol: O'Reilly and Associates, Inc.

Clements, P., Kazman, R., & Klein, M. (2002). *Evaluating software architectures: Methods and case studies.* Addison-Wesley.

Cockburn, A. (2001). *Writing effective use cases.* Boston: Addison-Wesley.

Dumas, A. (2002). *Select the best approach for your EAI initiative* (Sunopsis White Paper). Sunopsis, Inc.

*The European enterprise application integration market.* (2001). Frost & Sullivan Studies.

Flege, O. (2000). *System family architecture description using the UML* (Fraunhofer IESE Report No. 092.00/E). Kaiserslautern, Germany.

Fowler, M. (n.d.). *Patterns in enterprise software.* Retrieved November 2005 from http://www.martinfowler.com

Hohpe, G., & Woolf, B. (2004). *Enterprise integration patterns.* Addison-Wesley.

IEC TC 65/290/DC. (2002). *Device profile guideline. TC65: Industrial process measurement and control.*

Jackson, T., Majumdar, R., & Wheat, S. (2005). *OpenEAI methodology, version 1.0.* OpenEAI Software Foundation.

Keller, W. (2002). *Enterprise application integration.* Dpunkt Verlag.

Laitenberger, O. (2000). Cost-effective detection of software defects through perspective-based inspections. In *PhD theses in experimental software engineering* (Vol. 1). Stuttgart, Germany: Fraunhofer IRB Verlag

Lam, W., & Shankararaman, V. (2004). An enterprise integration methodology. *IT Pro.*

Martin, W. (2005). *Business integration 2005: Status quo and trends.* Proceedings of EAI Competence Days Road Show.

Murphy, G., & Notkin, D. (1995). Software reflexion models: Bridging the gap between source and high-level models. *Proceedings of the Third Symposium on the Foundations of Software Engineering (FSE3).*

*Process Family Engineering in Service-Oriented Applications (PESOA)*. (n.d.). Retrieved November 2005 from http://www.pesoa.org

*PuLSE™*. (n.d.). Retrieved November 2005 from http://www.iese.fhg.de/pulse

Schmidt, J. G. (2002). Transforming EAI from art to science. *EAI Journal*.

*Software inspektion*. (n.d.). Retrieved March 2006 from http://www.software-kompetenz.de

Themistocleous, M., & Irani, Z. (2003). Towards a novel framework for the assessment of enterprise application integration packages. *Proceedings of the 36th Hawaii International Conference on System Sciences (HICSS'03)*.

Wohed, P., et al. (2003, April). *Pattern based analysis of EAI languages: The case of the business modeling language*. Proceedings of the International Conference on Enterprise Information Systems (ICEIS), Angers, France.

# Chapter X
# Visual Environment for Supply Chain Integration Using Web Services

**Guillermo Jiménez-Pérez**
*ITESM – Monterrey Campus, Mexico*

**Javier Mijail Espadas-Pech**
*ITESM – Monterrey Campus, Mexico*

## ABSTRACT

*This chapter presents a software tool to simplify application integration among enterprises. The visual environment provides facilities to display in a graphical interface the structure of databases and applications. By clicking in the appropriate tables, functions, and fields, enterprises could specify the data sources needed for integration. The details of the applications are extracted from WSDL and metadata definitions. Once the fields for every access are specified, integration code is automatically generated in either Java or C#. The chapter describes the visual tool and its use for automatic supply chain integration from metadata or WSDL descriptions. It describes how users specify the data elements and the integration links and how the code integrating the specified elements is automatically generated. The generated code uses Web services standards to integrate the specified sources.*

## INTRODUCTION

Today, for a large number of companies, the information systems landscape is a complex mixture of old and new technologies. Traditionally, companies built heterogeneous applications for specific problems and took advantage of business opportunities. The normal approach was that enterprises developed their applications in an isolated way, with every department independently developing their own applications and data storages. Currently, these companies confront the necessity of sharing information among those heterogeneous applications (commonly named legacy systems) and the new systems being developed (Juárez Lara, 2001). The Internet brought about an increased necessity for companies: the needs of Web presence and the integration of their systems.

As more and more companies pursue the benefits of electronic business through the Internet, they face the need of enterprise integration (EI) as key technological enabler in transforming their business processes. A typical form of EI is Web integration. In this scenario, a company wants to offer its existing products and services over the Internet, so it builds front-end Web systems and integrates them with its back-end legacy systems. A more complex scenario involves enterprise application integration (EAI). In this scenario, the company links up its previously isolated and separated systems to give them greater leverage (Wing & Shankararaman, 2004). An emerging EI scenario is business-to-business (B2B) integration, which occurs when companies integrate their own business processes with those of their business partners to improve efficiency within a collaborative value chain. Examples of B2B applications include procurement, human resources, billing, customer relationship management (CRM), and supply chains (Medjahed, Benatallah, Bouguettaya, Ngu, & Elmagarmid, 2003). Effective and efficient supply chain integration is vital for enterprise competitiveness and survival, and with the emergence of the e-business era, supply chains needs to extend beyond the enterprise boundaries (Siau & Tian, 2004). The need for supply chain integration techniques and methodologies has been increasing as a consequence of the globalization of production and sales, and the advancement of enabling information technologies (Ball, Ma, Raschid, & Zhao, 2002).

This chapter describes the design and implementation of a visual tool for automatic supply chain application integration by using Web services technologies within the PyME CREATIVA project (explained later). The tool provides visual specification facilities and the capability of code generation for B2B integration of enterprises in supply chains.

The chapter is organized as follows. The following section describes the PyME CREATIVA project as an example of a project addressed to supply chain integration, and presents an overview of electronic commerce and supply chain integration. Then the chapter describes the core technologies used in implementing the visual tool—EAI, Web services, and code generators—and presents the analysis and design of a visual tool for enterprise integration to a supply chain. Next it describes a study case of supply chain integration using the infrastructure described. Finally, results and conclusions are given.

## THE PyME CREATIVA PROJECT

The creation of industrial networks to foster the competencies of small and medium enterprises (SMEs) needs integrated information services (e-services) that enable coordination and cooperation among the different SMEs, allowing them to share their technological resources, creating virtual organizations. These e-services should be integrated in an open technological platform that is easy to access, which is known as an e-hub (Molina, Mejía, Galeano, & Velandia, 2006). A main goal in the PyME CREATIVA project is to produce a low-cost management and operational infrastructure for value-added networks of SMEs. Several steps are being applied to build the infrastructure defined by a set of services. The first step is defining the business model of value-added networks. Secondly, one needs to design the IT architecture. The third step consists of designing the services supporting the operation of the network. Lastly, it is necessary to use a combination of open-source technologies to produce a low-cost implementation of the architecture. The e-hub tries to offer a variety of integrated e-services that are necessary for dynamic market competition. This e-hub can be seen as a marketplace platform, where SMEs can execute trading processes, purchase orders, supply chain management, request for quotations, and other types of e-services with other SMEs in the hub. For the PyME CREATIVA project, the following e-services are being developed and integrated within the e-hub architecture.

- **E-marketing:** Integrates different technologies for the development of customizable portals to allow SMEs to promote their products and services
- **E-productivity:** Incorporates technologies for the SMEs' diagnosing, planning, and monitoring
- **E-engineering:** Engineering collaboration that integrates design and manufacturing technologies for integrated product development
- **E-brokerage:** Integrates technologies to support the development of virtual organizations, enterprise industrial clusters, and so forth
- **E-supply:** Implements technologies for the integration of logistic services, industrial

plant configuration, client-supplier relation management, and supply chain tracking. This electronic service is used later as study case of visual application integration.

Figure 1 shows the architecture of an e-services hub. Many technologies are involved, but the main components are a service-oriented architecture (SOA) and business process management (BPM) to control the workflow execution within the e-services. The portal platform is accessible through standard Web protocols (hypertext transfer protocol [HTTP], Web services description language [WSDL]) and a servlet engine (J2EE). To access the e-services, SMEs deploy their own Web portal that acts as an access point. The following section describes the concepts and technologies involved in implementing the e-services hub. Open-source technologies are used to implement the collaboration services to keep cost low for SMEs.

*Figure 1. E-services hub architecture*

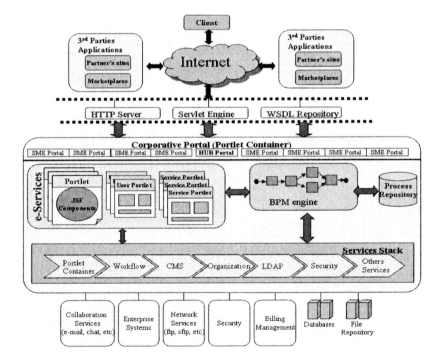

*Figure 2. Application integration problem domains (Linthicum, 2004)*

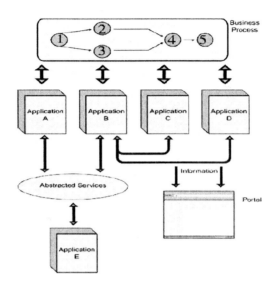

## INVOLVED TECHNOLOGIES

### Enterprise Application Integration

Application integration (AI) is a strategic approach for binding many information systems together, at both the service and information levels, supporting their ability to exchange information and leverage processes in real time. Application integration can take many forms, including internal application (EAI) or external application integration (B2B). AI involves a combination of issues. Each organization and trading community has its own set of integration issues that must be addressed. Because of this, it is next to impossible to find a single technological solution set that can be applied universally. Therefore, each AI solution will generally require different approaches (Linthicum, 2004).

For example, in Figure 2, a business process is represented as a set of steps. Each step may require information from different applications, and in some cases, applications need to interact and share information with other applications or Web portals.

Several technologies are available for application integration. For example, RPC (remote procedure call) was one of the first technologies used to build distributed systems with server-based applications. Another important technology is RMI (remote method invocation), which allows the communication between distributed Java objects. CORBA (Common Object Request Broker Architecture) and DCOM (Microsoft technology) are component objects models for distributed systems and application interoperability. A newer technology for integration is service-oriented application integration (SOAI), which allows enterprises to share application services. Enterprises could accomplish this by defining application services they can share. Application services can be shared either by hosting them on a central server or by accessing their interapplication (e.g., through distributed objects or Web services; Linthicum, 2004).

## Web Services

A Web service is a software system designed to support interoperable machine-to-machine interaction over a network (Haas & Brown, 2004). Services are described by interfaces using a machine-processable format (specifically WSDL). Other systems interact with the Web service in a manner prescribed by its description using SOAP (simple object access protocol) messages, typically conveyed using HTTP with XML (extensible markup language) serialization in conjunction with other Web-related standards (see Figure 3).

In a typical service-based scenario, a service provider hosts a network-accessible software module (an implementation of a given service). Figure 3 depicts a service provider that defines a service description of the service and publishes it to a client or service discovery agency, through which a service description is published and made discoverable. The service requester uses a find operation to retrieve the service description, typically from the discovery agency, and uses the service description to bind with the service provider and invoke the service or interact with service implementation.

The technology necessary for SOA implementation relies on a common program-to-program communications model built on existing and emerging standards such as HTTP, XML, SOAP, WSDL, and UDDI (universal description, discovery, and integration; Kreger, 2001).

The wide support of core Web services standards by major enterprise software vendors is a key reason why Web services technology promises to make application integration both within an enterprise and between different enterprises significantly easier and cheaper than before. Standards mean that not only applications can be implemented on different platforms and operating systems but also that the implementations can be modified without affecting the interfaces (O'Riordan, 2002). To build integration-ready applications, the service model relies on SOA. As Figure 3 shows, SOA is a relationship of three kinds of participants: the service provider, the service discovery agency, and the service requester (client). The interactions involve publishing, finding, and binding operations (Champion, Ferris,

*Figure 3. Service-oriented approach (Adapted from Papazoglou, 2003)*

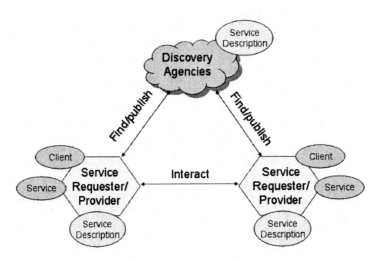

Newcomer, & Orchard, 2002). These roles and operations act upon the service artifacts: the service description and the service implementation (Papazoglou, 2003).

## Code Generators

Software system generators automate the development of software. Generators automatically transform specifications of target systems into actual source code, and rely on libraries of parameterized, plug-compatible, and reusable components for code synthesis (Batory & Geraci, 1997). The primary benefit of any code generator is to reduce the amount of repetitive code that must be produced, thus saving time in the development cycle (Whalen & Heimdahl, 1999). Another benefit to this approach is the ability to extend the services generated, enabling the code generator to act as a force multiplier for programmers. Having a code generator synthesize complex code dealing with concurrency, replication, security, availability, persistence, and other services for each server object will ensure that all servers follow the same enterprise rules. By using a code generator, developers can experiment more rapidly with different architectures. Another benefit obtained when using a code generator for the data layer of enterprise architecture may be its ability to deal with evolving technology (Whalen & Heimdahl).

Generators are among many approaches that are being explored to construct customized software systems quickly and inexpensively form reuse libraries. CORBA, for example, and its variants simplify the task of building distributed applications from components; CORBA can simplify the manual integration of independently designed and stand-alone modules in a heterogeneous environment by generating proxies for client interaction with a server. Generators are closer to tool kits, object-oriented frameworks, and other reuse-driven approaches because they

focus on software domains whose components are not stand alone, and that are designed to be plug compatible and interoperable with other components (Batory & Geraci, 1997).

## VISUAL INTEGRATION TOOL ANALYSIS AND DESIGN

### Target Overview

PyME CREATIVA requires at least four types of integration with the e-services: common legacy applications, .NET applications, J2EE applications, and hub internal services. The goal of a generator aimed at application integration is to produce code to integrate the e-services hub with legacy applications and other platforms through Web services technologies.

Figure 4 shows an overview of the visual tool and involved components. One component is a graphical user interface (GUI) for visual specification of the integration requirements and methods. Basically, the GUI consists of a set of forms or screens to introduce information related to application integration. Another component performs XML data mapping of the visual specification; this component creates an integration project for each enterprise. A third component is the code generator; this component will produce the necessary code to achieve the application integration through Web services technologies. The generated Web services code is deployed in the Web services container.

For implementing a code generator for application integration in a domain, it is necessary to analyze and understand the domain processes and operations. Once operations are identified, it is necessary to define access methods to the information involved. Visual integration provides the integration developer with the necessary elements to specify the information sources. In this way, the integration developer just needs to

*Figure 4. Visual tool overview*

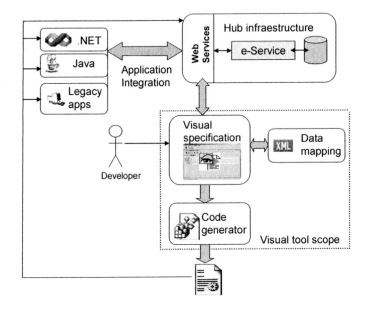

*Figure 5. Service-level integration*

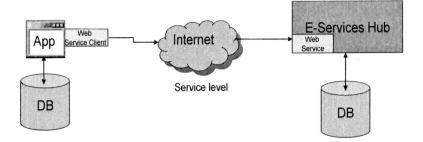

specify the operations in the domain processes through visual selection of information sources (remote operations).

The code generator produces two different code packages: one for the application to be integrated and another for the e-services hub; this is depicted in Figure 5. The generated code allows the e-hub and clients to interact through a Web services mechanism such as SOAP messages.

The main purpose of a visual integration tool is to provide integrators a simple configuration environment for application integration into the e-services hub.

Figure 6 shows the information sources to the visual integration tool: Web service descriptions and directives descriptions. From these and the visual specification provided by the integrator, the generated code is produced.

In the following sections, the visual tool is modeled using UML (unified modeling language) diagrams. The main user of the visual tool is an application integrator, in this case called the inte-

*Figure 6. Interfaces of application integration tool*

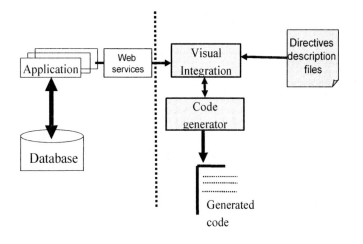

*Figure 7. Introduction of configuration data*

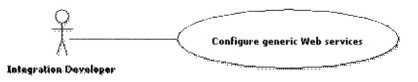

gration developer, who is responsible for creating the integration between the e-services and the external application.

## Use Cases

### Use Case: Configuration of Generic Web Services

In this use case, the developer configures the set of generic Web services, providing the required information, which is stored and updated in an XML file (webservices.xml).

Figure 8 shows the screen where this use case is executed. The screen has a tree of the generic Web services registered in the visual tool (1). In another area of the screen (2), it is possible to add a new generic Web service or edit one.

### Use Case: Log-In

At the log-in step, the developer requires authentication for a specific enterprise. The visual tool uses a Web service to retrieve the list of enterprises that belong to the hub.

Figure 10 shows the log-in screen, where the tool shows the list of enterprises (1) belonging the hub platform. The developer provides a project name (2) for this specific integration.

The developer also has to type a user name and password (3) for the selected enterprise. The authentication process is done by a Web service deployed in the hub platform, therefore the visual tool acts as a client of this Web service in order to secure the use of the information.

*Figure 8. Screen of configuration of generic Web services*

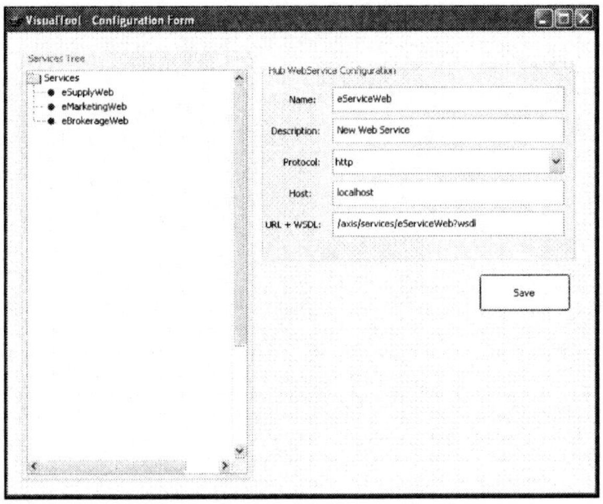

*Figure 9. Log-in*

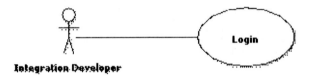

Integration Developer

Login

*Figure 10. Screen of log-in*

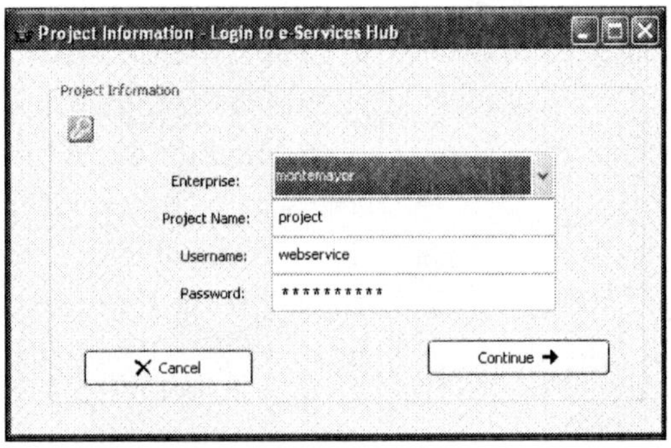

## Use Case: Web Service and Operation Specification

The following step to achieve application integration is specifying the generic Web service and the operation to be executed by the client program.

Figure 12 shows the screen for this input. The screen contains a tree of the generic Web services and its defined operations. The developer clicks on an operation in the tree and its definition is displayed (2). Then it is possible to select this operation for integration (3) and continue with the code generation step.

## Use Case: Code Generation

The last step consists of generating the necessary code for both the Web service and the programs at the client side.

The developer may independently generate the integration code for any operation by returning to the last screen (operation selection). The code generation component will take information from the generic Web service, operation description, and directives files associated with the operations linked to the Web service and generate the code for the Web service integration mechanism.

Figure 14 shows the screen for the code generation step. It is possible to generate clients in Java,

*Figure 11. Web service and operation specification*

*Figure 12. Screen of Web service and operation specification*

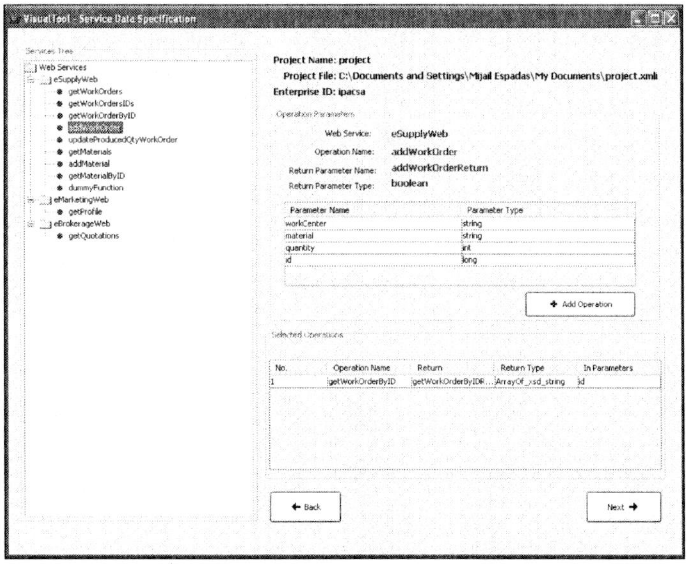

*Figure 13. Code generation*

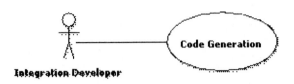

*Figure 14. Screen for code generation*

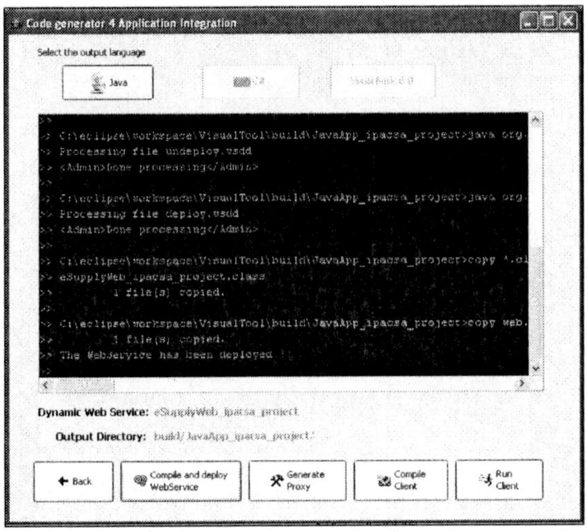

C#, or Visual Basic 6.0 programming languages (1). The selection of the language involves the template adaptations and the creation of the package with the generated source code: the Web service, the wrapper, and a client. Once the source code has been generated, the developer can compile and deploy (2) the generated Web service into the Web services container. The visual tool also generates the proxy classes (3) that implement the skeletons of the remote Web service and operation. Finally, the developer can compile and execute (4) both the wrapper and the client in order to test the integration between the client and the remote Web service just created.

## Class Diagrams

In order to simplify the description of the main classes of the visual tool, Figure 15 shows a package diagram containing the classes belonging to each component and other utilities.

As Figure 15 shows, five main packages implement the software for the visual tool.

- **Visual specification:** A set of classes that implement screens
- **Code generator:** Submodel classes for code generation
- **Data mapping:** Classes for mapping information to XML files

*Figure 15. Main packages of the visual tool*

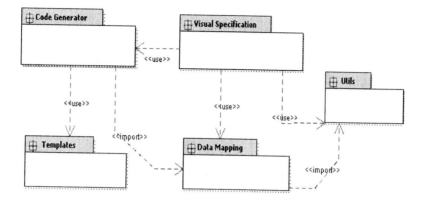

- **Utils:** Utilities for WSDL documents, project management, and I/O (input/output) classes
- **Templates:** A set of files that represents the templates assembled by the code generator

The following sections describe the class diagrams for the three main components of the visual tool: visual specification, data mapping, and the code generator.

## Visual Specification

The visual specification package comprises a set of forms or screens where the developer introduces information about the integration. Figure 16 shows the classes that implement the visual specification package from Figure 15.

The first screen showed by the visual tool is produced by the PrincipalForm class, allowing the developer to open or create an integration project. After that, the visual tool shows the LoginForm where an enterprise is selected; the project name, user name, and password are entered here. If authentication is successful, the visual tool shows the SpecificationForm screen, where the user can select the generic Web service corresponding to the e-service and the remote operation for this Web service. The last screen is the GenerationForm where it is possible to select the programming language for the client of the remote operation. This form also allows one to compile and deploy the generated Web service and compile both the wrapper and the client. For instance, in the case of the Java programming language, the GenerationForm uses Apache Axis tools to generate proxy classes (stubs and locators) from a WSDL document. Another screen related to the PrincipalForm is the ConfigurationForm where the set of generic Web services are configured using the visual tool.

As Figure 17 depicts, the process to specify the complete integration consists of three steps beginning with a log-in screen, followed by the Web service and operation selection, and finalizing with the code generation form. The log-in step uses some additional features that the visual tool provides like a Web services deployed in the hub platform to retrieve a list of enterprises that belong to the hub platform and to authenticate the user or developer who wants to integrate the enterprise. The operation selection step involves access to the XML configuration file (webser-

*Figure 16. Visual specification classes*

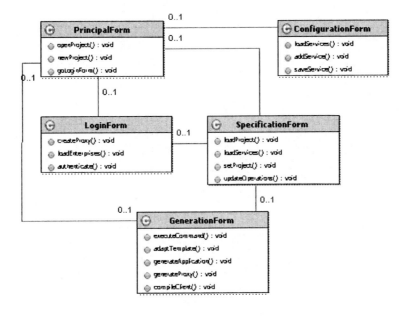

*Figure 17. Visual specification flow*

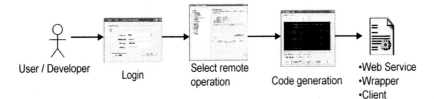

vices.xml) that declares where the generic Web services are deployed. Then, the screen shows a tree of the generic Web services from which the user can select a Web service and an operation through visual selection.

The following XML code is an example of the webservices.xml file, which contains the declaration of a set of URI to WSDL documents. The WSDL documents in Box 1 are definitions for the generic Web services mentioned before.

The last step, code generation, involves a code generation functionality to produce the necessary code, based on the specified operation in

the operation selection step. The code generator component is explained in the next sections.

## Data Mapping

The data mapping classes implement the functionality to map information using XML directive files. These classes define operations that both visual specification and code generation classes use to achieve the complete process of code generation for integration. Figure 18 shows the class diagram for the data mapping package.

*Box 1.*

```
<webservicesbean>
 <webservices name="eSupplyWeb">
  <description>Web service for e-supply hub integration</description>
  <WSDL>http://10.17.56.89:8080/axis/services/eSupplyWeb?wsdl</WSDL>
 </webservices>
 <webservices name="eMarketingWeb">
  <description>Web service for e-marketing hub integration</description>
  <WSDL>http://10.17.56.89:8080/axis/services/eMarketingWeb?wsdl</WSDL>
 </webservices>
 <webservices name="eBrokerageWeb">
  <description>Web service for e-brokerage hub integration</description>
  <WSDL>http://10.17.56.89:8080/axis/services/eBrokerageWeb?wsdl</WSDL>
 </webservices>
</webservicesbean>
```

*Figure 18. Data mapping classes*

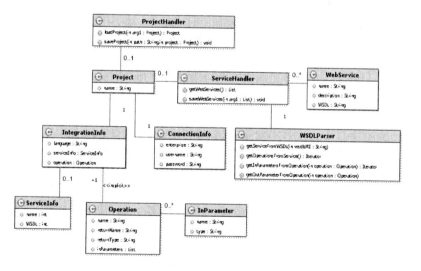

At the top of the class diagram is the Projec-tHandler class that allows retrieving and saving a project (directive files). Once the project is loaded or created, it is possible to manipulate its two attributes: the IntegrationInfo and Connection-Info classes. The IntegrationInfo class includes information for the integration process like the operation and Web service. The ConnectionInfo class contains the enterprise, user name, and password for authentication. In order to load the generic Web services stored in the webservices.xml file, the ServiceHandler class implements mechanisms to retrieve information from this file. The WSDLParser class reads information from each WSDL document (specified in the webservices.xml file) such as the set of operations, and their types and parameters. Then, the ServiceHandler class stores this information in a list of WebService classes.

The directive files in the visual tool are seen as project files because it is possible to open the XML files and recover or save directives. For a directive representing access to a remote parameter through a Web service interface, the following information is required.

- Project name
- Information for authentication and connection
  - o Enterprise to integrate
  - o User name
  - o Password
- Information for integration
  - o Language for client and wrapper
  - o E-service to integrate
- WSDL of the generic Web service by including the URI where the Web service is located; is composed by
  - o Transport protocol (HTTP, FTP [file transfer protocol], file, etc.)
  - o Host name or IP (Internet protocol) address
  - o Port

  - o Relative context of the WSDL (i.e., /axis/services/eSupplyWeb?wsdl)
- Operation to execute
  - o Type of data returned by the operation invocation
  - o Required parameters for the operation
    - ➢ Name
    - ➢ Type

Taking as example a service to retrieve work orders, the directive file for integration through Web services is as shown in Box 2.

This directive specifies an operation named getWorkOrders, modifying the date parameter releaseDate. The operation returns a vector parameter getWorkOrdersReturn. In this case, the operation accesses a Web service that resides within an application server implementing HTTP at the IP address 10.17.56.89 through port 8080 in the /axis/services/eSupplyWeb context inside the application server. The complete URL (uniform resource locator) specification is http://10.17.56.89:8080/axis/services/eSupplyWeb?WSDL.

These previous examples may give a general idea of the tasks that should be performed by a visual specification and integration tool using directive description files as input, and the output that should be produced for service level integration. In order to deal with the heterogeneity of data types, WSDL documents are commonly deployed with general types that the visual tool converts to specific types. Within the application systems, parameters can be of the following data types.

- **String:** Data that represent a succession of ASCII (American Standard Code for Information Interchange) codes
- **String[ ]:** Data that represent an array of string types
- **Integer:** Represents an integer quantity
- **Double:** Represents a floating point value
- **Date:** A time-stamp instance

*Box 2.*

```
<project name="project1">
<connectionInfo>
 <enterprise>ipacsa</enterprise>
 <username>webservice</username>
 <password>webservice</password>
</connectionInfo>
<integrationInfo>
<language>net</language>
<serviceInfo name="eSupply">
      <WSDL>http:// 10.17.56.89:8080/axis/services/eSupplyWeb?wsdl</WSDL>
</serviceInfo>
<operation name="getWorkOrders">
    <return-name>getWorkOrdersReturn</return-name>
    <returnType>vector</returnType>
    <inParameters name="releaseDate" type="date"/>
</operation>
</integrationInfo>
</project>
```

- **Vector:** An array of data objects

Directives represent the basic information necessary to execute operations and their parameters. For every access operation, it is necessary to define a set of directives to identify the parameters involved and the way in which the operation will be executed (Web service). In this way, directives are a very flexible mechanism easy to adapt when changes to parameter access are necessary.

## Code Generator

The package of the code generation represents a submodel for code generation facilities.

Figure 19 depicts an interface named Generator that defines general methods for all subclasses. There exist two subclasses of the Generator class:

WServiceGenerator and ClientGenerator. The first one defines methods for the generation of a Web service adaptation and a subclass named WServiceGeneratorJava that implements the generation functionality for a Web service in the Java language, including its deployment. The ClientGenerator subclass defines methods for the generation of clients. There exist three subclasses of ClientGenerator. The ClientGeneratorJava subclass implements functions for the wrapper and client generation in Java. Both ClientGeneratorCSharp and ClientGenerationVBasic implement functions for wrappers and clients in C# and Visual Basic languages, respectively.

The code generator component uses configuration wizards' techniques (Jiménez, 2003) to generate code based on specifications introduced through visual screens and stored in XML direc-

*Figure 19. Code generator classes*

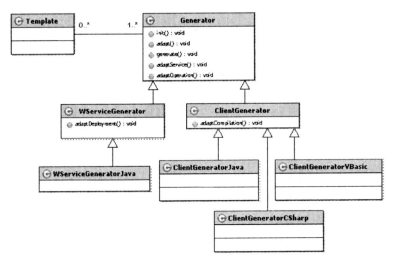

*Figure 20. Code generator description*

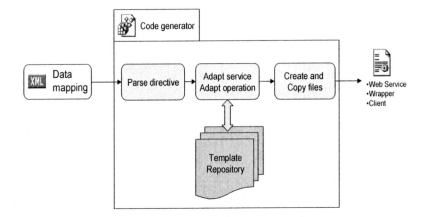

tive files. This code generator uses a set of generic templates (known as wizlets) that are adapted by the visual tool according to the visual specification of the operation to integrate. The code generator takes as input an XML file and parses the directive file to obtain both integration and connection information to adapt each template. Figure 20 shows the code generator of the visual tool, which uses a template repository containing a number of files, mostly pseudo-applications, and some deployment descriptors. They have

no-valid markup to compilation or execution, but once adapted by the visual tool, a new application with the functionality already implemented is created.

An example of a generic template is the JavaWrapper.temp template presented in Box 3.

In the code in Box 3, the no-valid markup (percent symbols) is underlined; it defines identifiers that will be replaced by valid code, according to the specification in the XML directive. Taking the example code and the work-orders example,

*Box 3.*

```
//JavaWrapper.temp
import javax.xml.rpc.ServiceException;
import localhost.axis.services.%webservice%.*;
import java.util.*;

public class JavaWrapper {
  static %Webservice% proxySOAP = null;
  static {

  try {
  %Webservice%ServiceLocator my%webservice% = new %Webservice%ServiceLocator();
  proxySOAP = my%webservice%.get%webservice%();

   } catch (ServiceException e) {e.printStackTrace();}
  }

  public static %Webservice% getProxySOAP() throws Exception{
    return proxySOAP;
  }

  public static void main(String[] args) {
    try {
  %Webservice% proxySOAP = JavaWrapper.getProxySOAP();
  %vars%
  //%type% = proxySOAP.%operation%(%params%);

      %show_results%
      } catch (Exception e) {e.printStackTrace();}
  }
} //end of class
```

a possible output of the code generator could be the adapted code in Box 4.

The code has been adapted and is ready to compile and execute, and the non-valid markups have been replaced by the appropriate code. The most important templates are the following.

- **Web Service.temp:** Template for the Web service that will be deployed
- **JavaWrapper.temp:** Template for a Java wrapper that makes the connection with the Web service created
- **JavaClient.temp:** Template that uses the JavaWrapper in order to present an example of how the integration is performed

- **Deployment Descriptors:** Templates that are files for the Web service deployment within the Axis container (in the case of Java)

There exist other templates such as wrappers for other programming languages, shell execut-

able files, and XML files. These allow the code generator to achieve the complete compilation, deployment, and execution of the new generated application (both Web service and the wrapper and client).

*Box 4.*

```
// JavaWrapper.java
import javax.xml.rpc.ServiceException;
import localhost.axis.services.eSupplyWeb_montemayor_project1.*;
import java.util.*;

public class JavaWrapper {
  static ESupplyWeb_montemayor_project1 proxySOAP = null;
  static {
  try {
ESupplyWeb_montemayor_project1ServiceLocator myeSupplyWeb_montemayor_project1 = new ESupply-
Web_montemayor_project1ServiceLocator();
  proxySOAP = myeSupplyWeb_montemayor_project1.geteSupplyWeb_montemayor_project1();

  } catch (ServiceException e) {e.printStackTrace();}
  }

  public static ESupplyWeb_montemayor_project1 getProxySOAP() throws Exception{
      return proxySOAP;
  }

  public static void main(String[] args) {
  try {
ESupplyWeb_montemayor_project1 proxySOAP = JavaWrapper.getProxySOAP();
Vector result0 = proxySOAP.getWorkOrders();
      for(int c=0;c<result0.size(); c++){String[] data=(String[])result0.get(c);for(int i=0;i<data.length;
i++)System.out.println(data[i]);}

          } catch (Exception e) {e.printStackTrace();}
  }

  } //end of class
```

## STUDY CASE FOR E-SUPPLY SERVICE INTEGRATION

The previous sections provide a general description of the technologies necessary for implementing an e-services hub and how a visual environment could be used to simplify interaction with external applications. Although explanations were biased toward PyME CREATIVA requirements, we believe a similar approach could be used to implement and integrate applications in other domains. This section gives more details specific to the e-supply service in the PyME CREATIVA context to clarify how the infrastructure described simplifies application integration.

## Process Definition

Supply chain integration (in the PyME CREATIVA context) is necessary when a supplier enterprise (member of the hub platform) wants to update the amount of work orders produced in their systems (could be legacy, .NET, or Java). This might be necessary because the clients of the supplier enterprise may want to track their purchase orders (a purchase order is a set of work orders) submitted through e-supply interfaces.

Figure 21 shows a UML activity diagram describing the steps involved. This process is very simple but will serve as a useful example to describe the functionality of the visual tool. This example involves two operations: a RetrieveWor-

kOrders operation and an UpdateWOQuantity operation. For this example, we will use service-level integration and the visual tool will generate Java code.

## Defining Service-Level Integration

Service-level integration is similar to data-level integration. In this case, it is necessary to define a basic data configuration that contains information about the Web service interface (as described before, the integration will use already deployed Web services). The use case of the introduction of the configuration specified the service-level integration with parameters such as the Web service, URL, context, and so on. In the same way as the data level, the Web service repository (in our case the Apache Axis container) is hosted at the IP 138.178.16.3 within the /Axix/Services context, accessed through the SupplyService interface that contains the integration methods. Then, once this configuration data are specified, we can link them with the UpdateWorkOrders process. The methods of the Web service interface are then retrieved, thus the appropriate selection of operation can be performed.

For the retrieve operation, the specification consists of the definition of the desired method provided by the Web service interface. The execution of this remote method will retrieve work-order information including the ID, requested quantity, and quantity produced, and can be stored in a

*Figure 21. A supply integration process*

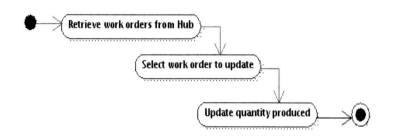

local variable (i.e., a vector of string arrays) for its manipulation.

For the update operation, using the retrieved work orders, it is possible to select the correct update remote method over the quantity produced for a work order based in its ID. Similar to data-level integration, it is necessary to specify two parameters for the operation: the ID as a filter and the new quantity for update.

## Generating and Executing AI Code

The specification of the code-generation use case is similar to that of data-level code generation. In this case, we need to specify the defined process, the location of the generated code, and the platform. The visual tool will generate the same two classes as those for data-level integration (UpdateWorkOrder.java and ExceuteCode.java), but with different implementation in the first class because this class uses Web services technologies to access the remote methods.

For example, a code snippet of the Update-WOQuantity method is as shown in Box 5.

The second file (ExecuteCode.java) contains the code to be executed for the instance of UpdateWorkOrder in its main method. This is

similar to data-level integration, but the integration implementation differs in the technologies used (database in the first case, Java code in the last case).

## CONCLUSION

The integration of SMEs to build virtual organizations of supply chains requires direct access to their databases or wrappers to link applications. Potentially every SME has its own database managers and platforms, thus integration could become a nightmare. PyME CREATIVA implements a set of e-services for operating virtual organizations of SMEs, thus standard interfaces could be defined. To simplify the interaction of SMEs into the PyME CREATIVA infrastructure, a visual tool to specify integration data and applications is very helpful. By extracting metadata from SME databases or using standardized interfaces, the integration code could be generated. This chapter described the aims of the PyME CREATIVA project, its architecture and supported e-services, and the specific interoperability needs. The goal is to integrate hundreds of SMEs into the hub, thus it is important to simplify the integration

*Box 5.*

```
import axis.services.eSupplyServiceAccess.*; //code generated by wsdl2java Apache Axis tool
public class UpdateWorkOrder{

//connection code
public boolean UpdateWOQuantity(int newQuantity, int id){
    ESupplyAccessServiceLocator myService = new ESupplyAccessServiceLocator();
    ESupplyServiceAccess mySOAP = myService.geteSupplyServiceAccess();
    boolean updateSucces = mySOAP.updateWOQuantity(100,3566121);
//exceptions handling
}
}
```

process as much as possible. The simplest way is obviously to have a tool able to generate the code for interoperation. Using the specification of the databases or applications, the tool generates the necessary code. A visual interface presents the relevant data, thus by selecting the appropriate parameters, the code for interoperation is directly generated.

Implementing visual generator tools is only possible in very constrained domains where all involved variables are known (Jiménez, 2003). This is the case in PyME CREATIVA as all services and the information interchange needs are known in advance. Other more open domains may present more difficulties for automatic code generation.

The examples presented here concentrate on generating Java code for application integration. However, the generator component is based on code templates to generate the integration code, so with appropriate templates, it would be possible to generate code for different Web services platforms, such as Microsoft .NET, for instance.

## ACKNOWLEDGMENT

The work described here was partially supported by a grant from the Interamerican Bank of Development for the PyME CREATIVA project and the ITEM-CEMEX research chair.

## REFERENCES

Ball, M. O., Ma, M., Raschid, L., & Zhao, Z. (2002). Supply chain infrastructures: System integration and information sharing. *ACM SIGMOD Record, 31*(1).

Batory, D., & Geraci, B. J. (1997). Composition validation and subjectivity in GenVoca Generatos. *IEEE Transactions on Software Engineering, 23*(2).

Champion, M., Ferris, C., Newcomer, E., & Orchard, D. (2002). *Web services architecture* (Working draft). W3C. Retrieved from http://www.w3.org/TR/ws-arch/

Guzmán Ruiz, F. (2001). *Mecanismos de comunicación externa para un sistema integrador de comercio electrónico de negocio a negocio.* Unpublished master's thesis, Instituto Tecnológico y de Estudios Superiores de Monterrey, Monterrey, Nuevo León, Mexico.

Haas, H., & Brown, A. (2004). *Web services glossary.* W3C. Retrieved from http://www.w3.org/TR/ws-gloss/

Jiménez, G. (2003). *Configuration wizards and software product lines.* Unpublished doctoral dissertation, Instituto Tecnológico y de Estudios Superiores de Monterrey, Monterrey, Nuevo León, Mexico.

Juárez Lara, N. (2001). *Integración visual de aplicaciones en sistemas de comercio electrónico de negocio a negocio.* Unpublished master's thesis, Instituto Tecnológico y de Estudios Superiores de Monterrey, Monterrey, Nuevo León, Mexico.

Kreger, H. (2001). *Web services conceptual architecture.* IBM Software Group. Retrieved from http://www-106.ibm.com/developerworks

Linthicum, D. S. (2004). *Next generation application integration.* Addison-Wesley Information Technology Series.

Medjahed, B., Benatallah, B., Bouguettaya, A., Ngu, A. H., & Elmagarmid, A. K. (2003). Business-to-business interactions: Issues and enabling technologies. *The International Journal on Very Large Data Bases, 12*(1).

Molina, A., Mejía, R., Galeano, N., & Velandia, M. (2006). The broker as an enabling strategy to achieve smart organizations. In *Integration of ICT in smart organizations.* Idea Group Publishing.

O'Riordan, D. (2002). Business process standards for Web services. *Web Services Architect.* http://www.webservicesarchitect.com.

Papazoglou, M. P. (2003). Service-oriented computing: Concepts, characteristics and directions. *Proceedings of the Fourth International Conference on Web Information Systems Engineering (WISE 2003)*, 3-12.

Siau, K., & Tian, Y. (2004). *Supply chains integration: Architecture and enabling technologies.*

Whalen, M. W., & Heimdahl, M. P. E. (1999). On the requirements of high-integrity code generation. *High-Assurance Systems Engineering: Proceedings of the Fourth IEEE International Symposium*, 217-224.

Wing, L., & Shankararaman, V. (2004). An enterprise integration methodology. *IT Professional, 6*(2), 40-48.

# Chapter XI
# A Framework for Information Systems Integration in Mobile Working Environments

**Javier García-Guzmán**
*Universidad Carlos III de Madrid, Spain*

**María-Isabel Sánchez-Segura**
*Universidad Carlos III de Madrid, Spain*

**Antonio de Amescua-Seco**
*Universidad Carlos III de Madrid, Spain*

**Mariano Navarro**
*TRAGSA Group Information, Spain*

## ABSTRACT

*This chapter introduces a framework for designing, distributing, and managing mobile applications that uses and updates information coming from different data sources (databases and systems from different organizations) for helping mobile workers to perform their job. A summary of the state of the art in relation to mobile applications integration is presented. Then, the authors describe the appropriate organizational context for applying the integration framework proposed. Next, the framework components and how the framework is use are explained. Finally, the trials performed for testing the mobile applications architecture are discussed, presenting the main conclusions and future work. Furthermore, the authors hope that understanding the concepts related to the integration of mobile applications through the presentation of an integration framework will not only inform researchers of a better design for mobile application architectures, but also assist in the understanding of intricate relationships between the types of functionality required by this kind of systems.*

## INTRODUCTION

Many workers in current organizations perform their activity in mobile environments. Sellers, architects, doctors, veterinarians, and so forth perform the most part of their work outside an office, many of them in cities or at rural and remote areas. Moreover, in many cases, the information required for mobile workers comes from different information systems and databases owned by different organizations or providers, so it is necessary to provide mobile workers with devices (handhelds, pocket PCs [personal computers], tablet PCs, etc.) with software systems that employ user interfaces appropriate for this kind of devices, and with the capabilities for accessing and updating several information systems.

In order to solve this problem, an integration framework, called DAVINCI, has been defined. DAVINCI is a framework for providing mobile workers with mobile software applications to query and update information coming from different and heterogeneous databases.

The DAVINCI project was first tested with the main aim of developing a solution to help veterinarians performing in-field sanitary inspections in cattle holdings across European Union countries; DAVINCI was tested by veterinarian services from Spain, Bulgaria, Latvia, and Czech Republic.

During these trials, we identified that one of the main advantages of the DAVINCI architecture is its capability to be integrated together with different European databases for animal health controlling (for instance, EUROVET in Bulgaria or SIMOGAN in Spain registering cattle census and movements). DAVINCI also permits the development of new data warehouse systems compliant with the previously cited regulation, providing large economic costs savings. On the other hand, DAVINCI is easily adaptable to procedures in different countries, each with a singular culture and organizational structure regarding the responsibilities for livestock sanitary control.

## STATE OF THE ART

Mobile computing devices (smart phones, PDAs (personal digital assistants), tablet PCs, or notebooks) increasingly include integrated wireless capabilities. Wi-Fi (wireless fidelity, 802.11) access points for wireless connectivity have appeared everywhere. Moreover, a growing number of complementary wireless networking standards, such as wireless personal area networks (802.15) and wireless metropolitan area networks (802.16), has evolved. In this sense, the users, who take their devices everywhere, expect their software applications to run as they do in the traditional network environment available at their offices.

To achieve such functionality transparently, however, these applications must meet a new set of requirements and support a specific set of capabilities related to the following.

- Provision of intelligent roaming capabilities to enable users to work without interruption, even when network connections are disrupted
- Exploitation of multiple network interfaces in a single device or the ability to select the most appropriate connection, for example, when two or more connections are simultaneously available
- Synchronization of databases by caching contents to local devices through asynchronous connections
- Access to data and applications on diverse devices through similar user interfaces
- Conservation of power at the operating-system level and maximization of performance

To implement mobile required functionality, application architects and developers have attempted to work around such problems without the benefit of development environments, application programming interfaces (APIs), or third-party middleware solutions that are tailored for mobile environments.

The tools that are available to application architects and developers to provide this functionality are as follows.

- Mobile-application architecture guides
- Mobile-computing application servers

## Mobile-Application Architecture Guides

The purpose of mobile-application architecture guides is to provide basic principles to design and develop mobile applications, facilitating the understanding of the high-level issues around mobility and mobilized software architectures.

This kind of guides identifies the primary capabilities required of mobile applications, such as efficient resource management, comprehensive context management, encoding, view consistency, extended policy and security functionality, durable storage, and reliable messaging.

Examples of mobile-application architecture guides are the following:

- *Intel® Mobile Application Architecture Guide* (Intel Corporation, 2006)
- *Mobile Applications: Architecture, Design, and Development* (Lee, Schell, & Scheneider, 2004)

## Mobile-Computing Application Servers

Mobile-computing application servers are software programs that run in a server and provide the following functionality.

- Application-level logic that handles business functions involved in a particular organization and its integration with back-end database or business application systems
- Presentation services for the mobile client device (handheld computers, notebooks, PDAs, etc.). This is also called GUI (graphical user interface) in some cases, though some handhelds are more like older text terminals than PCs. It includes breaking the messages into smaller chunks, filtering redundant information, and even logically compressing the data.
- Transaction services, in some cases including multithreading for heavy volumes and persistency
- Application programming-level interfaces with specialized communications protocols

Actual implementations of an application server vary from one vendor to another. Some application servers are generic Web servers with an SDK (systems development kit) or API capability to pull data from enterprise database systems and send them to browser-based client software on a handheld device.

Depending on the heritage of the vendor and its core expertise, you can categorize application servers in the following broad classes.

- Generic application servers with a Web-based SDK, for example, Netscape, Microsoft, Sun, SilverStream, and BEA, may have support for handheld devices and wireless networks strapped onto the basic application server
- Database vendors' application servers, for example, Oracle 9i and Sybase's iAnywhere application server
- Data synchronization vendors, for example, Puma, Synchrologic, and Extended Systems
- Specialized Mobile or Web computing application servers, for example, IBM's WebSphere
- WAP-centric application servers, like Nokia's WAP Server
- E-mail-centric application servers, for example, Microsoft's mobile Information server and the EdgeMail application server

## DAVINCI INTEGRATION FRAMEWORK'S MAIN CHARACTERISTICS

Our work applies many of the principles presented in the mobile-applications architecture guides to provide a framework for information systems integration in mobile working environments, paying special attention to some problems that are not well-solved by commercial mobile application servers, related to the following.

1.  Provide standardized ways (independent from the vendor) to accomplish the following:
    *   Define the mobile application's user interface and establish the data to be managed with this application.
    *   Ship the user interface definition and data required to manage mobile applications.
    *   Process the user interface definition and data to present mobile applications to the user.
2.  Facilitate the development of mobile applications that integrate data coming from the following.
    *   Different databases that are stored in different DBMSs (database management systems)
    *   Other existing systems that have been programmed in different languages and operative systems
    This reduces the time, effort, and cost required for the development phase.
3.  Optimize communications capabilities by accomplishing the following.
    *   Reducing the size and number of the messages interchanged between servers and client devices
    *   Increasing the possibility to adapt the mobile applications integrated to several different communication architectures

*   Recovering automatically the interrupted data messages
*   Selecting the optimal communication channel depending on the available bandwidth and the economic costs associated

Moreover, the integration framework should be deployed in different technological infrastructure coming from different vendors.

## INTEGRATION FRAMEWORK WORKING CONTEXT

Before beginning with the detailed description of the integration framework, it is necessary to analyze some of the basic concepts of the working context related to information systems integration in mobile and remote settings. These concepts are the geographic area, zone of work, protocol, campaign, and workers in the field.

*   **Geographic area:** This is a geographic zone that includes geographic locations (represented as geographic coordinate points) where the information management tasks are initiated. For example, in the business related to veterinarian control of pets and cattle, a geographic area could be a region of a country. In the business of the population censuses, a geographic area could be a town or a district of a large city.
*   **Work zone:** This is a concrete point within a geographic area that will be visited by the field workers to perform their concrete tasks. For example, in the business related to veterinarian control of pets and cattle, the work zones could be each one of the cattle holdings and farms to visit. In the business of the population censuses, the work zones could be the streets or buildings of a town or a district of a large city.

- **Mobile workers:** They are the people in charge of performing concrete tasks in the assigned work zones. For example, in the business related to veterinarian control of pets and cattle, the mobile workers will be the veterinarians employed to perform the cattle and pet inspection on a farm. In the business of the population censuses, the mobile workers will be the people employed to visit each family to fill in the census forms.
- **Campaign:** A campaign is the grouping of a set of tasks of the same type that will be performed by mobile workers in several work zones and geographic areas using the same procedure.
- **Protocol:** It is the work procedure used in a work zone by a mobile worker to satisfy the objectives established in the campaign definition. In the scope of this integration framework (DAVINCI), a protocol is implemented by means of a set of applications or forms that the worker should complete or execute.

Each DAVINCI application is composed of a set of tasks or activities linked using a precedence sequence. Each one of these tasks is implemented by a form that accesses (external and/or internal) databases and updates the pertinent information through the mobile devices used by mobile workers. The forms should be defined using the standard of XForms.

The sequences of forms that define the campaign protocol configure the mobile user interface. Using this user interface, the mobile workers will be able to manage (querying, modifying, inserting, and deleting) the data concerning each protocol task. These data will be stored in local databases, physically located in the mobile devices. In order to relate the data shown by the form and the data stored in the database, a file, named modelmapper, should be defined. These modelmappers define the relation between each one of the fields of the form and the concrete field of the local database, and the way (SQL statements) to access and update the data.

*Figure 1. DAVINCI mobile work concepts*

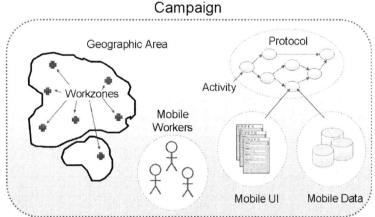

According to the concepts presented below, the integration framework works with general concepts, so this architecture is able to be applied to different business areas with very few enrichments and/or modifications.

The integration framework will only store generic data on geographic areas, work zones, and mobile workers: concretely, its internal identifier, name, description, and an identifier or code assigned in an external database that contains extended information related to the concrete business area related to the tasks to be performed by the mobile workers. For example, the data to gather by a veterinarian in a work zone (cattle holding) will be different than the data collected by a questioner in a family house related to a population census.

For access to these extended data that will be stored in external databases, DAVINCI uses a specific data access module denominated DBSync.

## INTEGRATION ARCHITECTURE

In order to provide support to this working context, the integration architecture suggested is based on client-server architecture, with a main module of communications and a message dispatcher, allowing the integration of any new service that is designed for a concrete area of business through its connection to the message dispatcher. Following this philosophy, DAVINCI'S integration architecture is organized in three levels, as it is shown in Figure 2.

a.  DAVINCI provides a user interface (Web interface) for defining the process that should be used by the mobile workers and the connection of this process with the mobile software applications to be used for its consecution.

b.  The core services are grouped in DAVINCI integration middleware, which permits the definition of mobile applications for per-

forming several types of mobile work without a strong effort in software programming. These applications are defined by means of the user interface specification using XForms language (a standard for defining multimodal and multidevice user interfaces), and the specification of the information sources (server, database, table, and fields) for each user interface element (labels, text boxes, combo boxes, etc.) and the policy for selecting and updating the concrete data of a form.

DAVINCI middleware also permits the use and adaptation of several software components for accessing and integrating heterogeneous databases that could be merged for providing the information necessary for a concrete mobile software application.

This middleware also permits one to distribute the assignments to each mobile worker available, send the application and data for using the software application to perform the work assigned, and receive the data obtained as a conclusion of the work.

c.  Moreover, DAVINCI provides some client components (currently developed for pocket PCs) that permit the following.

   •  The communication with the middleware for receiving the information by the applications and data used and updated

   •  The use of DAVINCI mobile applications with voice and pointing interfaces, providing access to additional capabilities related to location- and geographic-information-based services

The following sections describe these three levels in detail.

### Integration Middleware

The components that permit the definition of the user interface of the final integrated user applica-

*Figure 2. Integration framework overview*

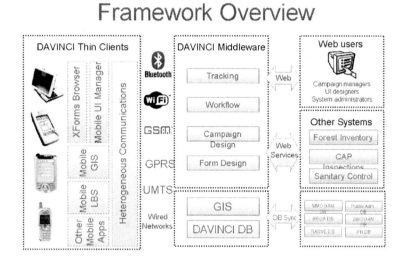

tions (Mobile UI Designer) and the components to access heterogeneous databases (DBSync) should be customizable; that is to say, in the standard version of DAVINCI, the core functionalities of these components are provided, but it is necessary to develop simple additional components to provide the full functionality needed in the final solution of the concrete sector.

Other components with full functionalities, related to the design of campaigns, the assignment of individual work to the mobile staff, and the synchronization of data and applications between the server and the mobile devices, are provided in the standard version of DAVINCI, so they do not have to be modified or enriched in any case.

## Customizable Components

These components should be applied to adapt and use the integration framework in a concrete business environment.

## Web Interface

The Web interface provides the required functionalities related to campaign design, the assignment of work (in terms of geographic areas or work zones) to each mobile worker, and controlling the advance of the work corresponding to a campaign.

Next, the main functionalities of the Web interface are presented in a detailed way.

- **Design of campaigns:** The design of a campaign consists of the specification of certain items.
  o   The tasks to perform
  o   The steps to follow for each task
  o   The selection of the geographic areas and/or work zones where the mentioned tasks should be performed

As we said previously, a protocol is implemented by an application (or set of forms) to complete or execute. Each application is composed of forms, each of them representing a task of the protocol. The forms are presented in sequence, representing in this way

*Figure 3. Campaign registration form*

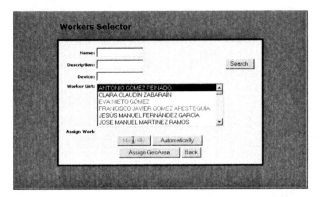

*Figure 4. Workers' assignment form*

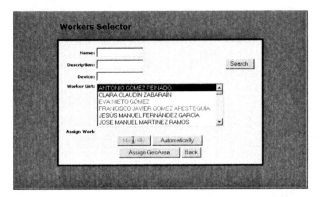

the precedence required among the protocol tasks. Once the forms are designed and the work assignments are planned, the forms and the related data are sent to mobile workers' devices. The above-mentioned forms are designed using the XForms standard. The query and update functions to process the data related to the forms are specified in the modelmappers files related to the forms.

- **Campaign work assignment:** Once the campaign has been designed, it is possible to begin the assignment planning. The work assignment consists of the determination

of the workers assigned to a campaign and selecting the geographic areas or work zones that mobile workers have to visit to perform the campaign task. Moreover, the Web interface permits the campaign manager to fix the concrete dates to perform the tasks, configuring the agenda of the mobile workers.

Once the work has been assigned to each available mobile worker, the campaign manager should send the assignments obtained to mobile workers; so, the integration framework sends, through a message-center

*Figure 5. Work-zone selection form*

*Figure 6. Campaign query form*

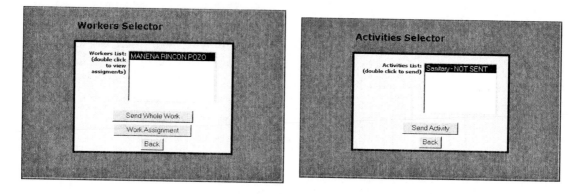

server component, to the mobile workers' devices the forms definition, model mappers files, and necessary data to perform the campaign tasks.

At any time, the allocations to the workers can be modified as long as the work in the work zone to reassign has not begun yet.

Also, the planning of a campaign could be consulted about at any time.

• **Control of the state of the work:** DAVINCI offers the required functionality to control the advance of the campaign work at any time. To achieve this aim, the work zones assigned to each mobile worker is consulted and, for each one of them, the advance degree is calculated through the information sent by

the mobile applications at the completion of each protocol task in every work zone and stored in the DAVINCI internal database. In addition, the integration framework allows the placement of the mobile workers in order to control their availability to receive new assignments. According to the accessories installed in the mobile workers' devices and the coverage of wireless communications in the zones where they are, the location processing will be based on GPS (Global Positioning System) positioning algorithms (without cell-phone coverage) or GSM provider positioning services (with cell-phone coverage).

*Figure 7. Work-zone selection form*

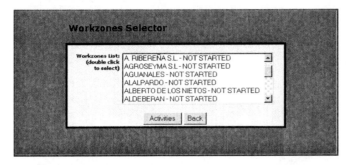

## Mobile UI Designer

The main purpose of this component consists of helping mobile-application integrators to design mobile applications using the XForms standard, preparing them for their introduction into the DAVINCI integration framework.

This component has not been developed yet. Temporarily, we are using any XML (extensible markup language) editor that is able to process XForm schema. Concretely, during DAVINCI's deployment in the European veterinarian sector, the editor used has been XMLSpy.

The activities required for this purpose are as follows.

1. Design of the forms of the mobile applications following the XForms standard
2. Creation of the modelmapper file containing the rules of insertion, modification, deletion, and consultation of the information of each form and each data item presented in it

## DBSync

This module is in charge of the communication and synchronization of data between the integration framework (DAVINCI) and any other system to connect.

DAVINCI works with a set of generic concepts that are easily adaptable to concrete concepts in external systems. For example, if we are processing work zones, which are the places where mobile workers perform their tasks, the DAVINCI integration system only stores the identifier of the zone and the extended information; the properties of the concrete business area are stored in an external system that is conveniently connected.

Moreover, DAVINCI applications will work with data stored in external systems and databases for recovery and modification of the business area information.

Next, DBSync's main functionalities are presented in a detailed way.

- **Get extended info:** In this use case, additional information is requested about any of the DAVINCI main subjects.
  - o   Work zones
  - o   Geographic areas
  - o   Campaigns
  - o   Workers

  The DAVINCI data model holds very little information about these elements in a generic way. Applications may need to show to the user more detailed data just for information purposes. The data obtained by using this use case is not modifiable by DAVINCI middleware, and DAVINCI does not handle them in any way but for showing it to the Web interface user in the suitable place.

*Figure 8. DBSync architecture*

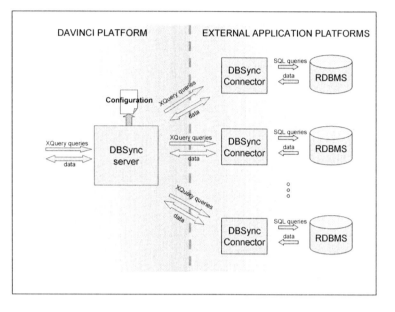

- **Query data:** Several DAVINCI middleware components will need to obtain data from the data sources integrated in order to send it to the clients for showing and updating. That data are queried in a generic way by using this use case.
- **Update data:** Several DAVINCI middleware components will need to update or insert data in the data sources integrated in the DAVINCI platform. That data usually come from data capturing applications used by workers. A certain format is specified to transfer data from clients through DAVINCI middleware to the original data sources.

The DBSync architecture is shown in Figure 8.

DBSync is composed of a server that interacts with the other DAVINCI modules needing data synchronization. The DBSync server uses different DBSync connectors to access different data sources integrated with the DAVINCI platform. The data sources can be located in geographically separated points, and can make use of different relational database managing systems (RDMS).

There is no limit on the number of DBSync connectors that can be used by the DBSync server.

- **DBSync server:** The DBSync server obtains the access data needed to reach the connectors from configuration, but the data can also be obtained dynamically.
- **DBSync connectors:** A DBSync connector is in charge of communicating with the DBSync server and performs the required operation regarding data querying and updating. Each connector handles one RDMS and can have its own policies about data updating and handling.

To integrate a new data source into the DA-VINCI platform, it is necessary to complete the following steps.

1.  *Define the data managing policies that will be used on the data source.*

The data updating and querying policies to be used on each data source are to be defined by the owner of the data source. The DAVINCI platform will provide new data and queries in the format specified above through the DBSync server, however it does not specify the way in which that data are to be updated and the queries are to be performed.

The following aspects are to be considered in order to design the data managing policies to be applied by the DBSync connector.

- Data may be overwriting when new data that has been captured by DAVINCI clients are updated in the data source. The DAVINCI platform does not perform any data storing or auditing, thus the DBSync connector should be able to manage data overrides in a consistent way with the data source. For instance, critical data should not be overwritten by new data if there is not a previous check, and historical data might be kept in order to roll back to previous states of the data source.
- Security procedures might be implemented in order to prevent unauthorized access to data or modifications. This depends on several factors such as the physical location of the data source, the connectivity of it with external threats, or application security requirements.
- There may be the need to dynamically modify the data access policies. If there is that need, configurable mechanisms can be implemented for being able to change data managing policies at run time.

2. *Implement and deploy a new DBSync connector to enable the DBSync server to access to data source.*

A new DAVINCI DBSync connector module must be implemented when a new data source is integrated into the DAVINCI platform.

The connector must expose a Web service in order to communicate with the DBSync server.

The data managing policies used by the connector to access its own data source are not restricted and are up to the particular implementation of each DBSync connector.

The connector Web service must be accessible from the DBSync server through HTTP (hypertext transfer protocol) or HTTPS. It can be deployed at any platform able to fulfill that requirement. There is no restriction about the technologies (programming language, platform operating system, applications server, etc.) used to implement and deploy the connector due to its Web service (SOAP, simple open access protocol) interface with the DBSync server.

An implementation using the same technologies as the rest of the DAVINCI middleware is provided for reference. It is built using the Java programming language and deployable on any J2EE-compliant application server; JBOSS was used during the development stage.

The DBSync module receives Xqueries in order to collect information from the database. That kind of queries achieves a high-level abstraction over SQL sentences in such a way that the same query could be applied over different types of databases (SQL Server, Oracle, etc.).

FLWR is the subset of the Xquery standard that will be used by DAVINCI modules. This standard defines five operations that could be defined in Xquery.

- *For* creates a variable that represents the table and associates it to a variable.
- *Where* filters the query using data from tables selected in *for* expressions.
- *Return* specifies the node set of the output document with variable references. After a *for-in* expression completes the iteration, *return* delivers the query result document.

There is one more operation offered by the FLWR standard, denominated *let*. However, this operation is not necessary to offer the functionality required by server modules of DAVINCI.

An example of a FLWR expression could be as shown in Box 1.

The result of this FLWR expression through a parser would be this SQL sentence.

**SELECT WR_ID FROM T_WORKER b,
T_DEVICE a WHERE b.WR_NAME
LIKE 'Fran' AND b.WR_ID != 0
AND a.DV_ID = 0**

In this case, the FLWR parser has been developed for MS SQL Server Database in such a way that the syntax of the SQL sentence is specific for that database server.

The DBSync module has an implementation of certain capacities of the FLWR standard. Through that implementation, a new connector developed for the DAVINCI platform will be able to transform a FLWR request into an SQL sentence for SQL Server Database.

## Fixed Server Components

These components should be installed in a server machine with a connection to the client devices. They provide full capabilities of the integration framework, so they do not have to be modified or enriched in any case.

*Box 1.*

```xml
- <q:query xmlns:q="http://www.w3.org/XQuery">
  - <q:flwr>
    - <q:for>
      - <q:forAssignment variable="b">
          <q:pathExpr>T_WORKER</q:pathExpr>
        </q:forAssignment>
      - <q:forAssignment variable="a">
          <q:pathExpr>T_DEVICE</q:pathExpr>
        </q:forAssignment>
      </q:for>
    - <q:where>
      - <q:binaryPrefixExpr name="AND">
        - <q:binaryPrefixExpr name="LIKE">
            <q:pathExpr>b/WR_NAME</q:pathExpr>
            <q:constant datatype="xsd:string">Fran</q:constant>
          </q:binaryPrefixExpr>
        - <q:binaryPrefixExpr name="AND">
          - <q:binaryPrefixExpr name="!=">
              <q:pathExpr>b/WR_ID</q:pathExpr>
              <q:constant datatype="xsd:decimal">0</q:constant>
            </q:binaryPrefixExpr>
          - <q:binaryPrefixExpr name="=">
              <q:pathExpr>a/DV_ID</q:pathExpr>
              <q:constant datatype="xsd:decimal">0</q:constant>
            </q:binaryPrefixExpr>
          </q:binaryPrefixExpr>
        </q:binaryPrefixExpr>
      </q:where>
    - <q:return>
        <q:variable name="WR_ID" />
      </q:return>
    </q:flwr>
  </q:query>
```

## Campaign Designer

This module encapsulates the functionalities for the management of campaigns and the protocols assigned to them.

Next, the campaign designer's main functionalities are presented in a detailed way.

- **Campaign design:** This functionality consists of one of the following.
  o Selecting an existing campaign to modify its geographic areas and protocols
  o Creating a new campaign and selecting the corresponding geographic areas related to the campaign in which one will work in that campaign.
- **Protocol design:** This functionality consists of one of the following.
  o Selecting an existing protocol, and adding and/or deleting activities. Moreover, for each activity, the application or form to complete is able to be changed
  o Creating a new protocol, defining the general information of each activity and the precedence between the protocol activities, and assigning a mobile application or form to each activity.

## Work Planning

Next, the work planning component's main functionalities are presented.

- **Assign and unassign mobile workers to geographic areas:** The first step in the work planning component consists of determining the workers who will perform tasks in each one of the geographic areas assigned to the campaign.
- **Assign and unassign work zones to mobile workers:** Once the workers are assigned to the different geographic areas, this function allows selecting from the mobile workers of a geographic area the concrete person who is going to visit each work zone.
- **Send the assigned tasks:** Once the tasks in the concrete work zones are assigned, the integration framework provides functions to send the assigned tasks to each mobile worker. This information shipment is composed of the mobile applications to run during the tasks in the work zones assigned, the list of the mentioned work zones, and the extended information that is required to run the applications correctly.

## MobileUI-Server

This component receives the orders (initiated by the user through the Web interface operations) sent by the campaign designer and work planning components, transforming them into the messages adapted for its shipment to the corresponding component in the mobile worker's device.

Moreover, MobileUI-Server processes the messages of data and requests sent by the corresponding component in the client side of this integration framework.

Next, the MobileUI-Server component's main functionalities are presented.

- **Application shipment:** This operation consists of looking for an application assigned to the worker, assembling the message to notify the client of the need of this concrete application, and sending the application to the client, which is in charge of verifying if this application is installed or not. If the application is not installed, the client will sent a new message asking MobileUI-Server for the files corresponding to the new mobile application to install.
  Moreover, this function collects the required extended data for running the application sent correctly, and sending them to the client side by means of the appropriate messages in order to be stored correctly in the corresponding mobile databases.

- **Forms request:** When a mobile device asks for the installation of a new application, MobileUI-Server sends a message with the description of the XForms documents that compose the required application.
- **Modelmappers request:** The process is similar to the comments for the forms request, being initiated by the same event of the installation of new applications in the client side, but in this case, the files sent correspond to the files that link the form fields to the local database fields.
- **Data request:** The process is similar to the comments for the modelmappers request, being initiated by the same event of the installation of new applications in the client side, but in this case, the information sent corresponds to information items to be inserted in the local database.
- **Data modification:** The purpose of this function consists of the synchronization of already updated information by the server to the client in order to permit the use of undeprecated information by the mobile workers to perform their tasks correctly.

In order to obtain this intention, whenever the server detects the modification of information shared by several mobile devices, a message to the affected devices, with the sentences to substitute the deprecated information with the valid one, is created.

## eSignatureServer

This module is responsible for verifying the information signatures generated in the mobile devices to check their validity.

## MessageCenter-Server

This module centralizes the communication between the different modules installed in the central integration server, considering that the communications server is the module that, in the server side, represents the mobile workers' devices.

When a server service sends a message to another server component, MessageCenter-Server redirects it considering that the information directed to a mobile device will be redirected to the communications server, which will send it to MessageCenter-Client (the message manager that is continuously running in each mobile device).

In addition, the shipment of messages between modules can be programmed, so in this case, the shipment is executed at a planned moment. In case of nonprogrammed messages, they will be sent as rapidly as possible.

MessageCenter-Server's main functionalities are as follows:

- Send a programmed message
- Send a list of programmed messages
- Send an instant message to other server component
- Send an instant message to a client component
- Start MessageCenter-Server
- Register new service

## Communications Server

This module is in charge of managing the communications between the client and server in the server side, handling a queue of messages to send, reconstructing the messages received from the mobile devices, and sending them to MessageCenter-Server, which is responsible for redirecting them to the corresponding server service.

The communications server's main functionalities are the following.

- Send a message to a client
- Receive message from client

## Fixed Client Components

These components should be installed in each mobile device to be used by a mobile worker. The client components provide full capabilities for using and managing the mobile applications that are in the scope of the DAVINCI integration framework, so they do not have to be modified or enriched in any case.

## Communications Client

This module is in charge of managing the communications between the client and server in the client side, handling a queue of messages to send, reconstructing the messages received from the server, and sending them to the message center, which is responsible for redirecting them to the corresponding component.

Next, the communications client's main functionalities are presented.

- **Send a message to the server:** When the message center has any message to send to the server, the mentioned message is introduced in a queue for its later shipment.
  The shipment will be made at any moment depending on the availability of the communication channels of the device (Bluetooth, GPRS, wired connection, etc.) and on the priority of the message to send.
  Internally, the message queue is stored in a database to avoid losses produced as a consequence of possible falls of the system.
- **Receive message from the server:** When the module of communications has received a message from the server, the messenger center is notified of this circumstance and is responsible for redirecting the message to the corresponding component.

## MessageCenter-Client

This module centralizes the communication between the different modules installed in the mobile device, considering that the communications client is the module that, in the client side, represents the central integration server.

When a client component sends a message to another client component, the MessageCenter-Client redirects it considering that the information directed to the server will be redirected to the communications client, which will send it to MessageCenter-Server (the message manager in the server part).

In addition, the shipment of messages between modules can be programmed, so in this case, the shipment is executed at a planned moment. In case of nonprogrammed messages, they will be sent as rapidly as possible depending on the amount of messages in the queue and the state of communications channels.

## MobileUI-Client

The MobileUI-Client component is formed by five differentiated subcomponents.

- Forms Viewer (XFormster)
- Forms Server (MobileWebServer)
- Forms Processor (XFormsModule)
- Data Access Control (MuiData)
- Applications Controller (MuiDataManager)

The dynamic collaboration between these subcomponents, with the rest of the modules of the side client, is shown in Figure 9.

The interaction begins when the DAVINCI server sends to the mobile device the applications to be installed. As it has been described previously, the messages interchanged among a concrete client and servers are provided by the communications client that is one of the services

*Figure 9. MobileUI-Client collaboration diagram*

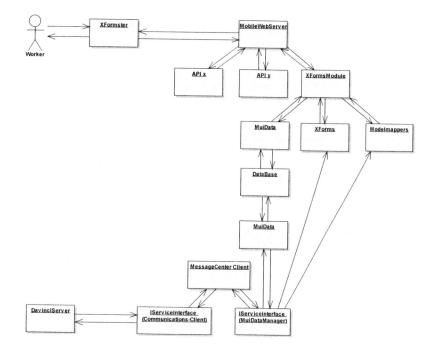

*Figure 10. MuiData collaboration diagram*

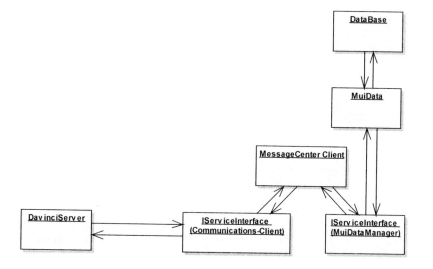

handled by the message center, as it is shown in Figure 10.

When a message of a new application is received, ClientMessageCenter redirects it to Mui-DataManager, which is in charge of managing the installations of new applications, the downloading of data from the server, and the modification of the updated data to the server. The forms (written in XForms) are stored in a specific repository, the data mapping files (modelmappers) are stored in another separated repository, and finally the data are stored in a local database.

For the management of the applications' data, an independent module called MuiData, which encapsulates the logic to manage data sources, is used, so it is possible to change easily the format of the data sources to another format (i.e., text files, XML files, etc.); replacing the MuiData component with another makes it able to process the new format.

On the other side, the mobile worker, when using DAVINCI mobile applications, interacts with the forms browser, called XFormster, to navigate between the forms that compose the mobile application. This navigation implies interaction with a forms server, called MobileWebServer, which is in charge of process the forms; this is illustrated in Figure 11.

The mobile Web server works with a set of APIs that processes different types of requests created and initiated by Web pages. In this case, the corresponding API to process requests of XForm forms will be the XFormsModule. This module is in charge of gathering the data sent from the Web form to the Web server, searching the modelmapper file corresponding to the mentioned form, and updating the data in the local database according to the rules specified in the modelmapper file.

Moreover, XFormsModule is also in charge of searching for the form that should be presented to the user as a consequence of any command processes being used for this purpose, displaying the new information according to the rules specified in the modelmapper file of the new form to show.

The forms browser has the user interface displayed in Figure 12.

Navigation between the different forms will depend on the specific design and the purpose of each mobile application implemented.

## eSignatureClient

This module is in charge of signing electronically the data generated by the mobile worker

*Figure 11. XFormster and MobileWebServer collaboration diagram*

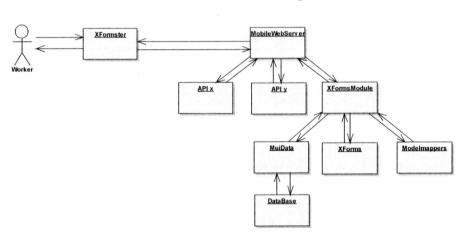

*Figure 12. XFormster user interface*

during the protocol application in a work zone and sending the signature to the central server for its verification and storage.

## VoiceUI

The VoiceUI component is in charge of providing the capability to handle XFormster forms by means of voice processing; that is to say, using the VoiceUI component, mobile workers are be able to handle the DAVINCI mobile forms using commands introduced by speaking.

The voice processing interface allows the user to introduce values for form fields and execute the commands provided by command controls provided by the form, even the activation of the main functionalities offered by XFormster.

## Print Manager

This module allows printing of any of the forms shown by the XFormster forms browser of the MobileUI-Client component. The print manager

uses the Bluetooth port to communicate with the printer in order to send the printing jobs.

When the user decides to print through XFormster, the print manager extracts the information of the labels and fields to be printed, and then prints and transfers them to a flat text document.

## Mobile Geographic Information Viewer (Mobile GIS)

This module allows the visualization of cartographic images and the presentation of information relative to parcels or enclosures defined on the shown cartographies.

Mobile GIS (Geographical Information System) is launched from the browser of MobileUI-Client's XFormster, but due to Mobile GIS's complexity and the size of the screen of some pocket devices, the cartographic information appears in a new window in order to not saturate the XFormster window; otherwise, the window to visualize GIS information would be totally unmanageable.

## Positioning Component (LBS)

This component is responsible for obtaining and handling the geographic position, in terms of coordinates, of DAVINCI mobile workers. This positioning component allows one to handle geographical, cartographic, and UTM coordinates, and to work with WGS84 and ED50 data.

A detailed diagram of DAVINCI integration components is shown in Figure 13 using UML (unified modeling language) syntax.

## STEPS TO USE THE INTEGRATION FRAMEWORK

If an organization wants to use the DAVINCI integration framework for the management of the applications used by its mobile workers, it is necessary to perform the following tasks.

*Figure 13. Integration framework architecture*

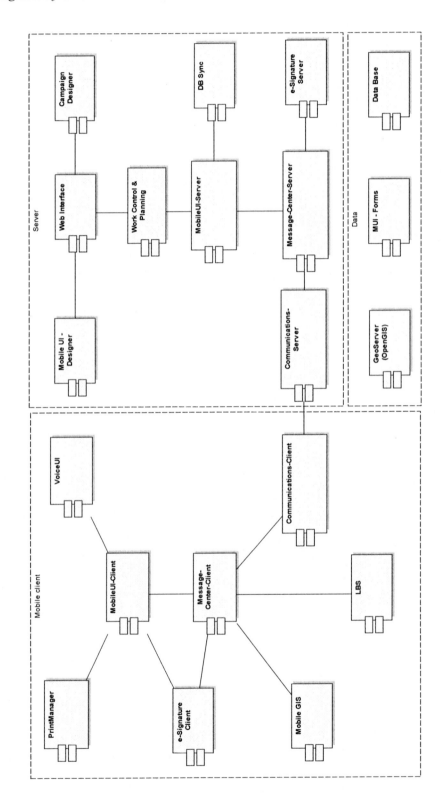

1. Install and configure the server components of the integration framework in a server able to be connected to client devices with the configuration. The server machine should have installed any J2EE-compliant application server and an RDMS. Our trials were done using JBOSS and Microsoft SQL Server (this database only manages the internal data required by the integration framework).

2. Install and configure the client components in the devices that will be used by the mobile workers. The configuration is different depending on the mobile device's capabilities. This is the most complicated task because there are several exceptions and special cases related to each device vendor and model. The effort of this task could be reduced by using the same model of mobile device.

3. Define the data managing policies that will be used on the data source used to obtain and update the data managed by the mobile applications.

4. For each mobile application to deploy, it is necessary to do the following.

   - Define, design, and implement the forms of the mobile applications using the XForms standard, preparing them for their introduction into the DAVINCI integration framework. This task could be done with any XML editor that can process XForms, for example, XMLSpy.
   - Define a modelmapper that contains the information of the graphical control used to show any item of the information, and the rules to query and update the mentioned data.
   - Implement and deploy a new DBSync Connector to enable DBSync Server to access the data source.

5. Design each company campaign using the Web interface of the integration framework.

6. Define the protocols that are assigned to the campaigns, and assign a form to each protocol activity.

7. Plan the work of each campaign, assigning the job to any available mobile worker.

Then, the applications and the required data are automatically sent to the mobile workers.

As each mobile worker performs his or her job, the integration framework automatically updates the information in the data source in accordance to the data managing policies established in Step 3 of this section.

When a mobile worker has new assignments, the integration framework sends the new data and, if required, a new mobile application to enable the worker to perform the job.

## SYSTEMS INTEGRATION IN EUROPEAN VETERINARIAN SECTOR USING DAVINCI

### Trials Purpose

The objectives that should be satisfied by means of the realization of DAVINCI trials and technology transfer activities consist of validating the improvements and technological innovations using DAVINCI in real working environments.

The main aspects validated were as follows:

- Capability of integrating the data provided by DAVINCI with Spanish and Bulgarian cattle exploitations and livestock-movement databases. This objective allows checking the effectiveness of DAVINCI for integrating information systems of different nature.
- Reduction of costs and effort spent in relation to the new mobile applications development.
- This validation was carried out by gathering related data on the necessary time to develop mobile applications to help veterinarians

perform information retrieval, and updating tasks related to the cattle sanitarian control program.

- Improvement of the ergonomics and veterinarians' comfort when performing the activities related to sanitary controls and clinical inspections. To fulfill this purpose, the effectiveness, efficiency, and ergonomic conditions for those carrying out the inspections were analyzed in detail. This analysis was performed by the veterinarians completing some evaluation questionnaires about the subjective estimation of these aspects.

## Trials Scope

Trials have been performed in four European countries (Spain, Latvia, Czech Republic, and Bulgaria) during the experimentation activities of the European-Commission-funded project called Advanced Management Services for Inspections and Sanitary Interventions in Cattle Holdings through Mobile Technologies (IPS-2001-42057).

The most efficient strategy for achieving the trial objectives was performing separate trials and technology transfer activities in Bulgaria, Spain, Latvia, and Czech Republic.

The reason for this separated approach is based on the existing differences between the infrastructures already present in each country.

In Spain, there is an information system for cattle-exploitations registering and livestock movements managed by each autonomous community and coordinated by the central government. The databases in the system satisfy the requirements in Directive Number 1760/2000 of the European Parliament. Also, in Spain there are standard GISs that are accessible by the project, so we will allow the checking of all of the project's required functionalities.

In Bulgaria, although there are databases for cattle exploitation and livestock movements, they are not standardized according to Directive Number 1760/2000 of the European Parliament. On the other hand, DAVINCI cannot use the standardized systems of the Bulgarian GIS to satisfactorily prove the GIS capacities provided by the project.

## Trials in Spain

A voice recognition system was tested in a small ruminant bleeding activity because the time saved with small ruminants is by far greater than the time estimated for large ruminants. The voice recognition system will be useful for some veterinary activities on the ground, but experience shows that the time saved in a bovine bleeding job using this technology is not as profitable as in small ruminant practices. For this reason, sheep control programs were selected for this trial.

The bleeding of large animals will be done using pocket PCs. There is a reason for this choice. In the case of large animals, more activities have to be performed and the use of a pocket PC, with all the functionality that it is able to provide, facilitates the job for these activities. The objective of this demonstration is not only the use of those devices on bleeding activities, but also to become conscious of the limits and capabilities of the use of technological devices on the ground in an unclean environment.

## Trials in Bulgaria, Czech Republic, and Latvia

Bulgaria believes that the approach to separate the demonstration from the prototypes is reasonable for a better evaluation of each one of them.

The use of a voice recognition system, no matter whether being used on small or large ruminants, is of significant interest for us, but due to some limitations on the languages that can be used in such a system, the demonstration of the technology could be limited to Bulgarian project coordinators reviewing the Spanish prototype.

*Figure 14. Trials execution in Spain*

*Figure 15. Trials execution in Czech Republic*

Although the mobile mapping system was tested in Spanish trials, we believe that this component will be of significant importance for us during the monitoring of culicids vectors on bluetongue disease, a disease with great epidemiological significance for Bulgaria and the region. Along with that, the application of restricting measures could be tested for a geographical area in case of a disease outbreak. All those activities could be done in a lab with all the data that we have—digital maps, data from culicid traps, and so forth—and we believe they are going to be sufficient for testing a prototype for the mobile mapping system.

## Trials Activities

In order for the testing routines to be able to provide a comparable set of conclusions, a common process for the trial development has been established by DAVINCI. The necessary activities to meet the established trial objectives and the desired technology transference are the following.

1.  Definition of the trials scope
2.  Determination of the region and holdings to be visited during the trials, looking for concrete parameters
3.  Selection of the participants in the prototype testing activity
4.  Configuration of the mobile devices' hardware
5.  Preparation of trial evaluation materials
6.  Performance of the cattle sanitary control activities
7.  Trials evaluation

*Figure 16. Trials execution in Bulgaria*

## CONCLUSION AND FUTURE WORK

The main success factor of DAVINCI does not lie in obtaining a system that provides a highly optimized performance, but in the automation of manual procedures, achieving acceptable performance. Moreover, it has been detected that the performance displayed is not optimal, but since it copes with minimal required performance, the fundamental objectives of the project are fulfilled.

As a summary, it may be said that the results obtained from the trials allow us to state that the main improvements brought about by DAVINCI are the following.

- The effort necessary to develop mobile applications to gather and/or consult livestock sanitary information has been reduced by up to 43.43%. The absolute value (in working hours) of this reduction will depend on the size (measured in function points) of the intended application.
  Nonetheless, it ought to be taken into account that the success mentioned has been produced in a limited context of a quite restricted set of mobile applications based on simple formularies, without complex displaying elements such as dynamic lists or tables, with simple access to databases not joining different data sources.
  Thus, this effort reduction is so far applicable to the development of simple mobile applica-

tions in terms of the insertion, modification, and consultation of data coming from a single data source.

- The economic costs needed to integrate mobile applications with existing systems and databases for livestock sanitary information management has been reduced by 9.03%.
- The average time to publish the data relative to a cattle holding has been reduced by up to 2.6 days, decreasing from 64.84 hours in the previous situation to 1.14 hours when using DAVINCI.
- DAVINCI easily adapts to the field working proceedings of different countries, with different cultures and organizational structures.
- The DAVINCI framework should adapt easily to technology changes in communications or to new mobile devices being used to perform fieldwork.
- The performance and easiness of mobile device configuration is acceptable, allowing the veterinarian to effectively carry out the field tasks.
- The communication components of DAVINCI are adaptable to several different communication infrastructures, allowing one to effectively resume interrupted messaging due to uneven communication coverage. Thus, the information transmitted through the available bandwidth is optimized.

However, it is necessary to perform, in the immediate future, an optimization process of the obtained solution, introducing, among others, the following improvements.

- To extend the set of available controls to be used by DAVINCI mobile applications, allowing the usage of more complex formularies with complex information, displaying elements such as dynamic lists and tables, and allowing simultaneous access to different data sources (joins)
- To carry out, keeping in mind the improvements of the available controls for mobile applications, a larger set of studies to assess the reduction of effort and time needed to develop mobile applications within DAVINCI
- To study and define algorithms allowing the broadening of coverage time, and thus lowering the number of messages that have to be resumed due to loss of communication with the coverage currently offered in rural environments by communication providers
- To improve the ergonomics and ease of use of the central components for in-field resource management
- To improve the ergonomics and ease of use of the mobile applications developed under DAVINCI, paying attention to those aspects related to input data validation and helping with the completion of less common tasks
- To provide the veterinarian with a mobile device that allows her or him to perform the fieldwork without the need to undergo configuration changes, such as battery changes, during the activities
- To improve the performance of DAVINCI's voice recognition system so that it consumes less resources and is less sensitive to noise while not lowering the speed for voice processing

# REFERENCES

Application servers. (2006). *MobileInfo.com.* Retrieved April 17, 2006, from http://www.mobileinfo.com/application_servers.htm

Baumgarten, U. (2004). *Mobile distributed systems.* John Wiley & Sons.

Boag, S., Chamberlin, D., Fernández, M. F., Florescu, D., Robie, J., & Siméon, J. (2005). XQuery 1.0: An XML query language. *W3C candidate recommendation.* Retrieved November 3, 2005, from http://www.w3.org/TR/2005/CR-xquery-20051103/

Boar, C. (2003). *XML Web services in the organization.* Microsoft Press.

Boyer, J., Landwehr, D., Merrick, R., Raman, T. V., Dubinko, M., & Klotz, L. (2006). XForms 1.0 (2nd ed.). *W3C recommendation.* Retrieved March 14, 2006, from http://www.w3.org/TR/xforms/

Intel Corporation. (2006). *Intel® mobile application architecture guide.* Retrieved April 17, 2006, from http://www.intel.com/cd/ids/developer/asmo-na/eng/61193.htm

Lee, V., Schell, R., & Scheneider, H. (2004). *Mobile applications: Architecture, design, and development.* Hewlett-Packard Professional Books.

Longueuil, D. (2003). *Wireless messaging demystified: SMS, EMS, MMS, IM, and others.* McGraw-Hill.

Mallick, M. (2003). *Mobile and wireless design essentials.* Wiley Publishing Inc.

Schiller, J. (2000). *Mobile communications.* Addison Wesley.

Siegal, J. (2002). *Mobile: The art of portable architecture.* Princeton Architectural Press.

# Chapter XII
# Enterprise Application Integration from the Point of View of Agent Paradigm

**Min-Jung Yoo**
*HEC (Ecole des Hautes Etudes Commericales), University of Lausanne, Switzerland*

## ABSTRACT

*The agent paradigm in general underlines the interaction phenomenon in a collaborative organization while respecting the autonomy and self-interested features of individual components. Relevant use of the agent paradigm will be one of the key factors to success in application integration projects in the near future. This chapter describes the basic notions of intelligent agents and multiagent systems, and proposes possible types of their application to enterprise integration. The agent-based approaches to enterprise application integration are considered from three points of view: (a) using an agent as a wrapper of applications or services, (b) constructing a multiagent organization within which agents are interacting and providing emergent solutions to enterprise problems, and (c) using the agent as an intelligent handler of heterogeneous data resources in an open environment.*

## INTRODUCTION

Today's business information systems are not isolated applications. Businesses are trying more and more to offer collaborative services, using an open communication infrastructure, such as the Internet, for the purpose of creating and operating worldwide services.

The early stage of integration efforts mainly addressed the problem of technical-level interoperability between heterogeneous applications providing application-to-application middleware. Nevertheless, the complex business context of nowadays requires higher level collaboration among applications in order to satisfy the properties of responsibility or agility, that is to say the following.

- Due to the responsibility, services required by clients should be safely offered

- Agility means continuously monitoring market demands and quickly providing new products

Concerning the first point, a business should try to satisfy its clients even when it is impossible to offer a service because of some failures. Searching for another service provider that is capable of offering a similar service may be a possible solution in that situation. In long-term vision, this strategy can better service clients. From the technological point of view, today's Web services standards may be useful for constructing integrated networks of applications among business partners. The environment of business information systems is highly volatile. For this reason, the creation, enactment, and management of such collaborative networks require special concerns. In many Web services composition projects, one of the hot issues is how to provide an effective means of searching for appropriate services and business partners.

For the second point, the possibility of reusing existing components will be a key factor for the successful development of new services within a short period of time. In that context, not only reusing one's own data sources and applications, but also benefiting from applications and services of other firms—through a certain form of collaboration and negotiation—will be helpful.

In order to confront these problems, CIOs (chief information officers) should think of progressively moving the abstract level of the interoperable medium from data handled by applications to services guaranteed by businesses. Rather than achieving integration, they should aim at constructing a collaborative organization that provides the dynamic just-in-time interaction of services. The agent paradigm in general underlines the interaction phenomenon in a collaborative organization while respecting the autonomy and self-interested features of individual components. The relevant use of the agent paradigm will be one of the key factors to success in the near future in application integration projects.

This is the reason why this chapter is dedicated to reviewing some principle concepts of intelligent agents and multiagent systems. This chapter is organized as follows. In the next section, principal concepts concerning intelligent agents and multiagent systems are briefly discussed. Then the chapter discusses the taxonomy of enterprise application integration in general. Next, it presents agent-based approaches to enterprise application integration. Finally, a short summary and concluding remarks are given.

## INTELLIGENT AGENTS AND MULTIAGENT SYSTEMS

First of all, it is necessary to give the definition of the term agent. The most widely accepted definition could be the one given by Wooldridge (2002, p.15): "An agent is a computer system that is situated in some environment, and that is capable of autonomous action in this environment in order to meet its design objectives." The kinds of intelligent capabilities of agents are then considered as follows.

- **Reactivity:** The ability to perceive the environment and respond in a timely fashion to changes
- **Proactivity:** The ability to pursue goals by taking the initiative in order to satisfy design objectives
- **Social ability:** The capability of interacting with other agents in order to satisfy the social goal

The studies of agents are multidisciplinary research results from the domains of artificial intelligence and logic, distributed computing, psychology, social science, artificial life, ecology, engineering and robotics, and so on. For this reason, it is always a debatable question of defining what an agent-based approach to a certain problem is. Meanwhile, if we are able to intelligently handle its diversity, there may be many ways of applying the agent paradigm to application integration and benefiting from it.

The most commonly accepted view of agent studies is regarding it as research on distributed artificial intelligence (DAI) or decentralized artificial intelligence (DcAI). DAI and decentralized AI have slightly different characteristics while sharing some common interests in the intelligent behavior of distributed entities.

In DAI, a global task is initially defined, and the problem is then to design the distributed entities to enable the performance of this global task. Therefore, the main issue is to study the distribution and the collaborative resolution of a given problem. The key problem to solve is how to efficiently distribute the entire problem among collaborative entities through tasks and resource allocation.

On the contrary, in DcAI, decentralized autonomous entities are initially defined. We then study how these entities are able to achieve tasks that may be personal or may interest several entities. Two important motivations in this field of research are the following (Demazeau & Müller, 1989).

- Designing more powerful and autonomous agents that are capable of processing incomplete and uncertain knowledge by communication with other agents. The DcAI approach can provide insights into improving and increasing the reasoning and decision capabilities of each agent. The agents may achieve their own goals in the society that are compatible within the range of a global goal. The key aspect is thus the collaboration among multiple agents. In the approach to collaborative distributed problem solving (CDPS) by multiagent systems, for example, cooperation is especially necessary when no single node has sufficient expertise,

resources, or information to solve a global problem.

- Understanding social behavior through collaboration between autonomous agents. DcAI can not only improve the intrinsic power of the capabilities of each agent, but also model social behavior emerging from the agent interactions. Since it is based on the framework of the interacting entities, it provides keys to understanding the interactions among intelligent beings that may belong to various groups of interests or societies.

As a result, the study of agents made its progress in the following issues providing related outcomes.

- The issue of intelligent agents
- The mechanisms of collaboration

In both cases, common features that characterize agents are the facts that (a) the entities are situated in an environment, (b) their rationality is expressed not only by their knowledge about the environment, but also through their actions in the environment, and (c) their collaboration is not hardwired.

## The Issue of Intelligent Agents

This approach is characterized by constructing intelligent autonomous entities, that is, intelligent agents, in a multiagent world. Constructing an intelligent program has been the main goal of traditional AI and logic-based approaches. The main difference between the traditional AI and the agent studies lies in how agents learn and handle conditions and knowledge necessary for reasoning. Intelligent agents try to solve problems situating in a partially known environment, whereas in traditional AI the environment or some mandatory conditions to problem solving are predefined. Another difference is that the agent's intelligence concerns not only their knowledge level, which is the case in traditional reasoning systems, but also the objective of their decisions. The reasoning purpose of an intelligent agent is thus to decide on an action, even though the agent has no sufficient information about its environment. This challenge resulted in many agent architectures, such as the BDI (belief-desire-intention) architecture (Georgeff et al., 1999) whose principle is based on practical reasoning systems.

The purpose of practical reasoning is in fact to direct the reasoning process toward action choices. The decision is based on two types of options: what the agent believes (the knowledge) and what the agent desires or cares about. These options can possibly be competing. In case there is competition or conflicting considerations between these options, the agent should weigh them to make a decision.

## The Issue of Multiagent Collaboration

This approach is characterized by studies on the collaborative organization of self-interested entities. In this approach, the relationship between agents and the environment, and the sphere of influence of agents to their environment are the primary considerations. The purposes of agent collaboration are as follows.

- Sharing knowledge and intelligence among collaborative agents. In the case of enterprises, the knowledge may concern the specialty, strategies, business intelligence, or some data in the enterprises' local databases.
- Benefiting from the capability of other agents, which is in fact the actions and services offered by other agents. It can be the enterprises' services.
- Exchanging goals that allow agents to achieve the objective of a whole organization while satisfying the individual goals of self-interested entities.

In any case, the agents have to communicate with each other for the purpose of collaboration. As for the communication mediums, agent systems may use a rich set of communication technologies for the purpose of data transfer between agent applications. The different communication mediums can be categorized into several types.

- **Concerning data transfer:** Shared memory, such as blackboard architecture (http://www.bbtech.com/), or message passing
- **Considering the types of communication infrastructure:** Wired or wireless
- **Considering the types of message distribution:** Point to point, multicast, broadcast
- **Considering the modes of message exchange:** Synchronous or asynchronous message passing

Particularly, the research issue of asynchronous message passing entailed several outcomes such as KQML (knowledge query and manipulation language) and KIF (knowledge interchange format; Finin et al., 1993), the FIPA (Foundation for Intelligent Physical Agents, 1997) standard for the agent communication language, various negotiation mechanisms for agent-based brokerage, and task allocation protocols such as Contract Net (Smith, 1980). A brief overview of some elements used for agent collaboration is given below.

### KQML and KIF

The most important contribution of KQML is the semantics of the communication protocols (domain-independent part), which is separated from the semantics of enclosed message content (which may be dependent on domain problems). The semantics of the high-level protocols, used by communicative agents, should be concise and have only a limited number of primitive communicative acts. The communicative acts of KQML, or "performatives" in other words, such as *tell, ask-if, reply, subscribe,* and so forth, enable you to construct a virtual knowledge base between heterogeneous software.

This part of knowledge is kept and managed by autonomous agents. The idea was that agents using KQML as a communication means might be implemented using different programming languages and paradigms. Any information that agents have may be internally represented in many different ways. Each agent is capable of the following.

- Understanding the high-level message semantics
- Handling the content of messages as it is required

While KQML is an upper language that permits us the knowledge-level data description, KIF was originally developed with the intent of being a common language for expressing properties of a particular domain. KIF is closely based on first-order logic in recasting its representation in a LISP-like notation.

## FIPA

The Foundation for Intelligent Physical Agents promotes technologies and specifications that facilitate the end-to-end interoperating of intelligent agent systems for individuals. Now, specifications proposed by FIPA become de facto standards for developing software agent platforms. The FIPA agent communication language (ACL) is one of such standards. The FIPA ACL is very similar to KQML, being composed of about 20 performatives. Apart from ACL, the FIPA specification also includes agent management technologies concerning agent management and agent message transport. The agent organization is managed by a standardized platform and agent directory facilities (yellow- or white-page services).

Already, a number of FIPA-compliant agent platforms are publicly available, such as JACK

Intelligent Agent (http://www.agent-software. com/), JADE (Java Agent Development Environment, http://jade.cselt.it/), ZEUS (http://labs. bt.com/projects/agents/zeus/), Agent Development Kit (http://www.tryllian.com), and so on.

## TAXONOMY OF APPLICATION INTEGRATION

An enterprise can achieve its integration purpose on its own terms, managing different technologies that are well-introduced in the other chapters. The integration model can range from a simple and rapid integration to a more complex form. The integration issues in general can be categorized as follows (Linthicum, 2003; Vinoski, 2002; Yoo, Sangwan, & Qiu, 2005).

- Common data transfer
- Database-level integration
- Interface processing
- Process-level integration
- Service-oriented integration
- Integration within a service organization

In this section, each type of integration is reviewed.

### Common Data Transfer

This is characterized by sharing text-based files between applications to be integrated, for example, using script languages such as Python or Perl in order to share information between different natures of applications (Vinoski, 2002). By the possibility of rapid prototyping and easy extensibility, this approach is often used by a large amount of communities for simple and short-term integration projects.

## Database-Level Integration

This approach attempts to interconnect databases by database-to-database integration or constructing a federated database in order to share consistent database content. Data replication is simply moving data between two or more databases. The databases can employ different models. Currently, many database-oriented middleware solutions provide the database replication service by transforming and adjusting heterogeneous database elements. The implementation is relatively easy and cheap.

Database federating tries to integrate multiple databases and database models into a single unified view by way of a federated virtual database. On the contrary to data replication, which physically replicates data contents, data federation constructs an additional virtual view of all databases that comprise many real, physical databases.

Considering that the ultimate purpose is to share data resources, the common data transfer and database-level integration can be compared with agent collaboration for sharing knowledge resources. Nevertheless, in the case of agent collaboration for sharing knowledge resources, the sharing decision is up to the agents. An agent can require some information and the other agent can decide whether to reject or accept it.

## Interface Processing

This approach uses well-defined application interfaces in order to provide a unified interface of all internal application services. The purpose of interfacing varies from the unification of information to the unification of services, or both of them. The interface layer can invoke local or remote applications. For example, JCA (Java Connector Architecture, http://java.sun.com/j2ee/connector/) can be regarded as this type of integration, which provides the interface of different ERP (enterprise resource planning) systems.

Concerning the agent paradigm, the possibility of providing a unified interfacing mechanism with the help of a certain agent communication language such as FIPA ACL can give interoperability among a high number of heterogeneous software and hardware systems. This allows integrating existing software systems as required, for instance, for operating the manufacturing enterprise, hardware modules such as machines, or sophisticated control mechanisms. FIPA particularly specifies as well the conditions to be satisfied for the agent software integration by a wrapper agent that is needed in activating software systems.

## Process–Level Integration

Process-level or business process integration deals with building enterprise-wide business processes incorporating existing applications into those processes. This approach is characterized by providing another layer of process management that resides on the top of other applications. This layer supports the flow of information and control logic between existing processes. A workflow management system (WFMS) is in this category of integration.

This level of integration allows the cooperative work of several processes while sharing the capability of processes. What will be the difference in the case of agent collaboration? The WFMS controlling mechanism is based on static process integration. Each process has no autonomous decision capability concerning its engagement with a certain type of collaboration, and all processes should be analyzed, including possible interfacing among them, during WFMS analysis time. On the contrary, the collaboration within an agent organization is strongly based on the agent's autonomy and self-interested strategy. The collaboration relationship might well change with time.

## Service-Oriented Integration

Service-oriented integration allows applications to share common business logic or methods. Web-based services are good examples of services to be integrated in this context. The main purpose is thus to provide end users with a unified service infrastructure that is composed of many service applications on the Web. By connecting to the composite application, a user can benefit from the services offered by existing applications. In this approach, the existing application logic is important from the point of view of the integrated services.

In fact, the notion of service is not a newly introduced concept. Meanwhile, the primary difference in the case of older service offerings is that human intervention was previously required in order to access and use services. The modern view of services goes beyond the old definition in terms of accommodating the openness of Web systems. This means describing services in a standard manner, arranging for them to be discovered and invoking them in a standard manner (Singh & Huhns, 2005).

## Integration within a Service Organization

Many researchers now predict that the future direction of business information goes toward global integration, and this should be guided by each entity's roles based on organizational analysis (Basu & Kumar, 2002; Zambonelli, Jennings, & Wooldridge, 2000). The ultimate purposes are twofold: constructing a unified view of an enterprise, and getting these different business services to collaborate to give each other value-added service stemming from synergy effects.

In that case, what is the difference between a service organization and service-oriented integration? Applications in a service organization know the semantics of other services as collaborative applications for the purpose of achieving its own

service. Applications in a service-oriented integration are not aware of the other applications. The characteristics of each service (i.e., the previously mentioned application logic) are visible only to the designer of an integrated service in the case of service-oriented integration. Contrarily speaking, in a collaborative service organization, each service has the capability of understanding the service semantics of others. The knowledge about the other services make it possible for a service to require a necessary service at run time for the purpose of achieving its own objective, or to find and bind dynamically to other services.

The above discussions are not only showing the variety of application integration, but also the direction of their evolution (Figure 1). Starting from static integration between applications, the evolution goes toward dynamic interoperability between services in an open environment. The service logic becomes more and more complex. As pointed out by Singh and Huhns (2005), the services are meant to be used in multiple contexts. Their argument continues as follows (p. 87):

That is, models for services apply both in terms of how the services are implemented and how they are used by others. For complex services implementation, we would need to model the databases and knowledge bases, applications, workflow, and the organizational roles involved.

At present, the integration means no more putting diverse concepts together, but rather matching the needs and preference of service consumers with the capacity and quality of services provided by service providers.

## AGENT-BASED APPROACH TO APPLICATION INTEGRATION: WHAT FOR?

Until now, we have seen the basic notion of the agent paradigm on the one hand, and the evolution of enterprise application integration on the other hand. It is vital to explain where the agents are

*Figure 1. The evolution of application integration*

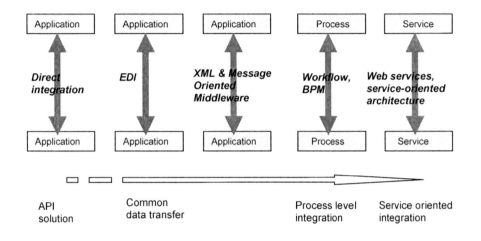

useful in the context of application integration. Our primary concern in this section asks the question, "What does it mean to use agents for enterprise application integration?" That is, we address the category of agent-based approaches to enterprise application integration. The discussions are based on three points of view.

- It means using an agent as a wrapper of applications or service execution.
- It means modeling enterprise applications using an agent organization within which agents are interacting and providing emergent solutions to enterprise problems.
- It means using agents as an intelligent handler of heterogeneous data resources in an open environment.

## Agent as a Wrapper of Heterogeneous Software

This means using a standard agent architecture in order to integrate the functions of applications (Figure 2). The interoperability problem addressed by Genesereth (1997) mainly concerns this issue.

The wrapper agent approach is comparable to interface integration or service-oriented integration considering the fact that it is to provide a unified access mechanism. Meanwhile, wrapper-based agent software integration not only offers a unified interface by means of agent communication languages, but also allows the service logic of an application to be meaningful in the integrated whole. This aspect enables an agent to understand the service semantics of other agents in the organization. The agent messages used for communication are more than the simple activation of other applications' internal processes. The messages in ACL are the real means of expressing the capacity or the preference of each agent application, and they ultimately influence the action of the other agents.

ExPlanTech presented in Pechoucek, Vokrinek, and Becvar (2005) can be taken as an example. ExPlanTech is a framework for agent-based production planning that offers technological support for various manufacturing problems. Users can assemble different components to develop a customized system for decision support in production planning. The platform is built on top of JADE. The approach helps encapsulate legacy software, such as those for scheduling, parts management,

*Figure 2. Wrapper agent approach*

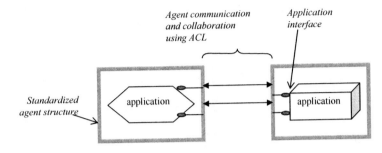

and stock management, as JADE agents. Developers can benefit from a predefined agent core with already implemented control and message transport protocols. This feature facilitates the easy integration and rapid development of new and third-party components.

With the help of using a standard agent architecture such as FIPA, the applications can also benefit from other services offered by the agent platforms in combination with Web services standards, ontology services, or security issues. Web Service Agent Gateway, for example, enables a connection between SOAP (simple object access protocol) messages and the FIPA-compatible ACLs. Already, several gateways have been developed for different agent platforms. Such a gateway service is accessible to both agents using FIPA ACL and Web service clients through SOAP messages. The gateway makes it possible to represent agent services as Web services, and the inverse, that is, to access Web services via the agent communication.

## Agents as Self-Interested Interacting Entities within an Organization

This means modeling enterprise applications using an agent organization within which agents are interacting and providing emergent solutions to enterprise problems (Figure 3). As it is underlined in Van Dyke Parunak (1999), if the characteristics of domain problems require the

particular capabilities of agents, agents can be best used for enterprise solutions. Some examples of such characteristics are that enterprise systems are modular, decentralized, ill-structured, and complex. Rather than functional decomposition that governs almost all enterprise applications, agent-based approaches follow physical decomposition. This aspect guarantees the modularity and ill-structured nature of resulting systems, which fits well to agent-based approaches to industrial application.

In this approach, access to existing applications or databases is often needed for the purpose of modeling the actions or internal resources of agents. As firms have their own legacy software, we must recursively solve the problem of application integration in general between agents and the legacy software. Middleware solutions and technologies for application integration, such as data transfer, message-oriented middleware, and XML-based standards, can be useful for the purpose of agent-to-legacy software integration.

In the ANTS (Agent Network for Task Scheduling) system (Sauter & Van Dyke Parunak, 1999), a traditional ERP approach, which is often used for supply chain management, is replaced by an agent-based system. In the organization of a multiagency, a single step in the process plan (operation), resources such as machines, operators, material, and customers or suppliers are separately modeled as small-grained agents. They are dynamically interacting using the contract net

*Figure 3. Organization of small-grained multiagency*

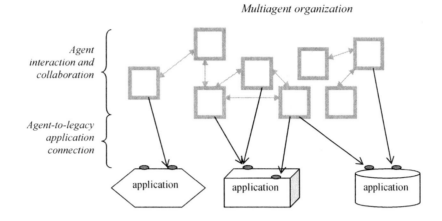

protocol (Smith, 1980) with a view to solving the negotiation problem in task and resource allocation among agents.

The main purpose of this approach is to encompass some limitations due to traditional centralized approaches that are too brittle in the face of the dynamic environment of real-world problems. Because the software environment becomes more and more complex, the successful implementation of such software requires a huge budget and effort. In order to solve this problem, small-grained agent-based systems offer a promising alternative to monolithic software modules by letting agents respond to their environment using simple reaction rules or directly interacting with other agents through predetermined protocols.

## Agent as an Intelligent Handler of Distributed Data Resources

This means using agents as an intelligent handler of nonintegrated and independently developed data resources that are used in different enterprises. This approach is helpful when enterprises need to reach distributed resources and understand them with a view to making a decision. The intelligent program, the agent on the semantic Web, is in this category. The aim of the semantic Web is to make the Web usable not only by people, but also by

software agents (Berners-Lee, Hendler, & Lassila, 2001). The agent is an intelligent consumer of these interpretable resources.

One of the mandatory conditions for using this approach is that the resources should be ready to be handled by a program, that is to say, written in particular semantic languages. Nowadays, semantic Web ontology languages such as RDF (resource description framework) or OWL (Web ontology language) allow the expression of the data resource semantics using XML. A more specialized ontology for expressing service semantics, OWL-S, is also available. OWL-S compliments current WSDL (Web services description language) by providing the possibility of describing the properties and capabilities of a service. Thanks to these features, agents can reason about the content of services and decide what services to use to achieve their own goals.

There are already many commercial products that enable us to create and manage such semantic documents. Oracle Spatial 10g, Altova's SemanticWorks 2006, or Cerebra Server are some examples of these products. Oracle Spatial 10g (http://www.oracle.com/technology/products/spatial/index.html) introduces an open, scalable, secure, and reliable RDF management platform for industries. Based on a graph data model, RDF triples are persisted, indexed, and queried in a

*Figure 4. Agents as semantic information consumers*

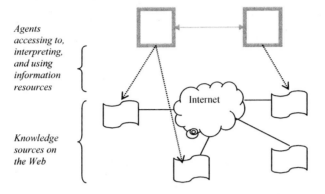

Agents
accessing to,
interpreting,
and using
information
resources

Knowledge
sources on
the Web

Internet

similar way to other object-relational data types. Altova SemanticWorks 2006 (http://www.altova. com/products_semanticworks.html) is a visual RDF-OWL editor from the creators of XML-Spy. The system enables semantic Web instance documents, vocabularies, and ontologies to be visually designed, and then outputs them in either RDF-XML or *N*-triples formats. Cerebra Server (http://cerebra.com/products/cerebra_business. html) is a platform that can be used by enterprises to build model-driven applications and highly adaptive information integration infrastructure. This platform also handles problems relating to security, user management, auditing, notifications, logging, and user interface services.

To summarize, the approach to intelligent agents on the semantic Web facilitates both inter- and intra-enterprise knowledge integration. Enterprises can reduce the cost and complexity of data management and systems development. Thomas, Redmond, Yoon, and Singh (2005) show an example of knowledge management using semantic documents and agents. Without changing the existing business process, which is described in a standard language, their approach proposes giving some additional information, described in OWL, concerning performance criteria for business processes and activities. These OWL documents provide semantic descriptions of the contextual requirements for agents to monitor business process performance. Different types of agents ultimately monitor the overall business process and the performance of individual business activities based on multiple criteria.

## CONCLUSION

This chapter described the basic notions of intelligent agents and multiagent systems, and proposed possible types of application to enterprise integration. As we have seen, agents are not panaceas that can completely replace one of the current application integration technologies. Agents are rather methodological metaphors that can be used with other technologies in order to achieve synergy effects.

Today's technologies can provide fast-to-achieve solutions to integration problems. However, they do not yet provide substantial architecture models or methodologies for open service organization. As for agent-based software solutions, in spite of the interest demonstrated, there are not yet many products available that are intended to encourage enterprise integration particularly using an agent paradigm. If companies prefer buying an integrated package for agent solutions, then it would be better to wait until the agent-based third-party products are much more mature.

In the meantime, we can stimulate the technology advancement in two directions. On the one hand, an agent-oriented organization model or modeling methodologies can guide third-party tool developers toward the effective use of the agent paradigm. In the agent-oriented modeling paradigm, a rich set of organizational structures and interaction protocols are well-studied. Concomitantly, enterprises can learn from the experiences in the agent domain and use the knowledge for developing their private solutions to application integration. These experiences will be helpful for modeling global business information systems in today's open environment.

## ACKNOWLEDGMENT

The author would like to thank Professor François Bodart from the University of Namur (Belgium) for providing helpful suggestions and discussions on the subject.

## REFERENCES

Akkiraju, R., Keskinocak, P., Murthy, S., & Wu, F. (1998). A new decision support system for paper manufacturing. *Sixth International Workshop on Project Management and Scheduling*, Istanbul, Turkey.

Basu, A., & Kumar, A. (2002). Research commentary: Workflow management issues in e-business. *Information Systems Research, 13*(1).

Berners-Lee, T., Hendler, J., & Lassila, O. (2001). The semantic Web. *Scientific American.*

Chen, Q., Hsu, M., Dayal, U., & Griss, M. (2000). Multi-agent cooperation, dynamic workflow and XML for e-commerce automation. *Fourth International Conference on Autonomous Agents.*

Demazeau, Y., & Müller, J.-P. (1989). Decentralized artificial intelligence. *Proceedings of the First European Workshop on Modelling Autonomous Agents in a Multi-Agent World.*

Finin, T., et al. (1993). *Specification of the KQML agent communication language: DARPA knowledge sharing initiative external interfaces working group.*

Foundation for Intelligent Physical Agents (FIPA). (1997). Agent communication language. In *FIPA 97 specification, version 2.0.* Geneva, Switzerland: Author.

Genesereth, M. R. (1997). An agent-based framework for interoperability. In Bradshaw (Ed.), *Software agents.* AAAI Press/MIT Press.

Georgeff, M., et al. (1999). The belief-desire-intention model of agency. In J.-p. Müller, M. Singh, & A. S. Rao (Eds.), *Lecture notes in artificial intelligence: Vol. 1555. Intelligent Agents V.* Berlin, Germany: Springer.

Goethals, F., Vandenbulcke, J., Lemahieu, W., Snoeck, M., De Backer, M., & Haesen, R. (2004). *Communication and enterprise architecture in extended enterprise integration.* Proceedings of the Sixth International Conference on Enterprise Information Systems, Porto, Portugal.

Inmon, W. H. (2000). A brief history of integration. *EAI Journal.* Retrieved October 2002, from http://www.eaijournal.com/applicationintegration/BriefHistory.asp

Jennings, N. R. (2001). An agent-based approach for building complex software systems. *Communications of the ACM, 44*(4).

Kishore, R., Zhang, H., & Ramesh, R. (in press). Enterprise integration using the agent paradigm: Foundations of multi-agent-based integrative business information systems. In *Decision support systems.* Elsevier.

Linthicum, D. S. (2003). *Next generation application integration: From simple information to Web services.* Addison-Wesley Information Technology Series.

Pechoucek, M., Vokrinek, J., & Becvar, P. (2005). ExPlanTech: Multiagent support for manufacturing decision making. *IEEE Intelligent Systems, 20*(1).

Sauter, J. A., & Van Dyke Parunak, H. (1999). *ANTS in the supply chain.* Workshop on Agent Based Decision Support for Managing the Internet-Enabled Supply Chain, Agents 99, Seattle, WA.

Singh, M. P., & Huhns, M. N. (2005). *Service-oriented computing: Semantics, processes, agents.* John Wiley & Sons.

Smith, R. G. (1980). The contract net protocol. *IEEE Transactions on Computers, C29*(12).

Thomas, M., Redmond, R., Yoon, V., & Singh, R. (2005). A semantic approach to monitor business process performance. *Communications of the ACM, 48*(12).

Van Dyke Parunak, H. (1999). Industrial and practical applications of DAI. In G. Weiss (Ed.), *Multi-agent systems.* Cambridge, MA: MIT Press.

Vinoski, S. (2002). Middleware "dark matter." *IEEE Distributed Systems Online.* Retrieved April 24, 2007, from http://www.iona.com/hyplan/vinoski/pdfs/IEEE-Middleware_Dark_Matter.pdf

Walker, S. S., Brennan, R. W., & Norrie, D. H. (2005). Holonic job shop scheduling using a multiagent system. *IEEE Intelligent Systems, 20*(1).

Wooldridge, M. (2002). *An introduction to multiagent systems.* John Wiley & Sons.

Yoo, M.-J. (2004, October). *Enterprise application integration and agent-oriented software integration.* IEEE International Conference on Systems, Man, and Cybernetics, The Hague, Netherlands.

Yoo, M.-J., Sangwan, R., & Qiu, R. (2005). Enterprise integration: Methods and technologies. In P. Laplante & T. Costello (Eds.), *CIO wisdom II: More best practices.* Prentice-Hall.

Zambonelli, F., Jennings, N. R., & Wooldridge, M. (2000). Organizational abstractions for the analysis and design of multi-agent systems. In *Lecture notes in computer science: Vol. 1957. First International Workshop on Agent-Oriented Software Engineering.* Springer-Verlag.

## ENDNOTES

[1] http://www.bbtech.com/

[2] http://www.agent-software.com/

[3] http://jade.cselt.it/

[4] http://labs.bt.com/projects/agents/zeus/

[5] http://www.tryllian.com

[6]

[7] http://java.sun.com/j2ee/connector/

[8] http://www.oracle.com/technology/products/spatial/index.html

[9] http://www.altova.com/products_semanticworks.html

[10] http://cerebra.com/products/cerebra_business.html, last accessed September 2006

# Chapter XIII
# Web Services Hybrid Dynamic Composition Models for Enterprises

**Taha Osman**
*Nottingham Trent University, UK*

**Dhavalkumar Thakker**
*Nottingham Trent University, UK*

**David Al-Dabass**
*Nottingham Trent University, UK*

## ABSTRACT

*Web services are used in enterprise distributed computing technology including ubiquitous and pervasive computing and communication networks. Composition models of such Web services are an active research area. Classified as static, dynamic, and semiautomatic composition models, these models address different application areas and requirements. Thus far, the most successful practical approach to Web services composition, largely endorsed by industry, borrows from business processes' workflow management. Unfortunately, standards subscribing to this approach fall under the static composition category, therefore the service selection and flow management are done a priori and manually. The second approach to Web services composition aspires to achieve more dynamic composition by semantically describing the process model of the Web service and thus making it comprehensible to reasoning engines and software agents. In this chapter, we attempt to bridge the gap between the two approaches by introducing semantics to workflow-based composition. We aim to present a composition framework based on a hybrid solution that merges the benefit of practicality of use and adoption popularity of workflow-based composition with the advantage of using semantic descriptions to aid both service developers and composers in the composition process and facilitate the dynamic integration of Web services into it.*

## INTRODUCTION

The last decade has witnessed an explosion of application services delivered electronically, ranging from e-commerce to information service delivered through the World Wide Web, to services that facilitate trading between business partners, better known as business-to-business (B2B) relationships. Traditionally, these services are facilitated by distributed technologies such as RPC (remote procedure call), CORBA (Common Object Request Broker Architecture), and more recently RMI (remote method invocation). Web services are the latest distributed computing technology. It is a form of remote procedure call like other distributed computing technology, but uses XML (extensible markup language) extensively for messaging, discovery, and description. The use of XML messaging makes Web services platform and language neutral. Web services use SOAP (simple object access protocol; Gudgin, Hadley, Mendelsohn, Moreau, & Nielsen, 2003) for XML messaging, which in turn uses ubiquitous HTTP (hypertext transfer protocol) for the transport mechanism. HTTP is considered a secure protocol, thus it allows Web services to be exposed beyond the firewall. The Web service messages and operations with invocation details are described using a platform-independent language WSDL (Web services description language; Christensen, Curbera, Meredith, & Weerawarana, 2001). Web services can be published and discovered using UDDI (universal description, discovery, and integration protocol). The Web services architecture centred on WSDL, UDDI, and SOAP is called a service-oriented architecture (SOA).

To take advantage of Web services features, network application services have to be developed as Web services or converted into Web services using a wrapping mechanism. Moreover, multiple Web services can be integrated either to provide a new value-added service to the end user, or to facilitate cooperation between various business partners. This integration of Web services is called Web services composition and is feasible to achieve because of the Web services advantages of being platform and language neutral and loosely coupled.

The logic for composition mainly involves two activities: (a) the selection of candidate Web services that fulfill the requirement in accumulation, and (b) the management of flow, which comprises control flow, the order in which Web services operations are invoked; and dataflow, the messages passed between the Web services operations. The level of automation provided in performing these two activities classifies composition into static, semiautomatic, or dynamic categories. Static composition involves prior hard-coding of the service selection and flow management. Performing selection and flow management on the fly, in machine-readable format, leads to dynamic composition. In semiautomatic composition, the service composer is involved at some stage (Thakker, Osman, & Al-Dabass, 2005).

## WEB SERVICES COMPOSITION APPROACHES

### Workflow-Based Composition

Workflow-based techniques approach Web services composition as a business process management (BPM) solution. Business processes can be considered as the group of activities that carry out business goals (Leymann, Roller, & Schmidt, 2002). Business applications represent such activities in the business processes; for example, a customer order fulfillment process will include individual applications for the activities of a customer placing an order, checking an account status, verifying an order, and dispatching an order. BPM deals with achieving the integration of these individual applications to achieve a business process view.

The main industrial standards to achieve such composition of Web services are WS-BPEL (Web services business process execution language; Andrews et al., 2003), WS-CDL (Web services choreography description language; Kavantzas, Burdett, Ritzinger, & Lafon, 2004), and BPML (business process modeling language; Van der Aalst, Dumas, ter Hofstede, & Wohed, 2002). These approaches use WSDL (declared operations and data-typed messages) extensively to compile the composition scheme. All these standards fall under the static composition category; that is, the service selection and flow management are done a priori and manually. However, commerce and industry remain loyal to the workflow-based composition for integrating services within the enterprise (enterprise application integration), and for forging B2B collaboration.

The prominent industrial standard, WS-BPEL, enhances and replaces existing standards XLANG from Microsoft and WSFL from IBM. Apart from being based on WSDL, it uses workflow management as the process model to achieve the formalization for control flow and data flow. The success of BPEL in the business community can be attributed to a number of factors. First, the standards are built on top of workflow management theory, making it ideal to model business processes' interactions. The second factor is that BPEL and its derivatives are now mature standards that provide a gamut of features for business processes, such as transaction processing, support for state management with the use of callbacks and correlation sets, provision for exception handling, and compensation fault processing features (Osman, Wagealla, & Bargiela, 2004) that are vital for the long-running and fault-vulnerable business transactions.

The adoption of BPEL as the Web service composition technology of choice has been reflected in the enthusiasm at large software houses in providing or including BPEL composition tools in their enterprise application servers, for instance, Oracle Application Server, Microsoft BizTalk Server, and the stand-alone tool from IBM, BPWS4J.

## Semantic Web-Based Composition

The fundamental premise of the Semantic Web is to extend the Web's currently human-oriented interface to a format that is comprehensible to software programs. Applied to Web services composition, this can lead to the automation of service selection and execution.

The automation is achieved by describing the Web services semantically, thus allowing software agents to reason about the service capability, and make all the decisions related to the composition on behalf of the user or developer. The decisions include the selection of appropriate services, their actual composition, and close examination of how they meet the criteria specified by the user. In contrast, in the static composition approach, the user or developer manually interprets the requirements for the required composition and the available service capability or functionality, and makes decisions regarding how services can be interweaved to make a value-added service.

DARPA's OWL-S (ontology Web language for Web services) is the leading semantic composition research effort. OWL-S (Ankolekar et al., 2001; Dean, Hendler, Horrocks, McGuinness, Patel-Schneider, & Stein, 2004) ontologies provide a mechanism to describe the Web service's functionality in machine-understandable form, making it possible to discover and integrate Web services automatically. An OWL-based dynamic composition approach is described in Sirin, Hendler, and Parsia (2003), where semantic descriptions of the services are used to find matching services for the user requirements at each step of composition, and the generated composition is then directly executable through the grounding of the services. Other approaches use the artificial-intelligence (AI) planning technique to build a task list to achieve composition objectives: the selection of services and flow management for performing the composition of services to match user preferences. McIlraith and Son (2002) use Golog, an AI planning reasoner, for automatic composition, while in a similar spirit some other approaches (Nau, Cao, Lotem, & Muñoz-Avila, 1999; Wu, Parsia, Sirin, Hendler, & Nau, 2003) have used the paradigm of hierarchical task network (HTN) planning to perform automated Web service composition.

Despite the enthusiasm of the research community about the semantic Web, there is still some way to go for creating a unifying framework facilitating the interoperation of intelligent agents or reasoning engines and attempting to make sense of semantic Web services.

## Summary

In large, efforts to facilitate automatic-composition Web descriptions through semantic descriptions have been progressing in parallel, but also in isolation, to developments in workflow-based standards preferred by the commercial organizations. These organizations prefer a here-and-now and practical, albeit static, composition technique that robustly supports their business needs to immature, research-biased, dynamic composition techniques that are more focused on the automation factor rather than on business-specific requirements.

In this work, we attempt to bridge the gap between the two approaches by introducing semantics to workflow-based composition. We aim to present a hybrid solution (Mandell & McIlraith, 2003; Traverso & Pistore, 2004) that merges the benefits of the practicality of use and adoption popularity of workflow-based BPEL composition with the advantage of using semantic descriptions to aid both service providers and composers in building the composition scheme and adapting new Web services to it.

# BRIDGING THE DYNAMIC COMPOSITION GAP IN BPEL

## The Implementation Scenario

We use the classic travel-agent problem (Thakker et al., 2005; White & Grundy, 2001) as the implementation scenario for our composition framework. The implementation is based on the composition of real-world services from the airline, hotel, and car-rental businesses. None of these applications interface to users through a Web service, hence, Web service wrappers were developed on top of their HTTP portals, and they were then subscribed to a local UDDI and made available for composition. For instance, wrappers were developed for three airline services: the EasyJet (http://www.easyjet.com/), WizzAir (http://wizzair.com/), and FlyBmi (http://www.flybmi.com/) portals. The parameters and field names of particular Web services are maintained the same as on the Web portals.

Our goal is to contribute to the automation of composing Web services using the BPEL approach. The automation should aid the service composer in modeling a generic composition scheme, and the service providers in adapting their application services to that scheme. The generic composition scheme contains the composition logic that integrates the functionality of two or more Web services to achieve a common goal. In order to reduce the complexity of automation, each composition scheme will be domain specific. This allows us to make the necessary assumptions about the details of the application domain, that is, its process model, ontology, and so forth.

In our approach, the service composer builds a BPEL-based scheme for the composition of services belonging to specific application domains; it is then the responsibility of the service providers to adapt their Web services, if necessary, to the domain interface of the composition scheme. One objective of this research is to make the adaptation process as seamless as possible to the service providers. The advantage to the service composer is the ability to recompile and fire the composition with different domain-specific Web services with minimal effort.

For instance, our travel-agent application composes services belonging to three domains: airline, hotel, and car rental. The travel agent prespecifies the functionality (domain interface) that it expects from each participant, for example, price quotation for the user-specified flight details. A large section of information engines and e-commerce services that integrate different Internet-based services through a unifying access interface fall under the same category, for instance, loan providers (loan assessor, banks, insurance companies) and shopping robots.

The following sections explain how the domain interface is specified and how it is exploited to facilitate the seamless, dynamic composition of Web services based on the BPEL approach.

## Specification of the Domain of Services

Central to the idea of grouping services in a domain is the presentation of a domain interface for the functionality expected from the service by the service composer in a standard, unambiguous format that is comprehensible by programs rather than humans. The Web services standard WSDL provides XML grammar to describe the working of the Web services. WSDL can be used for defining the expected functionalities from a participant Web service for a particular domain. However, the problem with WSDL is that it is a syntactical standard that is developed for human developers rather than program-based automation.

The recent trend in the area of Web automation is initiated by the Semantic Web. A fundamental component of the Semantic Web is the markup of the traditional World Wide Web to make it computer interpretable, user apparent, and agent ready (McIlraith, Son, & Zeng, 2001). The approach adopted by the semantic Web to achieve this is

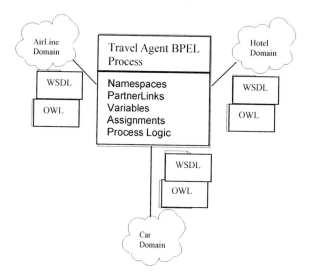

*Figure 1. Domain-specific composition*

formalized in terms of layers built on top of XML as XML alone provides only syntactical support and has no notion for the meanings required for achieving goals for the semantic Web. RDF (resource description framework), RDFS, and OWL are the specifications from the W3C (World Wide Web Consortium) to add semantics. These specifications provide language expressiveness and simulate human reasoning. OWL uses and extends RDF to specify ontologies. Ontologies define common specifications of domain-related concepts. Ontologies are like dictionaries, where the meaning of a concept can be described in the form of unambiguous semantic descriptions. Another aspect of ontologies is that the reasoner can be designed to interpret these conceptual meanings or derive deduction from the semantic description, making the solutions program based and computer interpretable.

In our suggested solution, we use ontologies extensively. OWL is the most expressive knowledge representation for the semantic Web so far. For designing the domain interface for expected functionality for a particular domain, WSDL files describing the domain functionality in XML grammar are accompanied with a semantic de-

*Figure 2. Domain-specific interface – WSDL file*

```
<wsdl:definitions  targetNamespace="http://travelagent.ntu.ac.uk/
AirLineDomainService">

<wsdl:types>

<complexType name="FlightQuery">

<sequence>
<element name="noOfAdults" type="xsd:int"/>
<element name="departure-date" nillable="true" type="xsd:dateTime"/>
…
</sequence>
</complexType>
```

scription of the service parameters expressed in the OWL ontology. This allows the description of expected functionality to be inferred in unambiguous form. Figure 1 illustrates the application of the above solution to our travel-agent example.

A snippet of such an OWL-WSDL domain interface for the airline domain is shown in Figures 2 and 3. The WSDL file complex-type FlightQuery of Figure 2 has been mapped into the OWL class FlightQuery of Figure 3; hence, an OWL reasoner can apply the class-relationship-based inference to verify that the mapped message type contains all required elements.

If a new domain-related Web service is to be created, the domain-interface files can be used to create a new Web service that adheres to the functionality expected by the service composer. Otherwise, the service provider needs to edit the ontology file to overcome any mismatches in the service descriptions (parameters and method names). In this case, the ontology can bridge the semantic mismatch provided that conceptual meaning remains the same. Figure 4 describes an ontology file provided by one of the candidate airline services to overcome semantic mismatches with the travel-agent domain interface. The ontology file in Figure 4 documents the fact that

the departureFlightDate element of this airline description is conceptually similar to the element departure-date in Figure 3.

## Dynamic Pool for Domain-Specific Web Services (DPDWS)

In the second phase of our framework, we attempt to integrate the domain-specific Web services into a dynamic pool, where the services can dynamically plug in and out of the composition scheme without the need to recode the composition logic. As explained in the previous section, the prerequisite for domain membership is the availability of a WSDL file describing the service functionality and an accompanying ontology file ensuring the compatibility of the service parameters to the domain interface.

### Domain Membership Verification

A module has been created for verifying the membership of Web services to a particular domain and ultimately the composition scheme. The module verifies the above-mentioned prerequisite according to the following steps (the airline domain is exemplified).

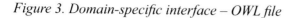
*Figure 3. Domain-specific interface – OWL file*

```
<owl:Ontology rdf:about="http://localhost/ntu/ac/uk/2005/
TravelAgent/AirLineDomain.owl">
</owl:Ontology>
<owl:Class rdf:about="http://localhost/ntu/ac/uk/2005/
onto/travelquery.owl#FlightQuery">
 <rdfs:subClassOf>
  <owl:Restriction>
  <owl:onProperty
rdf:resource="http://localhost/ntu/ac/uk/2005/onto/travelquery.owl#noOfAdults" />
   <owl:someValuesFrom>
    <rdfs:Datatype rdf:about="http://www.w3.org/2001/XMLSchema#
int"/>
   </owl:someValuesFrom>
  </owl:Restriction>
 </rdfs:subClassOf>
  <rdfs:subClassOf>
  <owl:Restriction>
  <owl:onProperty
rdf:resource="http://localhost/ntu/ac/uk/2005/onto/travelquery.owl#departure-date" />
   <owl:someValuesFrom>
    <rdfs:Datatype rdf:about="http://www.w3.org/2001/XMLSchema#
dateTime"/>
   </owl:someValuesFrom>
  </owl:Restriction>
 </rdfs:subClassOf>
```

1. Parse the OWL-WSDL files of the candidate Web services against the domain interface to check all the possible mappings between what is expected and what is provided by the candidate service. If the candidate service description file (WSDL) has a different format from the domain description file, the supplied ontology is searched for a mapping for this mismatch. If the ontology file has the required mappings, the mappings are stored for future use when the actual composition with this service takes place. For instance, the membership module stores the valid mapping departure-date-> departureFlightDate for the EasyJet service (see Figures 3 and 4).

2. If the service parameters match semantically, make the service available within the Airline DPDWS (Figure 5), that is, composition ready; this involves storing a reference for the service with the composition-necessary details: the target name space, mappings between required and provided elements, the operation name with corresponding

*Figure 4. Ontology file for EasyJet airline service*

```
<owl:Ontology rdf:about="http://localhost/
ntu/ac/uk/2005/EasyJet/easyjet.owl">
</owl:Ontology>

<owl:DatatypeProperty rdf:about="http://localhost/ntu/ac/uk/2005/ EasyJet/
easyjet.owl#departureFlightDate">
  <owl:equivalentProperty>
  <owl:DatatypeProperty rdf:about="http://localhost/
ntu/ac/uk/2005/onto/travelquery.owl#departure-date">
  </owl:DatatypeProperty>
  </owl:equivalentProperty>
</owl:DatatypeProperty>
```

*Figure 5. Membership verification module for DPDWS*

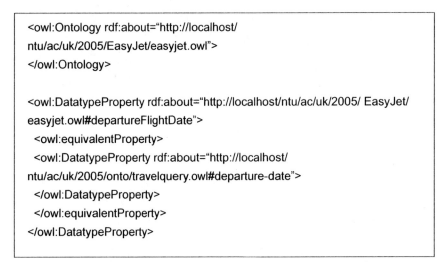

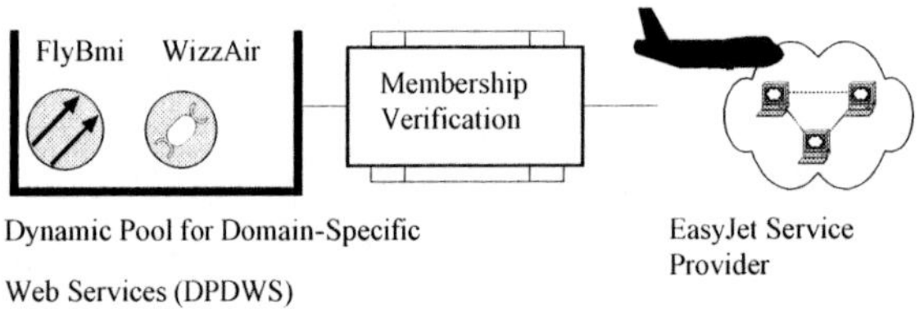

port types, message names, and message types. The verification module also creates the partnerLink name, partnerLink type, and partnerLink role based on the service name for this service. These details can be used when the actual composition is carried out.

Figure 6 is a snapshot of the implemented airline domain membership verification module, which implements this algorithm and is designed using Jena (2001), the Pellet (2004) ontology reasoner, the DOM XML parser, and Java technology. The only input required from the service provider is description and ontology files, and our composition takes care of making the service

composition ready by following the membership verification algorithm.

The next section details the mechanism for automating the dynamic selection of Web services from the dynamic pool and their integration into the composition scheme.

## Dynamic BPEL-Based Service Composition Facilitated by DPDWS

### Overview

In our framework, dynamically adding a Web service from the domain pool constitutes placing an instance of the service in the composition scheme file. For example, to add the functionality

*Figure 6. Airline domain membership verification*

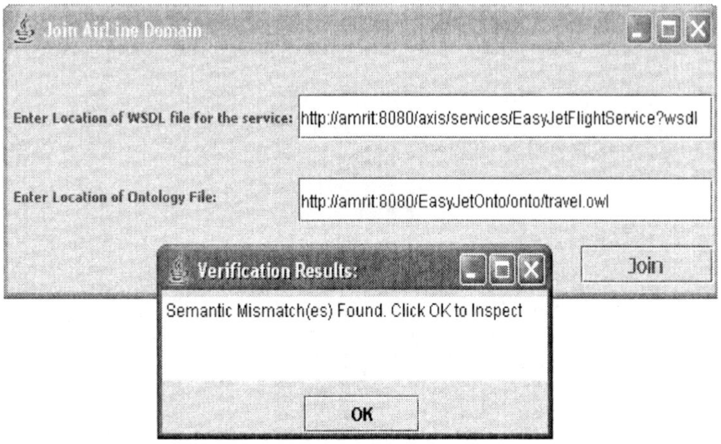

*Box 1.*

```
[<invoke name partnerLink="EasyJetPL"          portType="ejet:EasyJetPortType"
operation="checkReservation"
inputVariable="inputEasyJet" outputVariable="outputEasyJet"/>]
```

of retrieving a price quote for a specific journey by the EasyJet airline service, the travel-agent service composer will have to add the instance shown in Box 1 to the relevant execution segment of the BPEL composition file.

Such integration is automatically performed by our dynamic composition framework. Hence, the BPEL process file does not have to be manually edited and recompiled to integrate alternative Web services into the composition scheme. Table 1 shows how a BPEL process can be created with our programming-based framework.

This implies that the process file can be created dynamically with the inclusion of the new services from the particular domain. This programming-language-based tool can create the service references by reading the WSDL file

and can add them throughout the composition scheme, making the creation of the process file automatic and execution ready. This makes the scenario in Figure 7 possible, where services from the domain can be plugged in and plugged out automatically.

The target BPEL execution engine for our framework is Oracle's BPEL Process Manager (*Oracle BPEL PM*, 2005). It is worth mentioning that this particular implementation of BPEL also requires two additional files to be input with the BPEL process file: a service wrapper WSDL file that contains information to make the service a partner in the business process, and a BPEL configuration file that identifies the location of the wrapper file and binds it with a particular Web service partnerLink. For each new service

*Table 1. Process file creation with Java*

| Required Composition Function | Corresponding Framework Method |
|---|---|
| Add partnerLinks for the airline service with particular values for the new service | public String setParetnerLinks( Document bpeldoc, String prefix, String partnerlink_name, String partnerlink_type, String partnerlink_role) |
| Set the process logic for the airline service by placing partnerLink, which has the price-check operation | public void setPriceCheckInstance (Document bpeldoc, String invar, String outvar, String portType, String operation, String partnerlink_name) |

participating in the composition, the bpel.xml file is modified to include the new service. Our implementation creates the process file and wrapper file, and adds the entry in the bpel.xml file, making the process files composition ready and acknowledging the inclusion of the newly added service.

Following is the algorithm for DPDWS-facilitated composition, which creates the BPEL process file automatically, allowing the services to be dynamically selected from the domain.

## Algorithm for DPDWS-Facilitated Domain-Specific Composition

Our framework performs the following steps for facilitating the dynamic composition of domain-specific Web services.

1. Initialize the composition framework. This will result in the DPDWS populated with the services, which were verified by the membership verification module. An arbitrary Web service from the domain pool will be selected to create a skeleton BPEL process

file, reflecting the travel-agent composition logic.

2. On the selection of an alternative domain service, generate a new BPEL file and other configuration files required by the BPEL execution engine. This is achieved as follows.

    i) Retrieve reference for this service from the membership verification module. This will include all the details pertaining to the service and required by the composition module including partnerLink details. Also retrieve semantic mappings from the membership verification module and use them wherever applicable during the process logic.

    ii) Add the new service name space in the root element for the newly created BPEL file.

    iii) Add partnerLinks for the new service, generate partnerLinks automatically, and maintain uniqueness.

    iv) Map the messages of the Web service to the variables; the variable names are

*Figure 7. Travel agent composition facilitated by DPDWS*

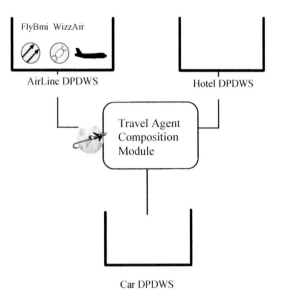

generated automatically. Steps ii to iv use the reference details created during membership verification.

v)  Build the process logic for the new service by placing all the service instances. This includes the addition of the service instance at all the places where the composition logic for a particular domain is defined in the default skeleton BPEL process file. Examples of such instances can be invoking the service, assigning responses to intermediate variables and passing them for particular operations, and so forth.

3.  Validate the newly generated BPEL file.

4.  Create the service wrapper file with the partnerLink information defined for the service reference and include a pointer to the location of the WSDL file within the wrapper file.

5.  Bind the partnerLink details with the service wrapper file location and modify the existing bpel.xml file to reflect the integration of the new service.

The composition module algorithm is implemented using Java technology and the DOM XML parser. Figure 8 illustrates the administration interface of our composition framework. The locations of process (BPEL skeleton files) and configuration files are necessary for the initialization of the framework. The list of available services to each DPDWS is dynamically populated with the membership verification module detailed earlier. The service composer can select any possible combination of services from domains for composition, and new process files with configuration files are automatically created; the composed service is fired if required.

*Figure 8. Travel-agent composition*

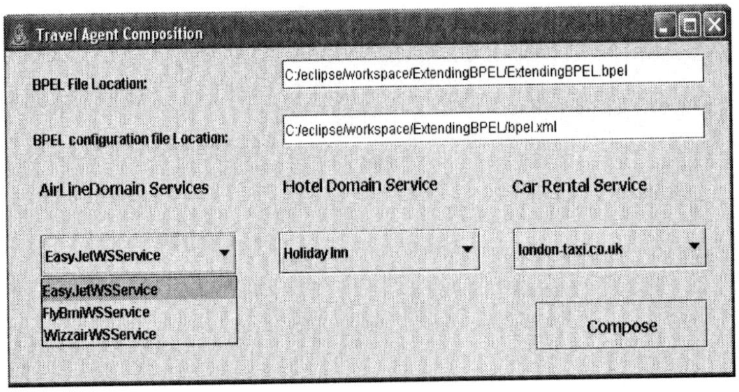

## RELATED WORK

In recent years, the research community has realized that the union of semantics with business standards can be helpful in mechanizing composition tasks.

Mandell and McIlraith (2003) propose a bottom-up approach for Web services interoperation in BPEL4WS (BPEL for Web services); they use OWL-S-based descriptions for the run-time binding of service partners. The implementation collects the OWL-S profiles into a repository and exploits the profile semantics to query partners for desired properties. This approach allows selecting partners at run time otherwise selected at design time according to the BPEL process model.

The approach uses semantic Web technology for automatic, meaningful service selection. However, the problem of making actual composition automatic is not addressed as the composition logic is built manually for the inclusion of partner services.

In our solution, we consider the composition from the service composer's perspective. The service composer categorizes the possible service partners into domains and makes the domain-specific interface (WSDL-OWL) available to the service providers. This interface serves as the prerequisite for joining the particular domain. Hence, we take a top-down approach by declaring expected requirements first and then populating domains with legitimate services (ones that fulfill the requirements), unlike Mandell and McIlraith (2003) who use OWL-S profiles for selecting service partners based on service descriptions. Our approach also allows creating a general reusable programming framework for selecting services from a particular domain and composing them automatically.

Traverso and Pistore (2004) represent an AI-planning-based technique to convert semantic (OWL-S) Web service process models into executable BPEL4WS processes. The implementation translates the OWL-S profile models into partially observable state transition systems, which are utilized for generating plans to reach the goals for composition. Their approach uses semantics at the composition level and takes advantage of the expressiveness and executable nature of low-level BPEL processes. The approach aims for the composition of services to be automatic, while service discovery and selection is manual.

In our approach, we also use semantics at the composition level; however, we exploit the

BPEL process creation mechanism combined with the domain concept to implement an automatic composition programming framework rather than using planning techniques. Our implementation also allows the selection and removal of service partners for the composition to be automatic.

## CONCLUSION

The aim of this work is to create a framework that alleviates the burden of dynamic Web services composition. We argue that despite the evident popularity of Web services as a secure distributed computing paradigm and the value-added dimension that composition adds to it, the practical adoption of the technology is still hindered by the knowledge and effort required for the compilation of the composition process and the manual adaptation of new and existing Web services to it.

After critical analysis of current approaches to Web services composition, we concluded that a practical and current solution should be based on a hybrid solution that merges the benefits of the practicality of use and adoption popularity of workflow-based (BPEL-based) composition with the advantage of using semantic descriptions to aid the composition participants in the automatic discovery and interoperability of the composed services.

The main premise of our approach is to aid the service composer in building a generic BPEL-based scheme for the composition of services belonging to specific application domains, and assist the service providers in adapting their application services to the composition scheme. Web services join the BPEL composition scheme by subscribing to a specific domain interface.

The domain functionality described in WSDL-XML grammar is accompanied by a semantic description of service parameters expressed in the OWL ontology, allowing the description of the expected domain functionality to be in an unambiguous form and catering for any mismatches in the Web services description. A domain membership verification module was developed that allows the service providers to adapt their application services to the domain interface and make them with minimal effort.

Once a domain Web service is declared composition ready, our dynamic composition framework transparently integrates the Web service into the BPEL process file; that is, it is automatically added to the pool of dynamic Web services for this domain. The chapter describes the algorithm for the dynamic population of the domain pool with Web services, thus allowing the service composer to effortlessly select any possible combination of services from the composition domains and fire the composed service.

Work under progress aims to extend the developed framework to facilitate the automatic matchmaking of Web services according to user criteria and the quality of the provided service.

## REFERENCES

Andrews, T., Curbera, F., Dholakia, H., Goland, Y., Klein, J., Leymann, F., et al. (2003). *Business process execution language for Web services, version 1.1.* Retrieved April 10, 2006, from http://www-128.ibm.com/developerworks/library/specification/ws-bpel/

Ankolekar, A., Burstein, M., Hobbs, J., Lassila, O., Martin, D., McIlraith, S., et al. (2001). DAML-S: Semantic markup for Web services. *Proceedings of the International Semantic Web Working Symposium (SWWS)*, 411-430.

Christensen, E., Curbera, F., Meredith, G., & Weerawarana, S. (2001). *Web services description language version 1.1* (W3C recommendation). Retrieved April 10, 2006, from http://www.w3.org/TR/wsdl

Dean, M., Hendler, J., Horrocks, I., McGuinness, D., Patel-Schneider, P. F., & Stein, L. A. (2004). *Semantic markup for Web services: OWL-S version 1.1*. Retrieved April 10, 2006, from http://www.daml.org/services/owl-s/1.1/

Gudgin, M., Hadley, M., Mendelsohn, N., Moreau, J., & Nielsen, H. F. (2003). *Simple object access protocol version 1.2* (W3C recommendation). Retrieved April 10, 2006, from http://www.w3.org/TR/soap/

Jena. (2003). Retrieved April 10, 2006, from http://jena.sourceforge.net

Kavantzas, N., Burdett, D., Ritzinger, G., & Lafon, Y. (2004). *Web services choreography description language (WS-CDL) version 1.0*. Retrieved April 10, 2006, from http://www.w3.org/TR/2004/WE-ws-cdl-10-20041217/

Leymann, F., Roller, D., & Schmidt, M.-T. (2002). Web services and business process management. *IBM Systems Journal, 41*(2), 198-211.

Mandell, D., & McIlraith, S. (2003). Adapting BPEL4WS for the semantic Web: The bottom-up approach to Web service interoperation. *Proceedings of the 2nd International Semantic Web Conference (ISWC2003)*.

McIlraith, S., & Son, T. C. (2002). Adapting Golog for composition of semantic Web services. *Proceedings of the Eighth International Conference on Knowledge Representation and Reasoning (KR2002)*, 482-493.

McIlraith, S., Son, T. C., & Zeng, H. (2001). Semantic Web services. *IEEE Intelligent Systems, 16*(2), 46-53.

Nau, D. S., Cao, Y., Lotem, A., & Muñoz-Avila, H. (1999). SHOP: Simple hierarchical ordered planner. *Proceedings of the International Joint Conference on Artificial Intelligence (IJCAI-99)*, 968-973.

*Oracle BPEL PM*. (2005). Retrieved April 10, 2006, from http://www.oracle.com/technology/products/ias/bpel/index.html

Osman, T., Wagealla, W., & Bargiela, A. (2004). An approach to rollback recovery of collaborating mobile agents. *IEEE Transactions on Systems, Man, and Cybernetics, 34*, 48-57.

*Pellet*. (2004). Retrieved April 10, 2006, from http://www.minswap.org/2003/pellet/

Sirin, E., Hendler, J., & Parsia, B. (2003). Semi-automatic composition of Web services using semantic descriptions. *Web Services: Modeling, Architecture and Infrastructure Workshop in ICEIS.*

Thakker, D., Osman, T., & Al-Dabass, D. (2005). Web services composition: A pragmatic view of the present and the future. *Nineteenth European Conference on Modelling and Simulation: Vol. 1. Simulation in wider Europe*, 826-832.

Traverso, P., & Pistore, M. (2004). Automated composition of semantic Web services into executable processes. *Proceedings of Third International Semantic Web Conference (ISWC2004)* (pp. 380-394).

Van der Aalst, W. M. P., Dumas, M., ter Hofstede, A. H. M., & Wohed, P. (2002). *Pattern-based analysis of BPML (and WSCI)* (Tech. Rep. No. FIT-TR-2002-05). Brisbane, Australia: Queensland University of Technology.

White, P., & Grundy, J. (2001). Experiences developing a collaborative travel planning application with .NET Web services. *Proceedings of the 2003 International Conference on Web Services (ICWS)*.

Wu, D., Parsia, B., Sirin, E., Hendler, J., & Nau, D. (2003). Automating DAML-S Web services composition using SHOP2. *Proceedings of 2nd International Semantic Web Conference (ISWC2003)*.

# Section IV
# Enterprise Integration
# Case Studies

# Chapter XIV
# Case Study:
## Enterprise Integration and Process Improvement

**Randall E. Duran**
*Catena Technologies Pte Ltd, Singapore*

## ABSTRACT

*Enterprise integration (EI) can be a major enabler for business process improvement, but it presents its own challenges. Based on the process improvement experiences of banks in several different countries, this chapter examines common EI challenges and outlines approaches for combining EI and process improvement to achieve maximum benefit. Common EI-related process improvement challenges are poor usability within the user desktop environment, a lack of network-based services, and data collection and management limitations. How EI affects each of these areas is addressed, highlighting specific examples of how these issues present themselves in system environments. The latter part of this chapter outlines best practices for combining EI with process improvement in relation to the challenges identified. Guidelines are provided on how to apply these practices in different organizational contexts.*

## INTRODUCTION

Business process improvement and highway traffic flow optimization have several similarities. For example, both try to improve the total volume supported, increase rates of flow, and decrease completion and journey time. Business processes benefit from reducing the number of exceptions, and traffic flow from reducing the number of accidents. Likewise, both are affected by less tangible factors. Drivers may be distracted by too many road signs or a rough road surface, whereas processes may be hindered by high levels of complexity or user-unfriendly systems.

Traffic planners and process improvement specialists have a wide range of techniques available to achieve improvement. Where a highway planner might designate a car-pool lane or change the speed limits, the process improvement specialist can eliminate handoffs or reallocate staff to reduce bottlenecks. Using these techniques, improvements can often be made without significant changes to the existing environment.

These minimal-impact approaches will improve performance, but their overall effect will be limited. Some advantage may be gained by using these techniques, but it may be the case that as much as 70% or 80% of the potential improvement still remains. Realizing these more significant gains requires changes to how the outside world interacts with these flows. In the case of optimizing highway traffic, it may require building new connecting roads, bridges, and on-ramps. For process improvement, it often requires the development of connectivity to the applications and systems that are used within the process. Enterprise integration (EI) can provide this connectivity and is critical to the overall success of many process improvement initiatives.

This chapter examines how EI can best support process improvement. The examples presented are primarily based on the experiences of three financial service institutions based in the United States, Japan, and Singapore. Each company had a different focus—call center, retail banking, and enterprise-wide processes—and is at a different stage of implementation: between 6 months and 3 years after starting to use EI for process improvement. However, the conclusions and lessons learned are not unique to the financial services industry; they are applicable to process improvement in many different industries. The objective of this chapter is to help those planning EI to better understand the context, perspectives, and challenges of process improvement initiatives, and to help those planning process improvement initiatives to understand how to best apply EI to help business processes achieve the significant productivity gains.

The structure of this chapter is as follows. The first part discusses process improvement approaches and considerations, and how they relate to EI. The second part reviews common process improvement problems and explains how EI can help. The third part outlines best practices for applying EI to support process improvement.

## PROCESS IMPROVEMENT

To appreciate the benefits that EI can provide, it is first necessary to understand the environment within which process improvement is achieved. The examples discussed are representative of situations that banking business units encounter across different functional areas in both the front office and the back office. Although the examples are drawn from the financial services industry, similar activities and processes are found in many other industries. Thus, the discussion of the process improvement environment will focus on common considerations rather than specific details that might only be relevant to a single industry or business area.

### The Process Improvement Environment

Processes that manage information usually make use of paper, software, or a combination of both paper and software. At one extreme, entirely paper-based processes may involve receiving handwritten or typed information and combining it with other printed information from paper-based files or printouts from software systems. These process inputs can result in stacks of paper that are physically moved between people who then check and verify information in the documents, make calculations, and eventually approve or deny the requests. While this type of processing may sound terribly inefficient and anachronistic to IT professionals, it is an environment that still exists in many business areas, such as loan application processing.

Given that it is now the 21st century, it is more often the case that at least some of the process information will be stored in and manipulated through software applications. Unfortunately, though, it is uncommon to find totally paperless business operations; both paper and software-based systems continue to drive business processes. For example, a call center operation may use software-based customer relationship management (CRM) and core banking systems to fulfill the majority of customer call requests. However, paper will still be abundant in the call center processes. For example, printed copies of reference information for recent sales promotions or tabular information that is used to support decision making may be tacked to cubicle bulletin boards, or paper forms may be used to initiate manual processes, such as special approvals that may be required when a transaction amount exceeds a specified limit.

In more IT-focused organizations, some processes are entirely paperless. While eliminating paper may sound like a panacea for process improvement, there is still a number of problems with which paperless processes often struggle. Two of the core problems are using too many different software applications within a single process, and using applications for purposes other than they were designed. The first problem relates to the fact that there is usually no single monolithic application that will perform all the functions required by a business process. The second problem is caused by existing application functionality not fully matching the business' requirements. Functional differences may result from using off-the-shelf software packages for more specialized functions; alternatively, existing software may not be enhanced to meet changing business requirements.

## Process Improvement Techniques

It is in this environment that process improvement initiatives must achieve goals such as reducing processing time, reducing processing effort, and improving resource utilization. Fortunately, there is a number of process improvement techniques that can help. Quantitative techniques use process execution measurements and statistical methods to help identify problem areas and verify improvement gains. Qualitative techniques review and analyze process flow steps and functions, search-

ing for the root cause of problems and potential areas for optimization.

The Six Sigma methodology (Pande, Neuman, & Cavanagh, 2000) has been used by many organizations to support process improvement. It provides a set of quantitative techniques that are used for process improvement. Six Sigma's define, measure, analyze, improve, and control (DMAIC) approach focuses on capturing and analyzing process execution information. Although significant effort may be involved with defining, analyzing, and improving processes, often it is the collection of process data for the measure and control stages that requires the most onerous effort. Moreover, the amount and granularity of the data gathered will affect both the quality of the analysis and scope of the collection effort.

Other, qualitative techniques, such as process flow analysis and value-added analysis (Harrington, 1991), focus on achieving performance gains by categorizing the process activities, identifying non-value-adding flows or activities, and then optimizing the flows to achieve gains. Improvements are often achieved by eliminating handoffs between people or groups, reducing delays, minimizing physical transport (i.e., moving stacks of paper around), reducing the likelihood of exception flows related to failure conditions, eliminating bottlenecks, and reducing overall process complexity.

In a business environment that is largely paper based, there is usually a number of areas that can be improved using these qualitative techniques. However, after the initial optimization is completed, further use of these techniques will yield limited results. Typically, to make further productivity gains, it will be necessary to automate the process steps using either business process management (BPM) technology (Khan, 2004) or customized software applications.

Alternatively, in business environments that have achieved a higher degree of automation, where IT-based systems are more prevalent than paper-based systems, the inefficiencies that qualitative analysis will uncover usually relate to the limitations of and incompatibilities between various IT systems involved in the process. In this situation, achieving productivity gains will depend on being able to take advantage of EI.

## CHALLENGES

Having reviewed some of the goals of and techniques for achieving process improvement in the previous section, this section will explore common challenges that process improvement initiatives face involving EI. For each area, we will review the scope of the problem, examine specific examples, and discuss how applying EI can yield material benefits for process improvement. The specific areas considered are poor usability within the user desktop environment, a lack of network-based services, and data collection and management limitations.

### Poor Usability within the User Desktop Environment

There are three main reasons why desktop computers do not help users execute processes as efficiently as they could. First, it may be necessary to use a large number of different applications with different styles of interaction at various times throughout the process. Second, the applications may not have been designed for the purposes that they are used within the process, and as a result, they are not user friendly. Third, all of the application functions and screens are often accessible to the user at any time without respect to the logical flow of process steps and the functions that are required at different stages. In this section, we will address the first and second problems; the third is an area that is best addressed using BPM tools rather than EI.

The imposition of disparate applications on business users is fundamentally an application architecture issue. However, this situation is prevalent in many business environments. While some CRM and enterprise resource planning (ERP) vendors promote their systems as the single monolithic source of all information and services, in practice this is rarely the case. The limitations of core applications, the need for specialized and customized applications, and the existence of legacy applications conspire together to prevent the single-application promise from being fulfilled.

A significant downside of using multiple user interfaces (UIs), from a process performance standpoint, is increased process complexity. Users must switch between applications within a single process. Also, when using multiple UIs, the user, that is, the process executor, may need to log in multiple times and consequently remember and manage multiple user accounts and passwords. In some cases, systems may automatically time out user accounts after 10 or 15 minutes. Thus, it may be necessary to log in multiple times each day. Another cost of having multiple UIs is that the user may have to enter the same data several times in different applications to execute a query or enter a data record that is maintained, in part or in whole, across multiple IT systems.

Beyond the relatively simple consideration of logging in, the user interfaces often differ dramatically in their look and feel and their manner of operation. Mainframe applications will be often run in 3270 emulation windows, using two- to four-letter typed commands as the primary means of navigating between screens. Thick-client Windows or Java applications are normally menu driven, and in some cases the applications will support drag and drop semantics and in other cases not. Thin-client, Web-based applications may have a click-through or tab orientation and may use bookmarks. Even if the applications in use are all of a single style, they will often use different terminology, lay out the same information in different ways, or provide the user with different idioms for interacting with the system.

243

It is not uncommon to have a combination of all of these types of applications as part of the desktop of a call center or back-office operation. Hence, when training staff on business processes, it is necessary to teach them the intricacies of each of the applications involved in the process. The more applications involved in the process, the greater the effort required by the users to learn and remember how to operate all of the applications. If a fully optimized user desktop environment were available, the user would interact with only a few applications, or possibly just one, to perform all process steps, and all user interactions would have a consistent look and feel.

It is also the case that applications are used for purposes other than designed: General-purpose applications are sometimes used for very specific functions, and vice versa. As a consequence, the user may have to perform many additional steps to complete a process task. For example, one of the steps of the process may require the user to check to see whether a customer account balance is above a specified limit. This single process step can then require the user to open the customer information system application, log in, navigate to the appropriate screen that holds the account balance information, and then scan the information shown, which may be quite dense, to find the amount. Then, the user must compare the balance amount to a limit that may be recalled from memory, read from a sheet of paper, or looked up in another application. In a fully optimized user desktop environment, only the relevant information—whether the account is above or below the limit—would be shown. The user would not need to open and navigate through the application to get one half of the information, and would not have to do a manual comparison of the balance and limit amounts.

Significant process improvement gains, through simplification and automation, can be achieved by the elimination of switching between different applications and by only displaying the data relevant to the current process step. A sim-

plified user environment will help reduce human error and, therefore, also improve process performance by reducing the number of exception cases that must be handled. However, streamlining the desktop environment requires that the required application data and services are readily accessible. The technology behind providing a consolidated UI application framework is not challenging; it can be implemented in a Web browser or thick-client application. It is providing the linkages from that consolidated UI to all of the other systems that is a major challenge. This is where EI provides an advantage.

Automation, another key process improvement aspect, is also highly dependent on EI. Decisions and calculations cannot be automated without being able to access the information that underlies the decision logic or arithmetic calculations in an elemental form. Often, these values are locked up: either embedded in screen displays, or stored in databases that are not readily accessible. EI is critical in unlocking these data resources, enabling the automation of process calculations and workflow to achieve process improvement.

In examining the desktop environment, we touched on the need to access multiple IT systems using EI. In the next section, we will examine the technical challenges encountered with this undertaking.

## Lack of Network-Based Services

The follow-on challenge to improving performance by streamlining the user interface is to effectively use EI to provide access from the desktop to the relevant IT systems. Ideally, a streamlined UI would provide access to both existing and new common network-based services using a common EI platform. Existing-system services encapsulate and expose functions in existing IT applications. Common services are synthesized functions that can be used by a number of different processes, contain business rules and validation logic, and may combine functions of multiple different IT systems.

In business processes, applications are usually accessed through their own proprietary UI. As part of a process, the application may be referenced through instructions such as "If the customer qualifies, go to the ZY screen and check the approved box," and "Then go to the QQQ application and initiate workflow #4 to send the request to the fulfillment department." The process focuses on the specifics of the applications rather than on the purpose and objectives of the business function. If the applications are changed, the processes must be redefined and users must be retrained. Ideally, in this case, process improvement would limit the user effort to verifying approval criteria and signifying approval or rejection. The specific applications that are used to record approval and initiate the follow-on workflow activities should be hidden behind EI in the form of network-based services to reduce the complexity.

In business process environments with limited EI, functionality that should be maintained in common services is often present as manual steps forming part of a workflow or embedded in existing systems as customizations. For example, process-specific logic for performing decisions and calculations may be manually executed, based on written instructions in the form of tables, formulas, or worksheets. This business logic may relate to transaction limits, customer profiles, or calendar schedules, and may frequently change and so cannot be hard-coded into the IT systems. From a process improvement standpoint, the centralization and automation of these activities will yield significant benefits. By making the decision logic or calculations automated, the process flow will be significantly simplified. The overhead of training users how to perform these sometimes very complex steps is eliminated. Furthermore, changes and updates to the business logic only need to be made in one place, so the effort of accommodating changes to the business logic is reduced. If an algorithm is changed, the common service can be changed rather than requiring new instructions to be distributed and the users to be retrained. EI can help provide flexibility to enable these common services to be implemented in many

*Figure 1. A typical software-driven process execution environment*

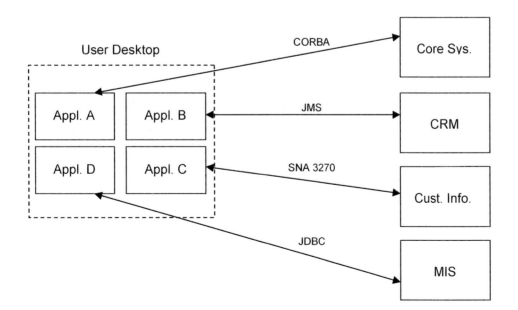

different forms, including some that are easily updated by the business users themselves.

Unfortunately, well-encapsulated, network-based services do not yet exist in many organizations, creating an impediment to the development of a streamlined UI and limiting overall process improvement capabilities. To do integration without such services, it is necessary to work with a mélange of different technologies to integrate multiple IT systems within the scope of a single process. These technologies may include screen scraping, SQL queries, messaging through Web services, JMS, COM, or CORBA, and mainframe SNA operations. Many of the IT systems may also be under the control of different departments, creating further challenges. The net result is that without a services model, even the initial investigation, definition, and specification of system integration requirements can be a long and daunting task for process improvement initiatives. As a result, sometimes the effort involved to integrate IT systems may outweigh the benefits that would be achieved for the process.

Having an EI strategy and making use of a service-oriented architecture (SOA; Krafzig, Banke, & Slama, 2005) can help get past these hurdles. Network services, provided through EI, can enable process improvement specialists to focus on the business abstractions and interfaces rather than on the details of the technical connectivity and application data structure. Additionally, services can often be reused across different processes, reducing the effort required to build new infrastructure as time progresses. If integration is done in an ad hoc fashion, there is normally little or no reuse of integration technology.

## Data Collection and Management Limitations

The growth of systems over time, both in number and complexity, often leads to difficulties in collecting data required for process improvement and management purposes. The root of the problem is that the information is dispersed across a number of different business applications and data stores. Typically, as the complexity of a process increases, so does the number of business applications and number of operational data stores involved. In some cases, the proliferation of data stores is unavoidable, particularly in cases where proprietary, third-party applications have databases embedded, which are not open for external access. More often, however, is the case where additional data stores, in the form of databases and spreadsheets, are created in an ad hoc fashion to meet tactical business or application needs. The availability of low-cost, easy-to-use databases has fueled this trend.

Common reasons for the creation of new data stores are the following.

- New, business-specific applications are developed by the business units without the coordination of a central IT planning function. For example, business users may want to keep track of the progress and/or results of a process for either management information purposes or so that other groups or customers can view the current state of the transactions.
- Limitations of existing, often-third-party systems require that information related to new business requirements be stored in separate tables or databases.
- Obstacles, either technical or security related, to accessing existing data stores make creating new databases the path of least resistance for business users.

Creating new data stores can provide tactical benefits, supporting the development of new systems, but it leads to strategic problems of increased complexity, duplicated information, and inconsistencies between different IT systems. While, fundamentally, this is a system and data architecture problem, the near-term benefits of data aggregation are hard to quantify. Thus, there are often difficulties gaining funding for such efforts.

Data warehousing projects tend to tackle this problem from the management information perspective. Data warehousing projects usually create an additional store that is a normalized and aggregated subset of the information from many different applications and data stores across the organization. The information available through data warehouses is normally suitable for management decision purposes, but is not adequate for process improvement. Likewise, information in a data warehouse is usually updated weekly or once a day at best, eliminating the possibility of using a data warehouse for real-time process monitoring. More importantly, the information captured in data warehouses usually does not provide the granularity required for process improvement analysis.

Carrying on from the call center example in the previous section, a process and improvement initiative may want to capture information regarding how long the user was accessing each application, which functions were accessed, how long they took to complete, whether the systems were always available online, and the patterns of system usage over time. While time and motion studies may help produce this information, they are time and labor intensive and only provide a sample of the true population. In this regard, there is a risk that a sample may not capture rare but costly exception cases. Likewise, time and motion studies will be limited as to the number of process variables, or dimensions, that can be monitored and captured concurrently.

If EI is used to provide connectivity to IT systems and data stores, the EI layer can also be used to filter and reroute information about events that are important to process improvement, such as the start and end times of certain process steps. When these types of events occur, relevant information can be routed to a process improvement database or a real-time monitoring application, enabling processes to be reviewed

when failures occur. The benefits of leveraging EI for process improvement data collection are that little or no human effort is required to capture the information, much larger sample sets can be captured, and more dimensions may be captured at the same time. The availability of a larger sample set improves the accuracy of statistical analysis done for process improvement purposes. The ability to capture more dimensions provides process improvement with more flexibility when defining the metrics to be analyzed.

Having examined some of the common process improvement problems and identified how EI can be applied to them, the next section will outline best practices for applying EI to achieve process improvement.

## BEST PRACTICES FOR COMBINING PROCESS IMPROVEMENT AND EI

Whereas the focus of EI is technical, process improvement relates more to the world of people and organizations. For this reason, best practices for combining process improvement and EI include both technical and organizational considerations. Just as the highway traffic planner has budget limitations and cannot close thoroughfares to tear up old roads and rebuild new ones, process improvement projects also have funding concerns and must ensure that progress is made in a way that is acceptable to all the relevant business stakeholders. In this section, we will review four best practices (BPs) for combining process improvement and EI to yield the maximum benefits. For each of the best practices, we will consider what it encompasses, why it is necessary, how it is implemented, and which organizational factors should be considered. General guidelines are provided on how best to apply these practices, but the details of how they are implemented will very much depend on the business, technical, and organizational environment.

## BP1: Combining Top-Down and Bottom-Up Process Analysis

When applying qualitative techniques to process improvement, the analysis often takes a bottom-up approach. That is, it focuses on the details, or micro aspects, of an existing process: the individual steps that make up the process. Bottom-up analysis can help identify logistical changes that can be implemented to achieve quick wins. However, bottom-up analysis may miss higher level, more comprehensive changes that could be implemented. These macro changes often have a more substantial impact on the overall process performance.

Alternatively, qualitative analysis can take a top-down approach. An existing process can be restructured and rationalized to produce a fully optimized process with little regard for the previous implementation. Top-down approaches focus on the macro considerations first, and only work down to the details in the final stages of analysis. This approach can be useful because it aligns the process structure with the strategic business objectives, ignoring the tactical details of the original process implementation, which may no longer be relevant. Top-down analysis is also useful for defining reusable, high-level process abstractions, which bottom-up analysis does not usually address. Top-down analysis has the disadvantage, though, of not leveraging past lessons learned that have been incorporated into the details of the existing process. Process steps that may seem superfluous from a top-down perspective may indeed be necessary for reasons that may be buried in the details.

Combining these approaches enables the process improvement to achieve the benefits of both. Top-down process improvement analysis identifies major changes and restructuring that can yield large-scale benefits. In parallel, bottom-up analysis helps discover any low-level considerations that conflict with the top-down analysis plan. Bottom-up analysis also identifies tactical process changes that can be implemented before, or in parallel with, more significant process changes.

One danger of only doing bottom-up analysis is that some of the most beneficial changes, such as the application of EI, may not be identified. Top-down process improvement analysis should review how IT systems are used within processes and how system access could be streamlined using EI. However, there is the risk that if only a top-down approach is used, fundamental limitations, such as the weaknesses of certain IT systems, may be glossed over. Combining top-down and bottom-up views helps balance out high-level visions with low-level realities. Finding a good balance between the two is critical.

To apply this best practice, process improvement teams should develop methodologies that incorporate both top-down and bottom-up approaches. Ideally, different people or teams should perform the analysis from different directions, and then combine their results and agree on a common set of changes that yield the greatest benefits. EI and systems rationalization objectives should be defined within the top-down process improvement analysis methodology. Top-down EI analysis recommendations should then be compared with the bottom-up analysis to verify the feasibility and benefits at the detailed level. The process improvement benefits of implementing specific EI requirements can then be compared to the cost so as to determine which are most critical and worth doing.

From the organizational perspective, it is beneficial to include an EI specialist as part of the process improvement analysis team. The EI specialist may be a permanent member of the team, or if there is an EI competency, an EI specialist can be provided on loan to assist with process improvement analysis.

## BP2: Develop an EI Competency

After the process improvement EI requirements have been identified comes the much larger task of implementing the EI. Note, however, that performing EI implementation is often outside the responsibilities of a process improvement team. Having a dedicated team for the implementation of EI allows specialists to focus on how best to provide access to existing systems and common services in a reusable form. This division of labor will allow the process improvement specialists to consider process improvement without having to worry about the often very complicated and technical details of how the EI will be implemented.

There are several benefits to this approach. While there will be some initial overhead incurred in developing the EI competency and putting an SOA in place for the first project, the benefits will increase over time as additional projects are able to reuse the SOA system connectivity and EI knowledge that has been amassed over previous projects. Beyond reusability, an EI competency will help ensure that EI is applied across all process improvement initiatives in a structured and orderly way. EI knowledge and experience can be centralized and consolidated rather than having to rely on the process improvement team members' EI knowledge, which may vary considerably.

There are several dangers of not developing an EI competency. First, EI may not be put into practice, even through it would yield significant benefits, because the process improvement team does not have sufficient skills to implement the EI. The EI effort may be deemed too difficult or risky. Second, if EI is implemented by process improvement initiatives on an ad hoc basis, there is the possibility of producing "spaghetti" integration. Loosely planned and managed integration can become a process hindrance itself, leading to unreliability, poor performance, and difficulties with maintenance. Lastly, without a centralized team specializing in EI, it can be difficult to de-termine what existing EI system connectivity is available and may be potentially reused.

Since many EI tools contain workflow facilities, it is often the case that simple workflows can be easily implemented within the context of EI without involving a process improvement team. Likewise, many BPM tools have built-in facilities for accessing databases and Web services, bringing into question the need to make use of a separate EI competency to integrate external data. These situations can create confusion and contention with regard to the interdependence of these two areas. While it is difficult to define hard and fast rules that will apply equally well in all situations, looking to the principles of consistency, simplicity, and reusability will usually provide the best guidance. Involving another group or infrastructure platform may not be appropriate if the integration or workflow effort is justifiably small, does not need to be reused, and can be adequately implemented and supported by the host group. However, it is critical to review these situations carefully to ensure that a proliferation of ad hoc solutions does not ensue.

Developing an EI competency is largely an organizational exercise; much of the work relates to determining the team structure that best fits with the organization's and the process improvement team's structure. One approach is to set up a dedicated EI group, which is sometimes referred to as the EI competency center. This approach facilitates specialization and provides focus, but usually requires that the group be instituted as a cost center that has sufficient funding to operate indefinitely. An alternative approach is to create an EI function within the process improvement team. This approach has the advantage of directly linking the EI focus to business-driven initiatives (and funding). However, it is important to recognize that EI is usually not a one-time exercise. Infrastructure and interfaces will need to be maintained over time. Process improvement teams may not be organized to provide long-term support and maintenance functions. Hence, a separate, dedicated EI competency group may be a better option.

One challenge for the EI competency is that the timing and volume of process-improvement-related implementation work can come in waves. This creates resource complications as many staff may be required at times to handle peak loads of integration work, while there may be periods when little EI work is required, especially at the start and end of process improvement projects and between them. One solution is to outsource the low-level implementation and testing work to ensure that EI implementation does not become a bottleneck for the process improvement projects. However, the EI competency should keep the process improvement EI analysis, API (application programming interface) design, and API specification work in house so that this critical, high-level knowledge can be accumulated and leveraged over time.

## BP3: Consolidate User Interfaces

As discussed previously, eliminating the need for users to navigate between different applications and matching screen layouts to process purposes offers major potential for improvements. Providing a consolidated UI that minimizes the number of applications with which the user must interact will also reduce the skills required and the training time for the process. Furthermore, the rate of human errors can be reduced by not displaying unnecessary information, and allowing the user to focus on exactly what is required at each step.

Besides streamlining data access and entry through a consolidated UI, this facility can also support data capture for process improvement and MIS purposes. For process improvement, the UI framework can be designed to record when and how many times different screens are displayed and how long they are viewed. MIS

*Figure 2. Example process execution environment with a consolidated UI and SOA*

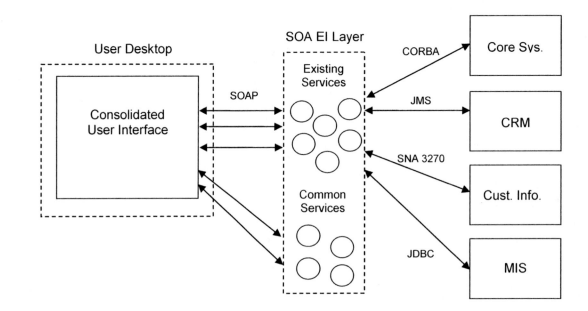

information can be collected through process-completion screens that prompt the user to record more qualitative information, as might be done manually with a tick sheet, about the transaction, such as the customer response rate to a recent promotion.

The risk of not consolidating the UI is that some of the largest process improvement benefits will not be realized. Worse, if the number of desktop applications is not decreasing, it may be increasing and introducing further complexity. Ultimately, there can come a point when business users revolt and refuse to accept new applications even if they are designed to improve efficiency, believing that that complexity of introducing a new desktop application will only make things worse.

There are a number of different technologies available that support the implementation of a consolidated UI. Thin-client Web interfaces are popular for their cross-platform compatibility and minimal desktop support requirements. Thick-client Java or native Windows applications, Web portals, BPM, and CRM applications are also viable options. The technology chosen to provide a consolidated UI will largely depend on the organization's general business requirements, long-term technology focus, previous technology investments, and expected use of BPM, CRM, or Web portal technology.

Aside from the technology chosen, the organizational considerations will also be substantial. Consolidating the user interface will probably strike a positive chord with many of the business users, but unfortunately, they may not be sufficiently motivated to invest the time and effort required to develop a new, fully optimized interface. The screens layouts do not define themselves, and only the business users can ultimately say what will work best.

Funding may also be a concern. Replacing multiple mission-critical application user interfaces can take years to complete, and at a significant cost. Delivering projects of such large scope is often beyond the capacity of an individual team.

Ideally, infrastructure-related efforts should be addressed at the departmental or enterprise levels. The consolidated UI infrastructure should be generic and applicable to a wide range of business processes across the organization. With respect to funding, it may be most practical to combine the development of a consolidated UI infrastructure with multiple process improvement projects. Rather than the infrastructure being viewed purely as a cost center, by combining it with process improvement opportunities, the development can be justified overall by the efficiency gains and cost reduction that is achieved.

Designing, implementing, and deploying the infrastructure required for a consolidated UI and SOA requires significant time, effort, and money. There are many challenges that will be encountered and that must be overcome. Few organizations have the luxury of allocating unlimited resources to a single project or process improvement initiative. However, it is possible to develop this infrastructure over the course of several projects while achieving tangible business benefits in the short and medium term. In BP4, we will discuss how this gradual development can be undertaken.

## BP4: Combine Process Improvement and EI Iteratively and Interactively

The waterfall project life-cycle model has been dominant for many years. This model expends extensive effort on defining requirements, analysis, and design with the goal of minimizing the development and testing work that is done subsequently. This approach, however, is often not necessarily ideal for process improvement, and especially not for process improvement that involves EI.

Process improvement streamlines activities for people to make them more efficient. Following the waterfall model, many assumptions may be made by either the process improvement analysts or the business users about how particular changes would affect the user experience and improve efficiency. However, sometimes these assumptions are not correct. Often, it is only when users try out a new approach can they definitively say whether the new implementation is an improvement, and what additional changes would yield even greater benefits.

Likewise, when process improvement is combined with EI, many assumptions are made about connectivity to various IT systems. These assumptions, as well as documented system behavior that is relied upon for the EI design, may be incorrect. With the waterfall model, it is usually near the end of the project, during system integration testing, that these faults are discovered. As a result, redesign and reimplementation may be necessary, resulting in significant delays and cost overruns. The waterfall model's failure to detect these problems early enough in the project life cycle makes the case for moving toward more agile alternatives.

A best practice for combining process improvement with EI is to start small and gradually build upon experience and success. For the initial delivery, identify the smallest scope that will provide benefits to the process owners and improve overall understanding of the problem domain. It is critical to set expectations regarding what will be delivered and what fulfills those expectations; scope creep can easily undermine the goal of producing results and gaining experience quickly. Identifying key challenges and rooting out invalid assumptions early in the project life cycle will enable further efforts to be successful.

Central to this approach is to demonstrate functionality early in the project life cycle with a proof-of-concept or prototype system to gain support and to quiet the naysayers. The end users must be involved with testing the proof-of-concept system so that they provide feedback regarding any fundamental weaknesses and/or potential improvements. The feedback thus derived will often require changes to be made to the original plan, but eventually result in a much more useful

system being delivered by the end of the project. This interaction will also help facilitate communication, get the business users' buy-in, and foster ownership. Success with the prototype and further real-world experience using the pilot system in a production environment will lay the foundation for larger, more comprehensive projects.

The risks of not moving forward incrementally can be illustrated as follows. The project might be going well until, near the end, integration testing identifies that the scope or the complexities of the EI were not fully understood. This added complexity requires redesign and/or rework, causing major delays to the project. Alternatively, the project may be going well, but the lack of visible progress could cause the business sponsors to lose confidence and support for the project over time. Another failure scenario is that the process improvement using EI may be delivered successfully, but unforeseen operational considerations may be discovered during testing or in production use that make the process improvements unusable as implemented.

When beginning a proof-of-concept project, it is best to "time box" the exercise to facilitate scope control. The elapsed time for a proof of concept should be between 1 and 3 months. First, determine the process improvement and EI functions that will demonstrate the greatest near-term results and test the most critical technical assumptions. Then, once the resource level for the project is known, determine which of those functions can be delivered within the allocated time frame. It is critical to underscope the initial exercise. Expect that the scope will grow of its own accord and that many things will be more difficult than expected, causing delays.

Most importantly, the proof-of-concept project should produce a system that can be rolled out as a limited pilot to production users. Use in a production environment should clearly demonstrate the benefits of both the process improvement and the EI. For the EI portion of the project, it is best to perform integration with one of the process's core IT systems, but only providing access to a small number of system functions. During the proof-of-concept implementation, short, successive iterations—2 weeks in length—should be used to define requirements, implement a working model, have users test the model, and then further refine the requirements. After a few cycles, the users should be able to verify that the improvements planned for full implementation will provide the intended benefits. Similarly, basic EI connectivity can be exercised and the behavior of IT system interfaces verified through a series of short iterations; performance considerations can also be examined.

At the end of the pilot, identify lessons learned and plan how to factor those considerations into future phases of the project. The challenges encountered in the pilot phase are unlikely to be unique; rather, they are more likely to be representative of common challenges that will be encountered again in the future. Leverage the success of the pilot to perform a second phase project that yields further process improvements and extends the EI. While the scope and time frame of this second phase should be greater, it should not exceed 6 months. The second phase provides an opportunity to deliver more significant business functionality, and, if possible, develop the initial SOA and consolidated UI frameworks. At the end of the pilot phase or at the beginning of the second phase, an overall road map of successive project phases should be developed. The road map defines the process improvements that will be delivered by each phase and the EI that must be implemented to support those improvements.

The iterative approach may be unfamiliar to some organizations. Many companies have long, drawn-out project definition and approval cycles that do not necessarily fit with a pilot-project model. Likewise, business users and technologists may not be comfortable with a more interactive, short-delivery-cycle approach. However, organizational, or personal, biases should not be allowed to get in the way of the process improvement gains that can be achieved using an iterative approach. For the pilot project, it is necessary to create a project team, of internal staff or external consultants, who have experience working this way and will make it a success. Then, the experience and success can be extended to other project areas in the future.

As a final note, it is important to understand that the iterative approach is not a substitute for extended requirements and design efforts, but rather is a tool for developing them in a more dynamic manner.

## CONCLUSION

For transportation planners, it may be simplest to add lanes to existing highways and change the speed limits: The addition of connecting roads and on-ramps may be ignored because they seem like too much work. Similarly, process improvement specialists may choose to reduce exceptions and reallocate resources to reduce bottlenecks, ignoring improvements that require EI. However, in both cases, the result of limiting the improvement effort could be to forego 80% of the potential gains.

In this chapter, we have highlighted what to expect when going after the gains that are possible by applying EI as part of process improvement. None of the challenges presented are extraordinary or insurmountable, but they are formidable and require consideration and planning to overcome. The best practices identified will help address these challenges. These practices may seem commonplace, but one or more are often not present in many organizations. While it is best if all four of the best practices can be combined, any one of them can stand on its own and provide benefits.

# REFERENCES

Harrington, H. J. (1991). *Business process improvement: The breakthrough strategy for total quality, productivity, and competitiveness.* New York: McGraw-Hill.

Khan, R. N. (2004). *Business process management: A practical guide.* Tampa, FL: Meghan-Kiffer Press.

Krafzig, D., Banke, K., & Slama, D. (2005). *Enterprise SOA: Service-oriented architecture best practices.* Upper Saddle River, NJ: Prentice Hall Professional Technical Reference.

Pande, P. S., Neuman, R. P., & Cavanagh, R. R. (2000). *The six sigma way: How GE, Motorola, and other top companies are honing their performance.* New York: McGraw-Hill.

# Chapter XV
# Case Study Implementing SOA:
## Methodology and Best Practices[1]

**Gokul Seshadri**
*Architect, TIBCO Software Inc., USA*

## ABSTRACT

*SOA or service-oriented architecture is an innovative approach to enterprise application integration that increases the benefits of EAI by means of standardizing the application interfaces. This chapter explores the gradual evolution of SOA through various phases and highlights some of the approaches and best practices that have evolved out of real-world implementations in regional retail banks. Starting from a step-by-step approach to embrace SOA, the chapter details some typical challenges that creep up as the usage of the SOA platform becomes more and more mature. Also, certain tips and techniques that will help the institutions maximize the benefits of enterprise-wide SOA are discussed.*

## INTRODUCTION

Service-oriented architecture or simply SOA is seen as the new face of enterprise application integration (EAI). By covering application-specific touch points with business-oriented interfaces, SOA is able to provide better design, agility, reusability, and maintenance savings, and has become the choice for the EAI approach.

This chapter details various steps involved in embracing the technology at an enterprise level, right from conception to enterprise-wide implementation. Throughout the chapter, we will be highlighting the approach using a standard example: A regional financial institution that attempts to embrace SOA in a methodical manner to realign its IT architecture with the business vision and goals.

International financial institutions have always been plagued with EAI problems. They are one of the earliest breed of business communities to quickly embrace SOA concepts and theology, and hence we chose this example. However, readers can quickly relate these examples and situations to any business community or sector they are associated with.

*Figure 1. Point-to-point interfaces for integration*

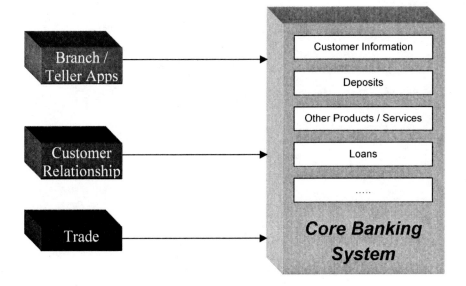

## APPROACHES TO INTEGRATION

### Problems with Traditional Approaches

Traditionally, financial institutions such as banks have been part of the earliest business institutions to adopt computerization because of the obvious benefits a mechanical computation machine brings to financial transactions. It is common to find traditional monolithic applications and programs built using procedural languages like COBOL and C in use in such institutions, even today.

Since these business applications could not function as independent silos, they had to communicate with each other. One required the data and semantics from another system in order to complete the desired business function. So, point-to-point communications interfaces were established between the applications. By point-to-point interface, we mean Application A directly hooking onto application B by means of whatever communication protocol that can be adopted.

With more and more applications seeking to communicate across one another throughout the enterprise, the number of point-to-point interfaces also increased dramatically.

The last decade saw the growth of professional prepackaged vendor applications meant for specific lines of business like loans, trade, or treasury. These boxed applications, once again, had to connect to existing applications and programs in order to achieve their functionality. The choice was, once again, point-to-point interfaces.

Substantial numbers of these point-to-point interfaces were batch programs, running at the end of the business day or end of the business week or month. Thus, when Application A updated a particular customer record, this change was available to Applications B and C only the next day or next week.

As banks expanded their business horizons into other related areas like selling credit cards and insurance products, the back-end applications and systems also grew in number. With the internationalization of business, every bank

*Figure 2. Communications clutter in a typical retail bank*

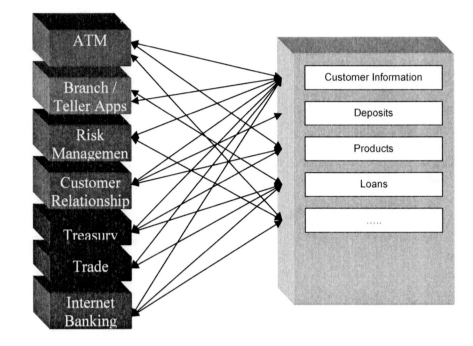

had to invest in their regions of interest by means of acquisitions and opening offshore branches. Needless to say, the net result was more point-to-point interfaces.

If we were to draw a logical representation of all these communications happening across different systems, we would end up with a huge communication clutter that is difficult to understand, comprehend, and manage.

With the growth of real-time computing and communication technologies like the Internet, batch interfaces were posing another challenge. When the latest information about a given business entity was not updated in all dependent systems, it resulted in a loss of business opportunity, decreased customer satisfaction, and increasing problems.

For example, let us say a customer has deposited $100,000 in his savings account and goes into the loans and mortgage division of the bank. If

the loans and mortgage application failed to get an immediate real-time update of the deposit the customer made just a few minutes ago, it would show an earlier record of the customer that might lead the lending officer to refuse the loan application of the customer. For the customer, who has no knowledge of how these back-end applications work, it is a frustrating experience because he has just deposited a large sum and yet the bank is not providing the facility he wants. This results not only in loss of opportunity for the bank (to sell a loan product), but also the threat of losing a good customer.

The challenges discussed above are quite common to all enterprises and financial institutions. And almost all of them are looking to adopt some kind of architectural strategy alighted with their business vision to solve these problem areas.

*Figure 3. Enterprise messaging*

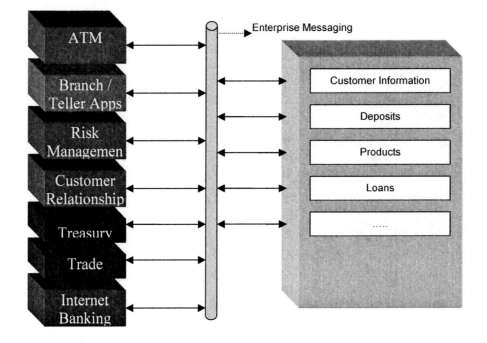

## Technical Solutions to Applications Integration

Technology companies around the world have been coming out with various solutions to address the problem of point-to-point communications and application integration issues.

One of the first solutions to appear on the horizon was enterprise messaging.

Messaging protocols are built on top of basic transport protocols, and they offer better features and flexibility. HTTP (hypertext transfer protocol), TCP (transmission-control protocol), and UDP are examples of basic transport protocols. IBM MQ Series® and TIBCO EMS® are examples of messaging protocols built on top of basic transports.

Messaging protocols advocated standardization in the way applications talked to one another. For example, when Application A wants to talk to Application B, both applications had to use

a common messaging protocol. This would mean that when Applications C and D join the bandwagon and are capable of talking the same protocol, then they could easily connect to Applications A and B that are already exposed to the common protocol.

The first and immediate problem with enterprise messaging were questions like, "What if Application A or B was not capable of talking in the enterprise messaging protocol?" Or, "What can be done in order to make Application A or B speak the common messaging lingo?"

There were two solutions to this.

- One was to enhance the existing application in one way or another so that it can communicate in the common enterprise messaging lingo.
- Another was to build a piece of software that would stand as a bidirectional translator, speaking in the preferred communication

protocol of various applications on one end and the common enterprise lingo on the other end. These eventually came to be known as application integration adapters.

Many enterprises accepted and started adopting this solution and to date it is very much in prevalence. The need for application integration adapters became an opportunity for software vendors to prebuild third-party adapters for standard business applications. For example, TIBCO has adapters for more than 100 such standard enterprise applications like Siebel, Peoplesoft, SAP, Oracle Financials, J. D. Edwards, and so forth.

However, messaging could not solve all problems. Enabling connectivity did not mean that Application A could talk to Application B straightaway. There was the data transformation and formatting issue. For example, Application A may impose that all its input and output data be fixed format strings, whereas Application B

may expect data as a delimited string. So, when data was sent from Application A to B, the fixed strings had to be converted to delimited strings and vice versa.

The other problem was business and logical transformation. For example, in Application A, the key to identify a customer could be a 15-digit customer numeric code, whereas Application B might expect an alphanumeric string. Thus, when a customer update was sent from A to B, key information had to be carefully managed in order to make the update successful. Such transformations are called logical transformations.

Initially, these transformations had to be hosted within the application integration adapters. But as the complexity of these transformations grew, a dedicated engine to design and host transformations came into being. Messaging became an integral part of this engine.

So far all solutions had been purely technical solutions. Architecture was built around tech-

*Figure 4. Semantic data transformation example*

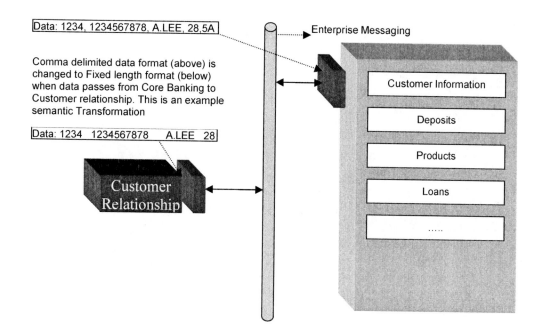

*Figure 5. Role of transformation engine*

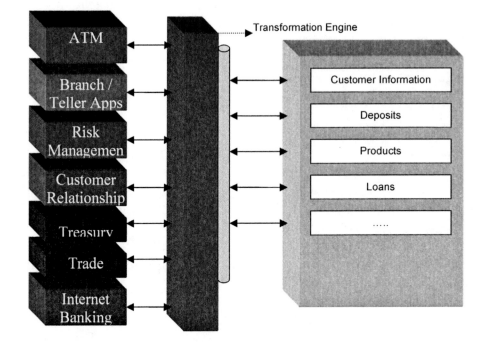

## SOA: CONCEPTUAL UNDERSTANDING

nologies advocated. At this stage, the inception of a new technology called Web services created a big impact on the traditional thought process that would change the landscape of applications integration forever.

Web services advocated a standard XML (extensible markup language) communication model and a simple HTTP transport. The simple object access protocol (SOAP) was a set of specifications that advocated a standard XML structure for all communications and standard methodology for binding transports. HTTP is the only ratified transport in the Web services world to date.

While Web services in itself was looking like another technical solution, IT architects around the world captured a nice concept that was looking beneficial beyond technology and specifications. This was the concept of SOA.

We will pause for a moment and understand what SOA is all about before proceeding further.

SOA is not a technology. It is an architectural approach built around existing technologies. SOA advocates a set of practices, disciplines, designs, and guidelines that can be applied using one or more technologies.

SOA encourages developing services around existing business functions offered by an application. Other applications that want communication with this application would make use of one or more services to accomplish the desired business task.

*Service-oriented architecture is all about building standard interfaces to access different business functions that are exposed by various core business backend systems. These functions could essentially be those that are frequently invoked*

*Figure 6. Bill-pay service example*

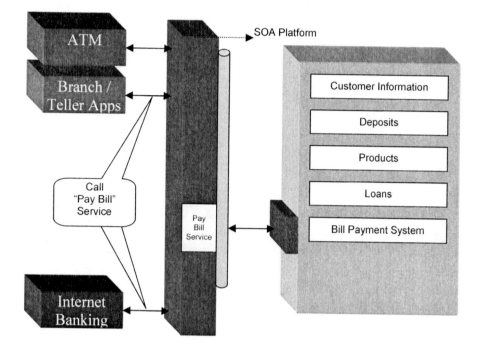

*by other business systems within the enterprise ecosystem.*

*When these standard interfaces are built for enterprise-wide usage and are used by many different backend and client applications spanning a wide variety of business and functional boundaries—we term such an implementation as Enterprise SOA.*

For example, let us say Application A is a bill-payment system. Applications B, C, and D need to invoke a particular business function called *pay bill*. This business function expects the following input parameters.

- The bill's customer reference number
- The service provider or the company whose bill is being paid
- The account number of the customer from which the amount needs to be debited

- The date and time of payment
- The payment amount

This function outputs the following parameters.

- The status of payment (success, failure, or pending)
- The payment transaction's reference number

For this situation, SOA advocates that Application A should build a service called the bill-pay service and make it available to all applications.

This service would consist of the following parts.

- A definition of the exact input and output data formats, preferably in the form of XML schemas, with details like mandatory and optional data elements

Figure 7. SOA implementation cycle from conception to reality

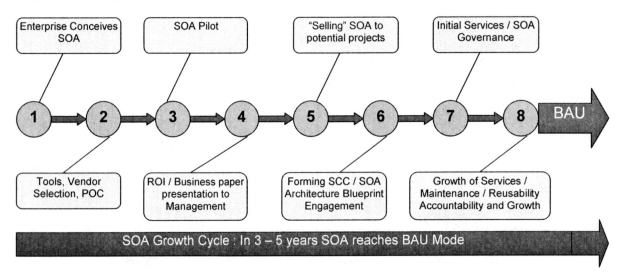

---

- A definition of the transport protocol(s) by the means of which this service could be invoked
- A service implementation, which is a piece of software that would accept the incoming XML, invoke the necessary back-end application function on behalf of the caller, get back the response, format the response in the service data format, and send it back to the calling client

In short, this service will be an abstract representation of a given business function offered by Application A without implementation details of how this function is being executed. It is the responsibility of Application Clients B, C, and D to provide the necessary data inputs and receive back the response.

## Sequential Steps in Approaching Enterprise SOA

An enterprise goes through various steps as it embraces SOA as the strategy for doing application integration.

The definitive milestones commonly observed in this journey are as follows.

- **The enterprise conceives SOA as a definitive strategy:** This is where all the high-level talks about SOA begin in all enthusiasm and earnestness. The enterprise usually formulates a task force constituting members from various related departments to do enquiries about SOA implementation and select possible approaches and benefits. The SOA task force begins to understand all the nuances of SOA from various books, vendors, and analyst reports. This phase is characterized by steep learning with few of the bank's staff emerging as champions in the race. These champions eventually pioneer the rest of the journey in SOA implementation.
- **Tools, vendor selection, POC, and so forth:** The SOA task force starts talking to various known and unknown vendors about tools and methodologies for SOA implementation. Naturally, the preference usually goes to vendors with whom the enterprise has

already been dealing with, but a careful analysis of known market leaders is advised and it is best to keep personal preferences aside and take a pragmatic look at who is leading the pack in terms of integration and implementation methodologies. This phase is challenging as it is marked by considerable confusion caused by various vendors and their interpretations about SOA. SOA, being an architectural concept, is flexible enough to lend itself to multiple definitions. Hence, it is the responsibility of the enterprise's task force to remain sane and look for things that are fundamental to SOA success. One of the best methodologies to evaluate tools and vendor offerings is to have a POC solving one of the organization's problems. This approach helps the vendors to showcase their strengths against real-life issues and helps the task force to see the weaknesses of the tools instead of solely depending on PowerPoint presentation data. More often than not, most enterprises face surprises when it comes to POC: Their initial conceptions, ideas, and thoughts on various tools and vendors usually undergo revisions.

- **SOA pilot or initial project(s) for SOA:** At the end of the vendor evaluation, the task force should be able to identify the vendor with whom the enterprise wishes to partner with. Having been convinced with the POC, now the task force is to host a small environment within the enterprise itself and do a sort of small-scale SOA implementation so that it becomes a benchmark, and so the implementation can be showcased to convince other people within the enterprise. The chosen pilot project should be a part of a business-critical, high-visibility project, but at the same time, it is best not to complicate or burden the pilot with too many activities. In some organizations, a full-blown enterprise project is set aside to try the SOA tools, whereas in others, only a

small portion of existing needs is made use of to test the tools. The chosen pilot should provide enough room to test the waters of SOA like the speed of development, reusability, business agility, and so forth. To the SOA vendor, it is a golden opportunity to highlight the strengths of the product and tools, and to showcase its professional services know-how. At the end of pilot, the task force should be in a position to decide whether SOA as well as the tools chosen for SOA can work for enterprise-scale implementations.

- **Return on investments and business paper presentation to management:** Once convinced about the pilot project and taking measurable outputs from project experience, the task force begins to prepare a business paper that puts forth the formal SOA proposal for management's approval. This is a complicated exercise as the task force needs to consider all the different aspects of enterprise SOA implementation. A rock-solid understanding of current architecture, transformed architecture (or the end goal), and various steps in achieving the desired architecture should be spelt out clearly. The overall budget of the enterprise-scale implementation in terms of hardware, software licensing, professional services from vendors, effort, and manpower required from the enterprise are all worked out in detail, and a maximum budget is projected. It is very important to project a correct figure and set the expectations right with the management in the first place. The ROI graph to be submitted along with the proposal should be able to project how long it will take for the enterprise to realize the amount that is being invested.

- **Selling SOA to potential project owners (or developing the SOA pipeline):** As enterprise SOA is approaching the stage of becoming a reality, it is important for the task force to develop a healthy pipeline of projects for which SOA can be made use of. It might require that the task force needs to do some sort of internal marketing campaign within the enterprise to convince the project owners to make use of incoming SOA for their projects. With management's approval, it is best to make it absolutely mandatory for all new projects to make use of the new SOA platform: Point-to-point connections or other proprietary means of connectivity requirements should not be encouraged unless there are extraordinary situations. It is natural to expect a lot of doubts, misunderstandings, and other complications on the part of various project owners about SOA; already, they have a host of risks to address, and SOA adds new doubts and risks on the overall project, in their perception. It is the duty of the task force to engage with them in a series of discussions so that these doubts are mitigated right in the initial stages. The idea is to convince them about the overall enterprise scope of SOA, why it is better than the current approach, and how management wants all projects to adopt this methodology. These risks are usually perceived to be very high on the initial set of projects that will be using SOA, hence it will be good on the part of the task force to provide some form of incentives or concessions to the initial set of projects (maximum of two) that decide to use SOA. This incentive can usually be charge-free development of the first 20 services or something similar to that. The overall idea is to provide a comforting feeling to those with incoming projects so that they do not feel they are scapegoats and the favors are one sided: The task force may achieve this by any means it deems fit.

- **SCC formation:** Once the management approval is obtained, the task force sets forth to form what is called an SOA competency centre, or SCC for short. The SCC is a body constituted within the enterprise, with fully dedicated and partially dedicated members, that develops, owns, and maintains services on an enterprise scale. This body should have a head who directly reports to top management. It should have developers, architects, technologists with in-depth know-how of enterprise internal systems, administrators, and infrastructure people. As far as possible, all task force members should eventually be integrated into SCC. The formation of the SCC, its role, and its constitution are broad topics and are beyond the scope of current discussions.

- **SOA architecture blueprint engagement:** The task force engages with the vendor's professional services team to deliver what is called an enterprise SOA blueprint. It constitutes various architectural deliverables like infrastructure design, services design, load balancing and scaling, fail over and fault tolerance, services monitoring, SLA measurement and management, and so on. It is this engagement that lays down the overall architecture for the entire SOA platform over which services will be designed, hosted, and maintained. It is critical to ensure all party participation in this architecture exercise as it has a long-standing implication on the overall success of SOA within the enterprise

- **SOA governance:** Along with the SOA architecture, the task force also needs to come out with a set of guidelines that will govern the development, usage, and eventual retirement of various services. SOA spans across the enterprise and each service could have a host of stakeholders. Hence, it is important to lay down a set of guidelines and principles that will govern what should be done and when. SOA governance includes topics such as services life cycle management, stakeholders of SOA and their responsibilities, guidelines for project owners who use

SOA, service-level agreements (SLAs), and so forth. Strong SOA governance prevents misunderstandings across various teams in the future and sets the expectations right across various stakeholders.

• **Delivery of services for initial project:** The SCC starts working on developing services for the initial set of projects. It is imperative that all service designs confirm to certain predefined standards and guidelines so that all services share certain common grammar and theology. The SCC should be careful to bestow enough attention to various details like how the new systems are going to connect to SOA, what protocols they will use, what will the XML format will be that they will use to connect, and so on.

• **Reusability, accountability, and maintenance:** Ongoing activities include providing a services catalogue for the entire enterprise to find what services have been hosted, what can be reused, and so forth. Every new requirement has to go through the SCC to ensure sufficient reusability. For each service, there should be a developer and a lead who should be made accountable for whatever services that are delivered. It is not possible to avoid perennial change requests to existing services, and this should be carefully managed because different projects are using the same service. It is best to go through some sort of version control for each service, and the services governance should dictate how many versions of a given service will be maintained, depreciated, and eventually retired.

• **Achieving BAU mode:** The eventual success of SOA depends on the increased usage of the new platform for all integration needs within the enterprise. More services and more reusability help to bring down the overall cost of integration. With the right strategies and support, the SOA platform becomes well-integrated within the bank and enters BAU, or business-as-usual mode, in 3 to 5 years time. Entering this mode signifies that SOA is already a part and parcel of the overall enterprise IT architecture and does not need to be treated as a strategic project anymore. Businesses already know that they need to set aside a portion of their project budgets for services development. A strong SOA platform provides room for the enterprise to think more about process modeling, business process management (BPM), and complex events management.

## SOA IMPLEMENTATION: TECHNICAL BEST PRACTICES

In the light of the topic under discussion, let us discuss certain recommendations, strategies, guidelines, and best practices that have emerged out of real-world implementation experience.

### Role of Enterprise Service Specifications

SOA defines a service specification for each service that details the data elements that are expected as request input parameters and what will be provided as the request output. SOA best practices indicate that this request and response structure should be inspired by the business function and business data models involved rather than the specifics of the application that is offering the service.

Coming back to the earlier example of the bill-pay service, we discussed the XML request and response formats that need to be defined so that calling clients can communicate with the service. SOA best practice advocates that these input parameters should not be influenced just by the input and output parameters defined by Application A. A bigger level view of what would be required for such a bill-pay service in the bank should be considered in designing the service. This ensures that what is required by the business is duly incorporated into the service specs rather than being led by the specifics of a vendor application. For example, let us say the bank's practice advocates providing a seven-digit requestor transaction reference for each financial transaction in the bank. Application A, which is providing the bill-pay service, does not require this parameter, but the service specs will incorporate the requestor transaction reference. This would ensure that if, at a later date, Application A is replaced by another application, H, then the overall business functionality of the service remains unaffected and the clients can remain transparent about this change of application.

Taking this a bit further, it will be realized that the enterprise data models that define the structure and entity relationships of various data elements of the bank should be duly considered while drafting the service specifications and contracts. In order to achieve this effectively, it is best to involve the expertise of the data management team within the bank. It is a good practice to have one or two representatives from the data management team to be involved right from the start of designing enterprise service specifications.

The enterprise service specifications define the service contract between the service clients and the SOA. It usually takes the form of XML requests and responses, with standard headers and varying body parts.

The following points are noteworthy:

- The enterprise service specification should not be confined by immediate project requirements at hand. The enterprise data models should help to understand the anatomy of underlying business entities. The specification represents an abstraction of the underlying business function: It is important that this definition is well-understood by those involved in writing out the specifications.

- It is best to adopt standardization while allocating name spaces to XML entities. Name spaces are to XML entities what addresses are to humans: They define a hierarchy in which the XML elements attach themselves. The top-level name space could be something like http://schemas.<enterprise-name>.com followed by other directory structures that need to be adhered to.

- Like the business requirements, enterprise service specifications also do not remain stagnant. As the requirements change, so do the specifications. Hence, it is important to version-control various schemas and specifications. A particular service can support one version of schema and so on.

- It is best to maintain some form of XML repository wherein all schemas can be hosted, uploaded, downloaded, and version-controlled.

## Support for Multiple Protocols and Data Formats

Not all systems within the enterprise can be expected to support XML data formats or Web service calls over HTTP. The SOA platform should be flexible enough to support clients with fixed, delimited, and other kinds of data formats, and allow services to be accessed via different protocols.

- The services should at least support the following messaging or transport protocols: JMS (Java message service), HTTP(S), TCP/IP (Internet protocol), and MQ. Depending upon the specific enterprise architecture, other protocols may be added or taken away depending upon the requirements.
- It should be well-remembered that all these different protocols have varying methodologies for load balancing and scaling. The service design should adequately take this into consideration.

- In terms of supporting fixed, delimited, and other non-XML forms of data, the normal methodology is to have a layer that will do the semantic transformation between the non-XML and XML formats. This, however, incurs additional overheads.

## Ensuring Reuse Across Multiple Applications and Projects

Services are originally conceived for specific projects and on top of specific applications. In order to ensure that these services are reusable for other projects in the long run, there are a number of disciplines that need to be enforced.

- The first and foremost discipline is to not let the service design on enterprise service specification be led too much by the scope of immediate project needs. Thought processes should stand outside what is immediately apparent at the time of development. It is good to consult other existing (or potential) users of the same service or business function.
- The other challenge is to align the enterprise specifications closer to the core business function offered by a given application rather than to the semantics of the application itself. This would ensure reuse of services, even if the boxed Application A is replaced by another boxed Application B that offers a similar function.
- It is equally important to build enough flexibility within the service itself so that it is able to accommodate growing changes so its reusability increases. With every other change in schemas and specifications, it is not possible to deploy one more version or instance of the service. That would be too expensive. Instead, service designers should build enough flexibility and adoptability within the core structure itself so that changes to a given service can be added on a continual basis without affecting the

currently running processes. One of the best methodologies to implement this is to attach a version number to a service or a schema. As long as a given client is using a service at a given version (which is already live), it should not be disturbed with future changes unless the service is going to retire. Newer versions of the service could be used by newer clients, but to existing clients this should be transparent.

- It is good to adopt a standard life cycle policy for services. Just like any other IT asset, services also go through a cycle of inception, growth, decay, and retirement. For example, when a newer and more efficient version of a service is in place, the old service may no longer be attractive or useful. The current users of the old service have to be given suitable time to migrate to the newer version, and then the old service can retire. Evolving enterprise-wide policies for service life cycle management is beneficial.

## Handling Varying Load Conditions for Different Services

SOA services are meant for enterprise-wide usage. Service ABC could be used by Applications A and B in the beginning, but over the passage of time, Applications C, D, and E might also start using the same service ABC. When Service ABC was originally designed, let us assume that it was designed to handle five transactions a second at its peak load. Now, however, with Applications C, D, and E also using the same service, the overall throughput of the service has come down to two transactions per second because of the increase in load. Hence, Service ABC needs more muscles or more process to achieve the original throughput desired.

More muscle does not necessarily mean more hardware. The overall capacity of the machine may still be sufficient, but Service ABC needs additional processes or handlers to handle the increase in load.

Couple this with another scenario. When the fate of Service ABC is detailed as above, Service DEF is facing a different kind of problem. It was originally designed to handle 20 transactions per second at its peak load, but in 2 years, it has been observed that the peak load has not crossed four transactions per second. Hence, the processes that have been attached to Service DEF need to be scaled down so that the machine resources can be saved and used elsewhere.

It is interesting to note that both of these services have been deployed in the same engine run time. Adding hardware to this run time will result in the linear scaling of resources, whereas what is desired is a unit-level (service-level) scaling manipulation.

The right solution to the above problem would be to have more than one process handler per service, and increase or decrease the process handlers with the changes in requirements.

## Services Monitoring

The term monitoring usually represents a mechanism that continuously monitors something against certain fixed rules and reports the outcome by means of alerts. For example, a monitoring rule could say, "Please alert me when the CPU (central processing unit) usage of the machine is above 95% for more than 5 minutes." Another rule could be, "Please alert me when the disk usage has crossed 80%."

Traditionally, monitoring tools have been focusing heavily on hardware. With newer technologies like SOA and BPM offering newer perspectives into the IT infrastructure, newer breeds of monitoring tools have also come to play.

In SOA, the basic service monitoring should cover the following aspects.

- Is the service and supporting components up and running? If not, start them. If there are any troubles in starting, please send an alert.

- Monitor the service log files for any errors that might have happened. If such error does occur, send an alert based on severity levels.

On top of this, SOA requires sophisticated monitoring covering the following aspects.

- What is the service level of a given service instance in the last 1 hour? If it is not within the SLA, how many requests have gone past beyond the agreed service time? If more than $N$ number of calls has suffered from delayed response, then it might mean that the service is experiencing load problems with insufficient resources for handling the load.
- What is the correlation between the overall resource (CPU, memory, etc.) usage on the machine, resources consumed by a given service, and the jobs that were being completed using those resources? For example, when a given service was consuming significant resources at some point in time, what was it really doing? It is because of a higher number of requests that had hit the service at that point in time? Is it due to some kind of housekeeping activity? Or is there any problem within the software (memory leak, etc.) resulting in this problem?
- Historically analyze overall service usage in, say, the last 6 months. How many services have experienced peak load and for what amount of time? Are the current hardware resources sufficient enough to cater for immediate projects or is new hardware required?

SOA offers a unique perspective into business that was previously unavailable: It offers a real-time view of what is happening in terms of transactions, usage, and so forth. This knowledge can be capitalized by business to increase customer loyalty, identify newer opportunities, and grow the business. The CEO (chief executive officer) of the bank may want to know what is happening today—right now—in the business, in the KPIs (key performance indicators) are

... area in computing ... ties monitoring. ... ion are provided

... ed in the pattern ... n customer's ac- ... withdrawn every ... nd this does not ... e pattern of the ... spects some kind ... d teller machine] ... ssword hijacked, ... and suspends the

... paign has been ... that displays a new ... r has logged in. It ... stomer is spending ... he or she normally ... e system suspects ... ding the promo in ... up and says, "Are ... oduct? We have a ... first 100 customers ... hong them."

... llateral and loans ... end from the past ... a sharp rise in the ... ho are applying for ... en period in a year. ... some social reason ... ecides to introduce ... oan product during ... year. The response ... omers for this new ... d with the previous ... rent product is fine- ... cess.

## Service-Level Agreements

In the SOA world, the term SLA represents the agreement between the SCC, and service consumers and providers in terms of the availability, load handling capacity, and scalability of a given service.

A given service could be used by different projects within the enterprise. Thus, the total load on the service per day will be a cumulative sum of all the individual loads imposed by the clients. The SLA defines the requirements of the client in terms of overall turnaround time expected and total volume of transactions that can be expected to be handled per day. The client should also project the overall increase in load expected (if any) in the next 3 to 5 years. The SCC takes these inputs while designing a given service and evaluates whether the current infrastructure and handlers are sufficient to handle the expected load. If not, the infrastructure or the number of handlers for a given service needs to be scaled up sufficiently.

The SOA platform depends on back-end systems to meet the SLA requirements. When a certain turnaround time is expected by the client, the SCC should evaluate what is the nominal time that will be consumed by the back end and add sufficient buffer for the time to be spent in SOA. Normally, the time spent in SOA should be around 5% and not more than 10% of the overall turnaround time.

The SOA platform's monitoring tools should help the SCC to ensure that all the promised SLAs are met. The monitoring infrastructure should continuously monitor the overall turnaround time taken by a given service and compare it with the normal or average turnaround time. If abnormal delays are encountered on a continual basis, then it means some investigation is necessary. The monitoring tools normally raise alerts in case such exceptions happen, and interested parties can subscribe to these alerts by means of e-mail, pager, and so forth so that they can be aware of what is going on in the SOA platform.

## SOA IMPLEMENTATION: ORGANIZATIONAL BEST PRACTICES

The following sections highlight organizational best practices that have emerged out of successful SOA implementation stories.

### Ensuring Widespread Adoption of SOA

An initiative as big as SOA cannot be a success if it is not understood and used by various business applications and projects across the enterprise. In big enterprises, it is common to find business and IT teams that are blissfully unaware of the existence of such a common infrastructure.

The following measures are recommended to ensure a high degree of SOA platform usage.

- The key is communication. Communicating the existence and relevance of SOA and ensuring that the knowledge of SOA is well-understood and used by independent project teams would be a key to ensure the successful adoption of the scheme. Seminars, technology update sessions, and other relevant means should be adopted on a continual basis to ensure that all the key stakeholders of IT are aware of SOA as well as the immediate services that are in development.

- Some enterprises have adopted the approach of putting an integration IT project council in place. The task of the project council, which mainly consists of members from the architecture and engineering team of the bank, is to review the integration requirements of the immediate projects in the pipeline and recommend the usage of relevant services.

Unless there are exceptional reasons, no project would be allowed to make independent connections to various applications: They have to go through the SOA platform and need to adhere to the standards and practices it advocates.

- The SCC team should proactively meet the project leads of incoming projects, dispel myths and assumptions if any, and ensure that the new project leads are comfortable using SOA for their integration needs.

### Services Development: Reusability Drives the Show

One of the biggest promises of SOA is the reusability of services across multiple business applications. It is one of the key drivers that will help the enterprise get proper return on its investments and help to reduce the budgets of individual projects.

For example, let us say the bill-pay service discussed earlier is initially built on top of Application A for Project 1. Later on when Projects 2 and 3, involving Applications B, C, and D, eventually need the same business function from Application A, they can leverage on the existing service and reuse the same one possibly without any changes. This would mean significant savings for the bank. As more and more services are built on top of the existing Application A, the bank would eventually reach a stage where most if not all integration requirements are met with existing services. This would drastically reduce the money spent on application integration. With more reuse, the amount spent to maintain a given service would also drive down.

The following points are noteworthy with respect to reusability.

- During the service request evaluation process, the SCC should evaluate the eventual benefit of building services for a given project in the long term. Let us say Projects 1 and 2 are requesting 10 services each to be built for their use. Let us assume the SCC has the budget for only one project. Now it is the responsibility of the SCC to evaluate what the level of reusability is between these two projects if services were built. If more applications within the enterprise are expected to use the services of Project 2, then it should be given precedence over Project 1 requirements.

- It will be noticed that at least a few services of a given project are not highly reusable. In the same example cited earlier, let us say out of 10 services, 8 are reusable and 2 are very specific to this project. It is the duty of the SCC to agree to build all 10 services because, just for the sake of 2 services, Project 2 cannot be allowed to establish point-to-point or other proprietary means of communications.

- Services would be developed against specific project requirements. For example, among the 100 business functions offered by Application A, if only 10 are required for Project 2, then only 10 services will be built. This would ensure that for every service that is being built and deployed into production, there will at least be one client. Later on, if more functions are demanded by other projects, more services would be developed and deployed.

## Managing SOA Infrastructure

Usually, a new hardware infrastructure is necessary to host the SOA platform and its associated tools. To support this new infrastructure, a dedicated team of system management people will be required. We will call this the SOA operation centre in our discussions.

The operation centre needs to support at least four different flavors of environment: development, SIT, UAT, and production. Disaster recovery could be added if required. UAT will closely

resemble production. Only UAT and production will have hardware fail-over capabilities, and only the SIT platform will have a connectivity option with back-end host systems while the development platform will remain as a unit testing environment.

It is the responsibility of the operations centre to ensure that the hardware and software required for various projects in the restructuring exercise are kept in tact. As various modules of a given project migrate from one environment to another, the operations centre would promote the piece of code after suitable testing.

The overall capacity of the hardware that constitutes the new platform would be reviewed from time to time, and additional hardware would be added when necessary.

## SOA Competency Centre

It is recommended that the task of developing, deploying, and maintaining services for various project requirements be allotted to a dedicated SOA competency centre, which will be a new constitution in the enterprise. This competency centre could initially be funded as a strategic initiative until it reaches a stage where it can be purely funded from project costs. In each project that makes use of SOA, a budget needs to be allocated for services.

The centre should be headed by a vice president. There should be three to four SOA architects, a few project leads, and a bunch of developers on the team. While initial projects can be developed by vendors themselves, the centre should slowly take ownership of services in a gradual manner.

## Geographic Expansion of SOA Platform

If the enterprise is a regional or global entity with major presence in more than one region, then it is natural to expect some form of regional expansion of the SOA platform. There will always be perennial expansions and increases in the scope of applications in newer countries and regions. This means that the SOA integration infrastructure has to cater beyond local requirements.

Three major strategies are adopted to cater this need.

• The lowest common denominator is to put up an SOA infrastructure that is similar to the current one in other regions.

• Having a common SOA platform to service multiple service clients and providers in different regions

• Having different infrastructures in different regions, but establishing some form of communication gateway between them and designing services that can talk to various service providers in different regions

It would be too expensive to put a dedicated infrastructure in every other region where the enterprise has a market. Rather, a careful evaluation of select regions where the customer base is significant would give way to fewer platforms. When the customer base in a given region is small, the integration needs can be taken care of by the nearest SOA infrastructure. Such SOA infrastructures are usually called SOA hubs.

One more measure that can be put in place is to have one common development environment to be shared across multiple regions. This would mean that the enterprise needs to put up only the production SOA infrastructure in various regions and control the development activities from a single hub. There are pros and cons to this approach: The obvious advantages are in terms of cost savings and the ease of maintaining one competency centre, but certain problems can be anticipated in terms of supporting multiple infrastructures. A nicer balance would be to maintain a skeletal SCC (of much smaller scale) in each region where an SOA hub has been established and maintain the core SCC in the main region where the SOA initiatives are being pioneered.

## ENDNOTE

[1] The content represents the views of the author and not necessarily the views of TIBCO Software Inc.

# Chapter XVI
# Application Integration:
## Pilot Project to Implement a Financial Portfolio System in a Korean Bank[1]

**So-Jung Lee**
*JLee Consulting, Singapore*

**Wing Lam**
*U21 Global, Singapore*

## ABSTRACT

*This case describes a pilot project to implement a financial portfolio system (FPS) within Jwon Bank. (The case is based on a real-life organisation, although the identity of the organisation has been disguised at the organisation's request.) A strategic IT review of Jwon Bank's IT systems and architecture revealed that a lack of integration between IT systems was hampering the bank's capability to meet its business vision. A key recommendation from the review was the development of an FPS that would enable customers to manage all their financial assets and access financial services from a single place. However, creating an FPS meant that Jwon Bank had to develop a strategic solution to meet the integration needs across the entire bank. Jwon Bank examined enterprise application integration (EAI) tools, and embarked on a pilot project to develop a prototype FPS and test the robustness of an EAI solution. The case highlights some of the management issues relating to integration projects of this nature, including strategic planning for integration, EAI tool selection and evaluation, and understanding of business process flow across divisional silos.*

## ORGANISATIONAL BACKGROUND

Jwon Bank is one of the fastest growing providers of financial services to consumers in South Korea. Out of all the banks in Korea, it is recognized as one of the most innovative and progressive. For example, it was one of the first banks to offer Internet banking and mobile banking services. The swiftness with which Jwon Bank has embraced technology to differentiate itself is acknowledged by industry analysts as one of the primary reasons for the significant increase in its customer base, which has grown by 15% in the last 5 years to about 5 million. The majority of Jwon Bank's new customers are young adults in the 20- to 30-year-old age group, precisely those who are most likely to adopt Internet and mobile banking services. In addition, Jwon Bank has been running a series of successful marketing campaigns that has given it a refreshingly youthful and vibrant image, unlike the conservative images associated with many of the other Korean banks.

Jwon Bank's competitive strategy is three-fold. First is to offer innovative financial services before its competitors. The introduction of new global trading services is one example of this. Second is to use technology to deliver financial services in flexible ways. The investment the bank made in new IT systems during the period of 2000 to 2002 is estimated at between $80 to 100 million. Furthermore, with the high IT literacy rates in Korea, the bank predicts that, within 5 years time, 80% of its customers will be conducting the majority of their banking through the Internet. Third is to provide exceptional standards of customer service to attract new customers and retain existing ones. A major service-quality initiative was recently launched by the vice president (VP) of customer services to monitor levels of service quality across the entire bank.

At the bank's 2003 annual conference, the CEO (chief executive officer) of Jwon Bank articulated his vision of being "Korea's preferred provider of financial services." The CEO also announced that Jwon Bank would significantly diversify its portfolio of financial services, enabling consumers to meet their entire financial needs, from investments to insurance, through Jwon Bank. The CEO believed that being able to provide customers with a holistic set of interlinked financial services would give Jwon Bank a significant competitive advantage in the industry.

## SETTING THE STAGE

### Jwon Bank's IT Architecture

Jwon Bank's IT architecture is divided into separate clusters of IT systems that are owned by individual business units (e.g., current accounts, investments, mortgages, trading, and insurance) and support the specific business needs of that business unit. Each cluster has between 5 to 20 IT systems. For example, there are a total of 14 IT systems in the credit-card cluster that collectively handle the management and processing of credit cards for the credit-cards business unit. In total, Jwon Bank has over 120 IT systems across all the different clusters.

Like many large organisations, Jwon Bank's IT architecture has evolved over a long period of time to include a diverse mix of IT systems running on different platforms and employing diverse technologies. For example, many of the IT systems in the current accounts cluster are legacy systems based on older, mainframe technology such as CICS and MVS. Many of the IT systems in the trading cluster are bespoke C++ applications running on the UNIX platform that have been custom built for Jwon Bank. On the other hand, the IT systems in the Internet-banking cluster are customized versions of packaged commercial off-the-shelf (COTS) systems running on the Windows platform. In addition, the IT system at the core of the customer services cluster is developed around the COTS Seibel system. The increased use of COTS reflects a general trend within the bank over the last few years of adopting a best-of-breed strategy to procure packaged COTS systems that were best in their class.

## 2003 Strategic Information Technology Review

In 2003, the VP of technology at Jwon Bank undertook a strategic review of the bank's IT systems. The main purpose of the review was to ensure that the bank's IT systems and technology architecture was well-positioned to meet future business needs and aligned to the bank's business strategy. The strategic IT review was carried out over a 4-month period by an independent team from one of the "Big Five" consulting firms.

The VP of technology was taken aback by the findings from the strategic IT review, which indicated that Jwon Bank had significant shortcomings in its IT systems and architecture. One of the key findings highlighted in the strategic IT review was the lack of integration between IT systems. The strategic IT review identified many cases of how the lack of integration between IT systems affected customers or potential customers of Jwon Bank. For example, if an existing Jwon Bank customer wished to purchase an insurance policy from the bank, he or she had to fill out a form and resubmit all their personal details again to the bank even though the bank already held such information in their IT systems. Also, if a customer informed the credit-card unit of Jwon Bank about a change in address, the change in address would not be automatically updated elsewhere in the bank's other clusters of IT systems. So, while credit-card statements would be sent to the customer's new address, bank statements would still be sent to the old address. In fact, the strategic IT review revealed that common customer data were duplicated in many different IT systems, thus leading to the possibility of data inconsistency.

The review also highlighted that the bank's IT systems and technology architecture were cur-

rently unable to fully support the CEO's vision of a holistic set of closely interlinked financial services. While certain financial services were interlinked, such as bank accounts and savings accounts, allowing the easy interchange of funds between the two accounts, a large number of other financial services were unlinked. A customer, for example, could not easily interchange funds between bank accounts and investment accounts. In many cases, there was no direct automation or real-time transfer of information between IT systems. Instead, back-end processes were conducted as daily batch jobs or, worse still, relied on some behind-the-scenes manual processing. For example, when a customer applied to open a trading account, information was captured in the IT system handling the applications. If an application was accepted, much of this information had to be manually rekeyed by the bank's back-office staff into the IT system that opened trading accounts. Not only was this a manual and intensive process, there was also the possibility of administrative errors.

Importantly, the strategic IT review highlighted that the lack of integration between IT systems hampered the development of customer relationships and opportunities for cross-selling. For instance, if a customer was arranging his or her mortgage through Jwon Bank, it was likely that he or she would also be looking to purchase home insurance. However, the current IT systems in the bank did not provide any automated alert to identify such situations, so cross-selling opportunities were lost. In short, each cluster of IT systems effectively operated within its own silo, reflecting in much the same way how the business units within Jwon Bank were operating. In many situations, however, the integration of IT systems is not only a technical activity, but a strategic enabler for new business models and processes (Sharif, Elliman, Love, & Badii, 2004).

## Recommendations

The strategic IT review team made a number of recommendations. One of the key recommendations was the development of a financial portfolio system (FPS). The concept of an FPS was to provide each customer of Jwon Bank with a personalised financial portfolio, that is, a single, consolidated view of his or her financial assets and commitments with the bank. Through an FPS, customers could manage their portfolios, for example, switch money in their savings account to a managed investment trust, through a single point of access rather than through many different interfaces. An FPS was seen as central to Jwon Bank's competitive thrust toward providing its customers with a means of holistically managing their finances. The FPS was also seen as an important customer relationship management tool through which opportunities for cross-selling could be acted upon. In fact, a few of the smaller banks in Korea had already begun to offer portfolio-centric services, thus providing

further impetus for the development of an FPS at Jwon Bank.

The strategic IT review indicated that the development of an FPS would require a significant degree of integration between the IT systems in Jwon Bank's IT architecture, which currently suffered from a lack of integration. Consequently, a further recommendation was that Jwon Bank should consider an enterprise-wide integration solution that would meet the strategic integration needs of the whole bank. An enterprise-wide integration solution would be a better alternative to having multiple localized integration projects that involved point-to-point integration between specific IT systems, as shown in Figure 1.

Although a point-to-point integration solution represents how organisations have traditionally integrated their IT systems, a major problem with this type of solution is that as the number of IT systems to be integrated grows, the cost of building and maintaining such a large number of interfaces becomes prohibitively expensive and often results in "spaghetti" integration (Linthicum, 2001). With

*Figure 1. Two integration solutions*

Point-to-Point Integration Solution

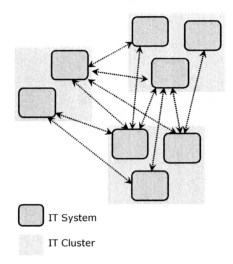

Enterprise-Wide Integration Solution

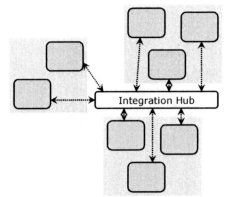

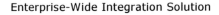

IT System

IT Cluster

an enterprise integration solution, commonly known as an enterprise application integration (EAI) solution, this problem is addressed by the deployment of an integration hub that serves as a central point for interapplication communication (Lee et al., 2003; McKeen & Smith, 2002). Because each IT system need only be integrated with the hub, an EAI solution significantly reduces the overall number of interfaces that need to be built and maintained. The IT systems in an EAI solution communicate with other IT systems by sending messages along the messaging middleware via the integration hub. Furthermore, an integration broker is able to support additional functionality, such as the translation of messages into different formats, transaction management, and the ability to apply business rules and logic.

The recommendations of the strategic IT review were taken with seriousness by the VP of technology at Jwon bank, who initiated a project to look into both the development of an FPS and an EAI solution within the bank. The FPS would be one of the IT systems that used the EAI solution to draw information from the bank's clusters of IT systems. So, although the VP of technology had initially considered treating the FPS and EAI solution as two separate projects, he felt that combining them into a single project with a clear business objective was more purposeful.

## CASE DETAILS

### Requirements Gathering

A project team was assembled within the bank comprised of a project manager, three business analysts, and three IT architects. The business analysts were tasked with the job of establishing requirements for the FPS. Although the strategic IT review had described the notion of an FPS, further analysis needed to be carried out to determine the specific functional requirements of the FPS. A series of interviews was carried out

by the business analysts with stakeholders from the different business units within the bank. Use-case scenarios (Cockburn, 2004) were developed to illustrate how the FPS would be used by the bank's customers; the following are examples.

- Changing the customer's address and personal profile
- Purchasing a life insurance policy
- Transferring funds between current, credit-card, trading, and investment accounts
- Viewing a single consolidated summary of financial assets and commitments
- Transferring funds to and from mortgage accounts
- Performing a financial health check
- Assessing the value of one's financial assets in a future period of time

Rather than develop the FPS from scratch, the business analysts recognized the potential to extending the bank's existing Internet banking system. This would have the advantage of building upon an existing IT system that was familiar to many of the bank's current customers. Furthermore, the Internet banking system was proven to be a reliable and stable IT system. The business analysts therefore proposed making major enhancements to the Internet banking system so that it became the FPS rather than developing the FPS completely from scratch.

### Feasibility Study

In parallel with the requirements gathering activity, the IT architects on the team were tasked with the job of evaluating different options for developing an integration solution, estimating costs, and making recommendations. Two external consultants with significant experience in large-scale integration from the consulting firm that had conducted the strategic IT review were also hired to work with the team as independent advice and guidance would be valuable.

To be as objective as possible, the team first looked at the feasibility of a point-to-point integration solution, where a set of programmatic interfaces would be built directly between the IT systems that needed to exchange information. In fact, several of the IT systems in certain clusters had already been integrated in this manner. However, after further examination, a point-to-point integration solution was rejected for the following reasons.

- A high level of custom development would be involved. With around 120 individual IT systems in the bank, the number of interfaces that would need to have been built was estimated at over 250.
- There was a high level of development risk. With so many legacy IT systems in Jwon Bank's IT architecture that used older technology, it was unclear that robust and reliable interfaces could be built from scratch.
- It would take a very long time to develop all the interfaces from scratch. Estimates suggested a maximum build capacity of 40 interfaces per year, resulting in a project that would take over 6 years.
- There was insufficient in-house capacity and expertise to undertake such a sizeable project.

The team agreed that a point-to-point integration solution had too many high-risk elements, and turned its attention toward an EAI integration solution as suggested in the strategic IT review.

As a starting point, the team identified the leading EAI tools currently on the market. This included EAI tools from vendors such as TIBCO, IBM, WebMethods, SeeBeyond, BEA, Mercator, and Vitria. EAI tools are generally comprised of an integration hub, messaging middleware, and adapters that connect IT systems to the messaging middleware. An important consideration in the selection of EAI tools is the range of available adapters (Lam, 2005). An adapter can be considered a kind of sophisticated gateway or wrapper (Brodie & Stonebraker, 1995), enabling packaged, mainframe, and other systems to be connected to the integration hub. Some adapters are prebuilt. For example, there is a prebuilt TIBCO adapter for the ERP application SAP R/3, which enables SAP R/3 to be connected to the TIBCO EAI tool. At a technical level, adapters typically make use of interfacing technologies such as application programming interfaces (APIs), SQL queries and even Web services. Where prebuilt adapters are unavailable, adapter development kits supplied by the EAI vendor can be used to create custom adapters, which is typically required for custom-built IT systems.

## RFP and Vendor Selection

A request for proposals (RFP) was written by the team and sent to 12 of the major EAI vendors. EAI vendors who were small in size or who lacked a strong Asian presence were excluded in the RFP. As a strategic mission-critical project for the bank, the VP for technology was adamant that the chosen EAI vendor would need to be sufficiently resourceful and provide strong local support. The RFP included information about the FPS and Jwon Bank's IT systems and architecture. In addition, a number of key technical requirements was highlighted in the RFP relating to reliability and resilience in the event of system failure. A high level of security was also stipulated given the sensitivity of financial transactions and information. EAI vendors were explicitly asked in the RFP to indicate the extent to which the EAI tool and adapters covered the diverse range of IT systems at Jwon Bank. Eight replies to the RFP were received that were considered complete, from which four EAI vendors were short-listed by the team based on the RFP response.

More in-depth face-to-face presentations and discussions were held between the remaining four EAI vendors. The VP of technology at Jwon Bank knew that selecting the right EAI vendor was as much a business decision for Jwon Bank as it is was a technological one. The EAI vendor would not just be a vendor; it would become a strategic partner who would need to understand Jwon Bank's business and work hand in hand with the team at Jwon Bank to ensure project success. It became evident to the team that only two out of the four EAI vendors appeared suitable partners for Jwon Bank. The other two EAI vendors had limited experience in working with clients in the financial services domain and did not appear to have as good an understanding of the domain as compared with the two short-listed vendors.

The EAI tools for the two short-listed EAI vendors were closely matched in functionality. However, a consensus was reached amongst the team to work with Mekon, an EAI vendor with a strong Asia-Pacific presence and an established track record working with financial institutions on the implementation of large-scale EAI solutions. The VP for technology at Jwon Bank agreed with Mekon that as a trial, they should conduct a pilot project based on a selection of the use cases identified in the requirements gathering phase. If the pilot project was successful, it would lead to the continued engagement with Mekon on the full project.

## Pilot Project

The main objective of the pilot project was to develop an FPS prototype with partial functionality based on a selected number of use cases. It was decided that the FPS prototype would focus on the specific issue of the customer profile, ensuring that consistent customer profile data existed across the different IT systems at Jwon Bank. The FPS prototype was considered sufficiently challenging in that it involved integration between a significant number of IT systems across different business units with Jwon Bank, and was nontrivial in nature. The intention was that the FPS prototype would not be made immediately available to Jwon Bank's customers. Rather, the

FPS prototype would enable the team at Jwon Bank to develop a proof of concept and, importantly, evaluate the utility, performance, and reliability of the EAI solution that would address the bank's enterprise-wide integration needs. Furthermore, the pilot project would provide an opportunity for Jwon Bank and Mekon to work together and assess how well they partnered with each other. A 6-month time frame was agreed for the pilot project to see what could be achieved.

Mekon fielded a team of four of its consultants to work with the team from Jwon Bank to develop the FPS prototype. Mekon followed their recommended methodology for implementing EAI solutions, which involved the following five steps.

1. Identification of business goals
2. Business process modeling and improvement (BPMI)
3. Mapping of business information flows
4. Systems connection
5. Business process execution

The teams from Jwon Bank and Mekon agreed on a project plan based on the above steps and worked together to execute the project. As part of Step 1, further interviews were held with stakeholders in the various business units at Jwon Bank both to clarify the scope of the pilot project and identify the ways in which each business unit handled customer profile data. Further sessions followed during BPMI (Step 2), in which business process models were created with each business unit. These described, for example, how customers changed their address and contact details, and the workflow that was involved in each business unit. What became apparent during BPMI was that although each business unit had an intimate understanding of the business processes within their own business unit, the understanding of business processes across business units was less well-understood. In other words, intrabusiness unit processes were well-defined, but not interbusiness unit processes. Such interbusiness unit processes, however, were central to the development of an FPS.

The mapping of business information flows (Step 3) involved establishing the flow of customer data from one IT system to the next. The business information flows were modeled in the EAI tool and it was determined how information would be sent via the integration hub from one IT system to another. In some cases, data would be sent to multiple IT systems. For example, if a customer changed his or her address, this information would need to be propagated to the IT systems that held address information in other business units. Importantly, this step in the methodology also highlighted where customer data were duplicated across multiple IT systems, leading to possible data inconsistency. Furthermore, information was stored in individual IT systems where the primary key was the account number or policy number. Hence, it was difficult for Jwon Bank to identify all the accounts or policies that any particular individual held. This holistic view of the customer was central to the development of the FPS and Jwon Bank's business strategy in general. The team therefore designed the FPS around a database that captured a centralized profile of each customer.

The systems connection step (Step 4) involved the installation of the EAI tool and the adapters that were needed to connect the bank's IT systems to the integration hub. For some of the bank's IT systems, such as the call-centre system, which was based around the Seibel packaged system, prebuilt adapters already existed. However, for several of Jwon Bank's custom-built IT systems, prebuilt adapters were unavailable and so custom adapters had to be developed. For example, the Internet banking system was custom developed using Java. The EAI tool included an adapter software development kit (SDK) with a Java API that enabled a custom adapter to be created for the Internet banking system. The IT staff at Jwon Bank did not have the necessary technical knowledge to undertake the development of

custom adapters, so this work was outsourced to the development services group at Mekon who had done many such custom adapter development projects before and so were well-versed in this type of specialized work. A development team at Jwon Bank worked on the development of other aspects of the FPS prototype in conjunction with the work on the installation of the EAI tool and custom adapter development. Extensions were added to the Internet banking system to create the FPS prototype.

After 6 months, the original planned duration of the project, it was clear that the prototype FPS was someway off being complete. The main reason for this related to the custom adapters that needed to be specifically developed to integrate the bank's custom IT systems. Unlike prebuilt adapters, custom adapters were like mini applications in their own right and required their own life cycle of specifying, developing, and testing. The development services group at Mekon had not originally anticipated that they would be so involved in the FPS prototype, so the manpower shortages on their side were also a constraining factor. Without the custom adapters ready, the team could not proceed to conduct full end-to-end testing of the FPS functionality. This delayed the project a further 3 months, and it was only after 9 months that the FPS prototype and EAI solution were ready. Although the user interface of the FPS prototype was rudimentary, the main objective was to test the functionality of the FPS prototype and the robustness of the EAI solution. A suite of functional tests were run with satisfactory results. Performance testing also revealed that response times of the FPS prototype under high loads was satisfactory for a system of this kind that would be accessed over the Internet.

## Evaluation of Pilot Project

The VP of technology at Kwon Bank considered the pilot project a mild success. The EAI solution proved to be robust and the prototype FPS was a system that could be taken further into full development. Furthermore, the teams from Jwon Bank and Mekon appeared to work well together and complemented each other. The team at Jwon Bank possessed the domain knowledge associated with financial services and Jwon Bank's business processes, while the team at Mekon provided the EAI solution implementation know-how.

However, the pilot project was not without problems that the VP of technology knew posed potential risks in moving forward with a full project. The prototype FPS implemented only about 10% of the full functionality of the FPS, yet took 9 months to develop, so an FPS with full functionality could conceivably take over 90 months. It was clear, therefore, that the development of an FPS was a significant undertaking and not something that could be delivered quickly. Rather, it was better to view the FPS as a project that would need to be divided into stages, with functionality released incrementally over a long period of time. Such an incremental approach would also allow Jwon Bank to better manage project risks. The team had already agreed that they should extend the Internet banking system to become the FPS. However, a challenge remained in deciding what functionality of the FPS should be implemented first. Although this was essentially a matter of business priorities, each of Jwon Bank's business units would claim that the FPS functionality pertaining to their respective areas was high on the priority list.

One of the most worrying aspects of the pilot project for the Kwon Bank team was the time and effort needed to develop custom adapters. Many of the IT systems in the IT architecture at Kwon Bank were custom developed, so it was clear that a significant number of custom adapters would need to be developed. The pilot project, however, had demonstrated that the development of custom adapters was probably the area that represented greatest risk and effort. On the other hand, the pilot project had been an extremely useful exercise as it had given Jwon Bank some firm idea

of the resources and costs required to deliver an FPS with full functionality. In particular, where custom adapters needed to be developed, costs would increase significantly. The team estimated that every custom adapter would require about $60,000 in development, testing, and installation costs. In comparison, the costs associated with the licensing of the EAI tool were relatively minor.

## CURRENT CHALLENGES AND PROBLEMS FACING THE ORGANISATION

Jwon Bank faces a number of challenges in going beyond the pilot project and developing an FPS with full functionality, and implementing a full EAI solution to address their enterprise-wide integration needs. One challenge is bridging the organisational silos that have emerged as a consequence of having separate business units responsible for different sets of financial services. Sawhney (2001) recognizes this as a significant issue in many large organisations, where "islands of applications" tend to emerge from business units creating their own IT systems without thinking about the broader needs of the organisation as a whole. In addition, the team discovered that intrabusiness unit processes, that is, business processes within a particular business unit, were well-understood, but not interbusiness unit processes. However, it is the interbusiness unit processes that are central to the development of the FPS because it involves taking a customer view across multiple financial services. Until now, there has been no specific unit or team within Jwon Bank that has examined interbusiness unit processes. Jwon Bank may consider addressing this void by reviewing the organisational structure of the bank and creating a working group comprising of representatives from each of Jwon Bank's business units.

A further challenge is managing a project of this sheer scale. It took the team 9 months to create a prototype FPS with 10% of the overall functionality required. It has already been recognized by Jwon Bank that a broader plan will need to be devised in which the functionality of the FPS will be rolled out in stages over a period of time. Jwon Bank must determine what functionality should be rolled out in which release, and the basis upon which this is to be determined. For example, is some functionality considered more urgent than others in terms of business need? Should functionality that is simpler to implement be rolled out before functionality that is more complex? Or is some functionality dependent upon other functionality already being implemented? Jwon Bank must consider all such issues in the formulation of a broader rollout plan for the FPS. There are also many other factors that need to be taken into account, including staff availability at Jwon Bank, adapter development capacity at Mekon, the impact of other IT projects in the business units, and business priorities for Jwon Bank. Furthermore, there are also potential internal political sensitivities to navigate, where the choice of some FPS functionality over others may be construed as favourtism toward a particular business unit.

It is clear that the development of the FPS will require input and involvement from many different stakeholders, including representatives from the respective business units. There are essentially many internal customers for the project, all of which have their own requirements to be satisfied within the FPS. The management of stakeholder expectations is therefore critical to the success of the project. Clearly, the amount of planning involved is significant, and a project of this nature will require a significant level of coordination. It may be sensible for Jwon Bank to treat the development of an FPS as a program, within which separate projects are organised and initiated. Within the umbrella of a program, some overall level of control can be maintained across projects and common standards applied. Jwon Bank might consider establishing a program steering committee for this purpose and developing a suitable governance structure.

Jwon Bank must also pay close attention to the management of project costs. IT projects are notoriously known for cost overruns (The Standish Group, 1994). The experience from the pilot project has shown that significant costs (and risks) are associated with the development of custom adapters. Unlike prebuilt adapters that can be installed and configured, custom adapters must be specifically developed for a custom IT system and therefore treated like a mini project in its own right. Jwon Bank will need to identify which adapters need to be custom developed and estimate development costs to formulate an appropriate budget. If the functionality of the FPS is to be rolled out incrementally, then a budget will need to be formulated for each release. Given its lack of in-house expertise, it would be sensible for Jwon Bank to outsource custom adapter development to Mekon as an integral part of the overall partnership between the two organisations.

The partnership with Mekon is crucial to the success of the FPS project for two reasons. First, the EAI tool of Mekon is the centerpiece of Jwon Bank's enterprise-wide integration solution. Second, Jwon Bank lacks the in-house skills and expertise to undertake such a project alone. Hence, Jwon Bank is very much dependent on Mekon to provide much of the required expertise. While this may be acceptable in the short term, Jwon Bank needs to think about building up its in-house skills and expertise and becoming less dependent upon Mekon in the longer term. In addition, although the pilot project suggested that both teams at Jwon Bank and Mekon worked well together, it remains to be seen whether this good working relationship is one that will be maintained over the course of time. One strategy that Jwon Bank may consider is to institute a staff development and training plan to build up in-house expertise. Another strategy is to consider a knowledge management initiative to capture, share, and disseminate knowledge and best practices with the organisation (Davenport, De Long, & Beers, 1999).

## ACKNOWLEDGMENT

The author wishes to thank employees at the real Jwon Bank for their support and assistance in the writing of this case.

## REFERENCES

Brodie, M., & Stonebraker, M. (1995). *Migrating legacy systems.* San Francisco: Morgan Kaufmann Publishers.

Cockburn, A. (2004). *Writing effective use cases.* London: Addison-Wesley Professional.

Davenport, T. H., De Long, D. W., & Beers, M. C. (1999). Successful knowledge management projects. *Sloan Management Review, 39*(2), 43-57.

Lam, W. (2005). Exploring success factors in enterprise application integration: A case-driven analysis. *European Journal of Information Systems, 14*(2), 175-187.

Linthicum, D. (2001). *B2B application integration.* Reading, MA: Addison Wesley.

McKeen, J. D., & Smith, H. A. (2002). New developments in practice II: Enterprise application integration. *Communications of the Association for Information Systems, 8*, 451-466.

Sawhney, M. (2001). Don't homogenize, synchronize. *Harvard Business Review.*

Sharif, A. M., Elliman, T., Love, P. E. D., & Badii, A. (2004). Integrating the IS with the enterprise: Key EAI research challenges. *The Journal of Enterprise Information Management, 17*(2), 164-170.

The Standish Group. (1994). *CHAOS report.* Retrieved from http://www.standishgroup.com/sample_research/chaos_1994_1.php

## ENDNOTE

[1] The case is based on a real-life organization, although the identity of the organisation has been disguised at the organization's request.

# Chapter XVII
# Case Study:
## Service–Oriented Retail
## Business Information System

**Sam Chung**
*University of Washington, Tacoma, USA*

**Zachary Bylin**
*University of Washington, Tacoma, USA*

**Sergio Davalos**
*University of Washington, Tacoma, USA*

## ABSTRACT

*The primary objective of this case study is to discuss a retail business information system illustrating an e-business integration example among a biometric attendance system, a surveillance system, and a point-of-sale system. Using a service-oriented architecture allows businesses to build on top of legacy applications or construct new applications in order to take advantage of the power of Web services. Over the past years, Web services have finally developed enough to allow such basic architectures to be built. Each of the components in the system will be designed and developed using a service-oriented architecture that clearly illustrates how such cutting-edge systems can be put together. By designing these components in such a fashion, this example will focus on applying service-oriented development and integration techniques to the retail sector. The result of this project will be an integrated system that can be used by businesses everywhere to learn how their organizations can benefit from service-oriented architecture. Also, the application of the service-oriented development and integration to systems that were previously stand-alone and heterogeneous is discussed. All previous experiences in object-oriented architectures and design methodology are naturally streamlined with the service-oriented architecture, supporting the loose coupling of software components.*

## INTRODUCTION

The purpose of this case study is to discuss a retail business information system (BIS) illustrating a service-oriented development and integration example among a biometric attendance system, a surveillance system, and a point-of-sale (POS) system. Various software systems typically used in businesses can be developed and integrated using SOA (service-oriented architecture; Alonso, Casati, Kuno, & Machiraju, 2004; Erl, 2004). An SOA is a way of designing a software system to provide services to either end-user applications or other services through published and discoverable interfaces.

The three heterogeneous applications used in this case study are a system for keeping track of how long an employee works called EAS (employee attendance system), one using surveillance systems to monitor an establishment called 3S (Smart Surveillance System), and a POS system. They are typical applications that are found in a majority of retail BISs. Unfortunately, each of these applications is almost always used in isolation and is difficult for users to interact with since they have been developed by different software companies, and integration with other systems is not considered at the beginning of their designs. In many instances, these components are constructed using different programming languages and are very difficult to upgrade and integrate with newer systems. For example, many POS systems need to manage employees and need a secure access mechanism through a fingerprint recognition system. Although EAS supports this functionality, EAS cannot be easily integrated into the existing POS system.

This case study demonstrates some of the advantages of SOA-based development and integration, which we call service-oriented development and integration (SODI), to construct a loosely coupled information system that can be effectively altered and upgraded as time passes. In order to demonstrate how the SODI can be

applied to the development of software systems for a retail business and the integration of the software systems, this case study considers two problems. The first problem is to propose how the SODI can be defined without losing the valuable experiences of the developers in object-oriented analysis and design. The second problem is to propose how the SODI can be applied to the development and integration of a retail business information system.

Web service technologies are still relatively immature in comparison to other networking protocols and technologies. At this time, the only relatively stable aspects of Web service technologies are the simple object access protocol (SOAP; World Wide Web Consortium Extensible Markup Language [W3C XML] Protocol Working Group, 2006), the Web service description language (WSDL; W3C Web Services Description Working Group, 2006), and universal description, discovery, and integration (UDDI; Organization for the Advancement of Structured Information Standards [OASIS], n.d.). Other standards such as WS-Addressing (W3C Web Services Addressing Working Group, 2006), WS-Policy (IBM, 2006b), WS-Security (OASIS, 2006), and semantic Web services such as the Web ontology language for services (OWL-S; W3C Semantic Web Services Interest Group, 2006b), Web services modeling ontology (WSMO; W3C Semantic Web Services Interest Group, 2006c), semantic Web services framework (SWSF; W3C Semantic Web Services Interest Group, 2006a), and Web services semantic (W3C Semantic Web Services Interest Group, 2006d) are still being researched and are not mature enough to be applied to real BISs yet by using industry-supported platforms such as .NET or Java.

SODI is a software development and integration approach that is architecture centric, integration ready, evolution based, and model driven. SODI is architecture centric and integration ready since three architectural patterns—three-layered architecture (Alonso et al., 2004), multitier architecture (Alonso et al.), and SOA—are employed for design and deployment, and the loosely coupled software components using Web services allow future integration to be ready even before the integration demands. Also, since a software development methodology called rational unified process (RUP; IBM, 2006a) is used, which has also been used in the development of object-oriented BISs, SODI is an evolutionary development and integration where all experiences in object-oriented software development can be employed. In addition, since the RUP has been used in traditional object-oriented software development with the 4+1 view model (Kruchten, 1995) and unified modeling language (UML; Object Management Group [OMG], 2007), the SODI approach using the RUP is model driven.

A retail BIS that consists of a biometric attendance system, a surveillance system, and a point-of-sale system is developed and integrated by employing the SODI approach. Based upon the case study, the SODI approach is analyzed and discussed in addition to the integrated system that can be used by businesses everywhere to learn how service-oriented architectures can benefit their organizations. The discussions of the SODI approach on this case study shows to Web services practitioners that Web service technology brings a BIS in which software components can easily integrate with others since they are loosely coupled through Web services. Also, the BIS developers can adopt this emerging technology of Web services naturally since their experiences of the architectural patterns and the design methodology in object-oriented development can be seamlessly reused for service-oriented development and integration.

Next we will discuss some of the previous approaches to integrating enterprise applications. Then, the three system components to be developed and integrated are discussed in more detail. After that, the chapter provides the SODI approach, which consists of three architectural patterns and the SODI design methodology using the 4+1 view and visual modeling. In the next section, the application of the SODI approach to the retail BISs is illustrated with visual models. The application of the service-oriented development and integration to the case study is analyzed and discussed in the last section, which also provides some conclusions in regard to the case study.

## PREVIOUS INTEGRATION EFFORTS

What were businesses using to integrate their internal applications before Web services came about (Linthicum, 2003a, 2003b)? The first generation of application integration techniques mainly included writing adapters that would take data from one application, adapt them for use in another application, and then send these data off to be used by another system. This first approach was very tedious to develop, and the complexity skyrocketed as new components were added. These troubles led to the second generation in application integration paradigms. This new generation utilized third-party components that already had adapters available for purchase. While this may have saved organizations some time, it still required quite a bit of money. Piecing third-party components together using adapters saved developers a large amount of time over first-generation techniques; however, the resulting system was still very complex to use. Trying to integrate applications developed in house with third-party components could be quite troublesome at times. While some of these problems have been ironed out today, many still exist and remain a concern for developers working on non-service-oriented integration projects.

Up to now, the efforts for enterprise application integration (EAI) and business-to-business (B2B) integration have been much more costly and time consuming than it would have been if a service-oriented integration strategy were used. An integration approach using the electronic data interchange (EDI) format does not share operations but only data, so the bandwidth between heterogeneous software components is higher than for SODI. Integration approaches using distributed object-oriented computing paradigms (Coulouris, Dollimore, & Kindberg, 2002) such as remote method invocation (RMI) or Common Object Request Broker Architecture (CORBA) need more effort from system developers and integrators compared to the service concept since a tightly coupled lower level abstraction concept called distributed objects are used.

## HETEROGENEOUS APPLICATIONS TO BE INTEGRATED

Three heterogeneous applications are developed, and some of their functions are integrated as Web services in other systems: EAS, 3S, and a POS system.

EAS is a system that allows small businesses to monitor and record an employee's attendance at work (Kerner, 2000). Businesses need some way to keep track of how long an employee has worked for a given day or week. There are many solutions to this problem, and some are much better than others. By *better*, it is implied that some systems are much less vulnerable to being manipulated by unscrupulous employees.

Instead of relying on arbitrary digits or strings for identification purposes, EAS in this case study uses an employee's unique fingerprint to identify an employee for purposes of signing in or out from work. While forging an identification number may be trivial, forging a fingerprint is exponentially more difficult. An individual's fingerprint consists of various loops and whorls intermingling with each other on the surface of our finger pads. Within these elaborate designs, one notices certain points that in combination can be used to uniquely identify a given person. To forge a fingerprint, one would have to make a replica of the surface of a person's finger, or as a slightly more drastic measure, sever off a person's finger.

Once employees begin using EAS for fingerprint recognition, managers can begin using the more advanced features of the system. Employee schedules can be created and entered into EAS. Then, at any time, managers can generate reports that outline an employee's actual work activity and compare it to the scheduled activity. These reports can also be used for tracking overtime work, people disobeying corporate policies, or payroll systems. The employee information is a good example of data that can be shared with other information systems.

Many businesses use surveillance systems for some reason. Typical uses are for monitoring customers and employees to (hopefully) prevent theft. Another use is for monitoring property to record break-ins. The major problem with these conventional security systems is that they often run on relatively ancient hardware, storing footage on VCR cassettes. The digital storage capabilities of computer hard drives have increased greatly over the years. Today, it is even possible to see systems with storage capacities of up to 1 terabyte or 1,000 gigabytes in size. 3S will take advantage of these large storage capacities. In addition, the use of Web service technologies will enable businesses to know anytime, anywhere when any unwanted activity takes place.

The final component is a newly designed point-of-sale system. Virtually every retail store now uses some kind of computer-aided POS system. These systems allow products to be scanned using some kind of barcode device, allows the product prices to be totaled, and provides mechanisms for cashiers to perform various operations such as selling or returning goods. There is a wide spectrum of different POS applications on the market today. There are off-the-shelf solutions that can be purchased and customized, and there are in-house solutions for a specific business. The problem with the first set of systems is that they are often too generic where the latter systems are too tightly coupled.

## SERVICE-ORIENTED DEVELOPMENT AND INTEGRATION

SODI is the software development and integration approach that is architecture centric, integration ready, evolution based, and model driven. The architecture-centric property means that multiple architectural patterns—three layered, multitier, and service oriented—are used for design, deployment, and integration. In the three-layered architectural pattern, a software application is designed by a set of three separate horizontal logical layers in which presentation, business, and data logics are independently developed and maintained, which is shown in Figure 1a. The presentation logic layer contains user-interface-related constructs and features that end users of the application may wish to know about. The business logic layer contains the core business logic for the applications being built. Lastly, the data logic layer contains database-related logic that manages and permits access to the database system of the application.

In the multitier architecture, *n* subcomponents of an application cooperate with one other using some protocols. Each tier is deployed on a separate host in a distributed network for security and reliability. Figure 1b shows both the four-tier architecture for a typical Web application and the three-tier architecture for a typical Windows application using client-server computing. Although a tier does not mean a host, multiple hosts are used in developing such distributed applications. Since the functions requiring similar execution environments can be located at the same tier and the location of functions to be deployed is known in advance, the multitier architecture has been used along with the three-layered architecture. For Web applications, client-side presentation logic in HTML (hypertext markup language) can be downloaded from the second-tier Web server and then executed on the first-tier client machine. At the second tier, the server-side presentation logic in JSP or ASP can be located and executed.

The business logic in JavaBean or Code Behind is executed at the third tier, while accessing the database is executed on the fourth tier.

In the service-oriented architectural pattern of Figure 1c, software can be considered as multiple interactions of the service consumer, producer, and broker if a software component can be declared as a Web service whose interface can be represented in a standard language, and can be discovered and invoked through a standard protocol. Currently, a service producer can generate the interface of a software component in WSDL by using a SOAP engine, and manually publish the Web service information onto a UDDI service registry that is managed by a service broker. If a service consumer needs the service, the service consumer can discover the service from the service registry manually or programmatically through a client program, although this autodiscovery function is very limited. The discovered service is invoked by the client and bound to the service through the interaction protocol SOAP that is located at the service producer's host.

The integration-ready property means that the integration of software applications using the multi-architectural patterns has built-in interoperability since components of each application are logically separated for better design, properly distributed over the network, interfaced in a standard interface language, and invoked through a standard interaction protocol. Some objects, which may be reused internally or externally in the future, are designed, implemented, and deployed as Web services that support interoperability for integration. Since interoperability is supported in those components through Web services before the actual integration demands occur, it is possible to plan for contingencies in the design phase of the applications.

The evolution-based property means that the knowledge and skills that have been adopted in the traditional object-oriented software development can still be used with Web service technology. Since a Web service can be considered a class

*Figure 1. Multiple architectural patterns*

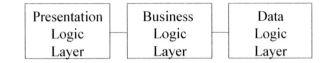

*(a) Three-layered architectural pattern*

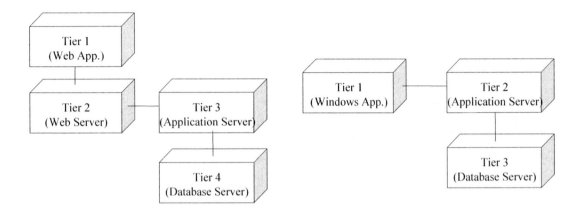

*(b) Multitier architectural pattern*

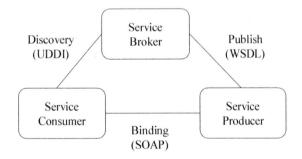

*(c) Service-oriented architectural pattern*

with standard interface descriptions in terms of object orientation, the methodology for object-oriented software development can be naturally used. RUP is a process framework that has been used in traditional object-oriented software development (Booch, Rumbaugh, & Jacobson, 1999; Jacobson, Booch, & Rumbaugh, 1999). The RUP is an iterative, incremental, concurrent, and visual model-driven design process based upon Kruchten's (1995) 4+1 view. The 4+1 view model describes the architecture of software-intensive systems based upon multiple and concurrent views—the design view, implementation view, process view, deployment view, and use-case view—representing logical or physical, and static or dynamic properties of a software application that are interconnected by user requirements. Since Kruchten's model was later incorporated into a famous computer-aided software engineering (CASE) tool, IBM's Rational Rose Modeler that supports RUP and UML, SODI using RUP is naturally model driven.

In SODI, while a software project is being developed through the inception, elaboration, construction, and transition phases, the activities for analysis, design, implementation, and deployment are done iteratively, incrementally, and concurrently. The activities are modeled based upon Kruchten's (1995) 4+1 view using UML diagrams. These visual models are used for modeling, that is, visualizing, specifying, constructing, and documenting the service-oriented software development and integration. Table 1 shows SODI using the RUP framework and 4+1 views.

At the inception phase, SODI practitioners bring the idea of using three architectural patterns to all workflows. During the inception phase, the use-case view is mainly modeled based upon the given or collected case scenarios. The use-case view model consists of use-case diagrams along with activity diagrams for more detailed coverage of use-case specifications. Use-case diagrams are generally used to describe the requirements present throughout the design of the system

while activity diagrams help describe the basic flow between application components. The user requirements are viewed in terms of the three-layered architecture, and each layer is represented as a subsystem in the use-case diagrams. If a use case or a set of use cases needs more explanations, activity diagrams are drawn for them.

During the elaboration phase, the design and process views are modeled based upon the use-case or activity diagrams. The design view model consists of class diagrams for a business logic layer and entity-relationship (ER) diagrams for databases (Connolly & Begg, 2005). Class diagrams are used to model classes and their relationships.

When Kruchten's (1995) 4+1 view model was proposed, the object-orientation concept was dominant at that time. Also, the concept of software development within an organization was the main concern instead of the integration of heterogeneous software applications across organizations. Additionally, the concept of process could not be clearly represented with the conceptual building block object. Therefore, the process view was limited to methods in a class, the states of an object, and interactions among objects. If a method of a class needs more explanation, an activity diagram can be used to explain the method. Also, the state change of an object can be explained in a state-chart diagram. The interactions among several objects can be described in an interaction diagram that may be a collaboration or sequence diagram.

Since a Web service can represent a business process, and a special class called an interface can represent the service, a class diagram can be used to show the relationship between the Web service and a set of classes to implement the service. Class diagrams mainly focus on the structure hidden behind the Web service interfaces. Using an activity diagram, we can represent the collaboration among services.

During the construction phase, the implementation view model is built by using component

*Table 1. SODI using the RUP framework and 4+1 views*

| Activities | View and Phase | Inception | Elaboration | Construction | Transition |
|---|---|---|---|---|---|
| **Requirements Workflow** | **Use Case View** | ✓ Use case scenarios<br>✓ Three layered architecture<br>✓ SOA<br>  ✓ Use Case diagram<br>    ✓ Activity diagram for a use case within a layer | | | |
| **Analysis/ Design Workflow** | **Design View Process View** | ✓ Three layered architecture<br>✓ SOA<br>  ✓ Class diagram for classes and services<br>    ✓ Activity diagram for methods<br>    ✓ Statechart diagram for classes<br>  ✓ Entity Relationship diagram for database<br>  ✓ Interaction diagrams fro object and atomic service collaboration<br>  ✓ Activity diagram for composite service collaboration* | | | |
| **Implementation Workflow** | **Implementation View** | ✓ SOA<br>✓ Multi-tier architecture | | ✓ Component diagram | |
| **Deployment Workflow** | **Deployment View** | ✓ SOA<br>✓ Multi-tier architecture | | | ✓ Deployment diagram |

*\* No service composition was employed at that time. A service consumer, which is either a Windows or Web application, only invokes Web services located at remote platforms.*

diagrams. These diagrams are closely linked to class diagrams but are a bit more physical in their existence. Component diagrams primarily focus on defining dependencies between components of any type for a layer on a tier.

During the transition phase, the deployment view is modeled using deployment diagrams. Deployment diagrams clearly illustrate how your system components should be deployed, and exactly on what system they should be deployed to. These deployment diagrams are used to outline the deployment of executable components onto a platform equipped with various software engines.

## CASE STUDY: SODI AND RETAIL BISS OF EAS, 3S, AND POS

Each of the individual applications has its own set of unique diagrams that describe the requirements of the system, the structure of the system, and how the system should be deployed. While some of the Web services and classes are shared, for the most part, each application is fairly different from one another in terms of implementation. The diagrams that appear in this chapter are just a handful of the many different diagrams developed to describe the three applications: EAS, 3S, and POS.

### Use-Case Scenarios

The first task is to collect the use-case scenarios. Although there are many other functions offered by EAS, the employment enrollment with a fingerprint is discussed in this chapter. The image in Figure 2 depicts a manager enrolling an employee's personal information in EAS. The upper half of the form contains personal details while the bottom half contains job-related details. Once the employee's personal information is entered, his or her fingerprint must then be enrolled. Once the fingerprint is enrolled, managers can begin creating schedules.

There are various problems associated with the existing surveillance solutions, namely, the systems can react to streamed video and perform some action. However, they are not proactive in nature. The 3S implementation provides a better control layer for managers to use when interacting with their security camera. 3S allows managers to schedule images to be taken at any interval from any period of time throughout the day. This fine-grained control allows managers to see only what they want at any time. Providing faster access to security recordings improves productivity and allows for more accurate archiving of footage. These features could be used for being notified with various images from around your store when the store is supposed to open or close. This would be a very handy feature for small business owners who want to make sure that their employees open their shop in the mall on time and do not skip out early at closing time. Figure 3 is of the start-up screen for 3S. The screenshots have been selected as images a manager would see if he or she wanted to schedule a set of snapshots to be taken and then view them remotely.

The new POS system takes advantage of the features of an SOA to decouple the application interface from its implementation. To enhance the robustness of the system, its database back end has been built to be fully compliant with the specifications outlined in the proven Association for Retail Technology Standards (ARTS, 2003) model standard. The ARTS model is an industry standard database schema that can be used for enterprise-level retail database applications. By using this standardized database back end, a pluggable interface, and a Web service middle tier, the POS is constructed in such a way that it is capable of being upgraded, extended, and integrated as new technologies and new business demands emerge. The main focus in designing this POS system was to create a basic interface and a set of Web services that the interface can use to handle most of its functionality. The screenshots in Figure 4 show one example interface that could be

*Figure 2. Employee enrollment in EAS*

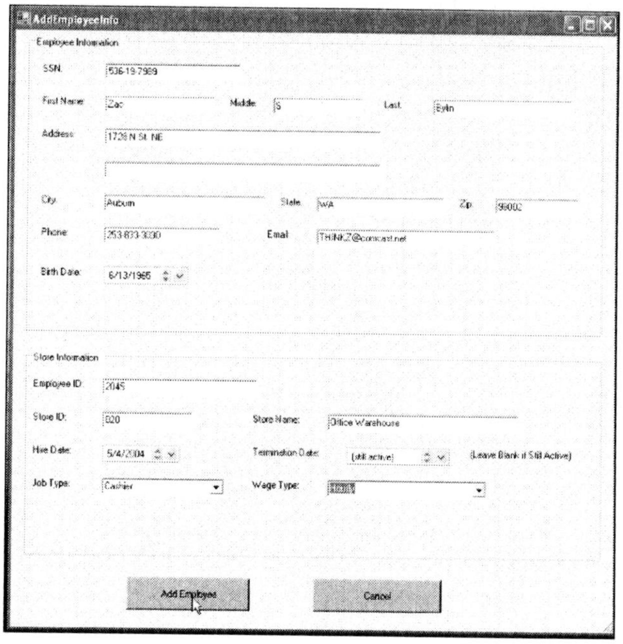

*Figure 3. 3S start-up screen*

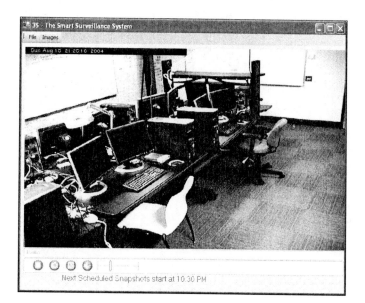

*Figure 4. POS start-up screen*

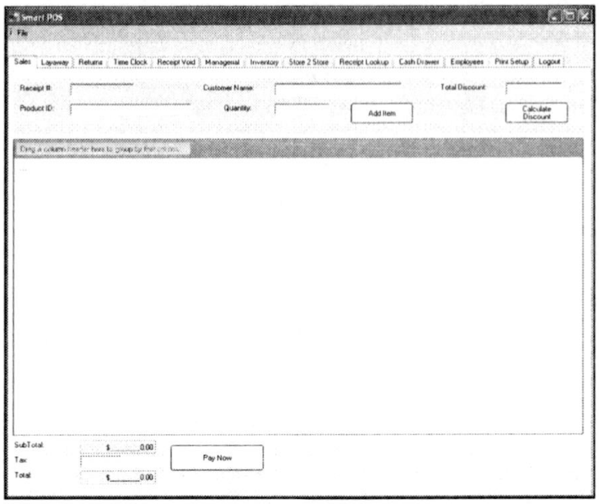

used in conjunction with the Web services being developed. These are the screens that a cashier would see after logging onto the system using a fingerprint when wanting to ring up a purchase. This screen shows the main interface that cashiers use. Cashiers enter the product ID and the quantity, and then click the *Add Item* button. The item is then added to the sales grid.

## Use-Case View

The core requirements for each application have been described in use-case diagrams. Actors and system components are first modeled using the use-case diagrams. Next, the primary functionality of each system is described in the three logic layers in the diagrams. Figures 5 and 6 show the use-case diagrams for EAS and 3S.

Figure 5a illustrates the main functionality of the EAS system, as well as the various subsystems that have been identified and modeled. Within EAS, there are five main subsystems: four that contain system functionality, and a database system. The recognition system is responsible for matching employee fingerprints. The enrollment

system is responsible for enrolling employee fingerprints. The management system is designed to add, update, and delete employee information as well as employee schedules. The report system is responsible for generating the various reports that can be created once schedules have been entered. The final system is the database.

The use-case diagram in Figure 5b illustrates a store manager controlling employee schedules. There are certain options that are presented to the manager in the presentation layer. These options such as updating schedules use the schedule management service in the business logic layer to accomplish its task. The schedule management service in turn uses the main EAS database and accesses the appropriate scheduling information.

The use-case diagram in Figure 6a illustrates a security person and a manager wanting to retrieve a snapshot recording by using 3S. Security personnel can both delete and view snapshots from the snapshot management service. The activity diagram in Figure 6b depicts the process for viewing and deleting snapshots. First, the snapshot service ensures that the snapshot actually exists, and then the service either retrieves

*Figure 5. Use-case view of EAS*

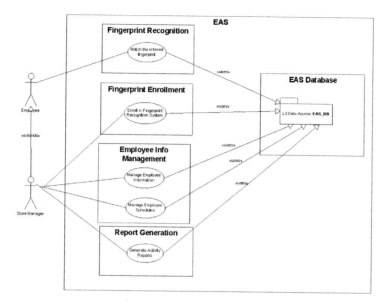

*(a) Use-case view with a use-case diagram for EAS*

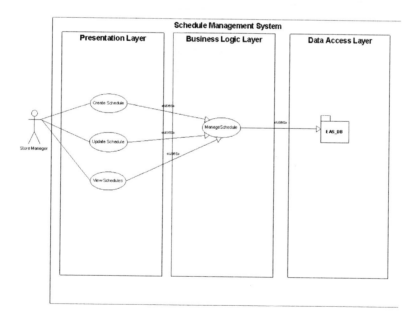

*(b) Use-case diagram with three-layered architecture for EAS*

*Figure 6. Use-case view of 3S*

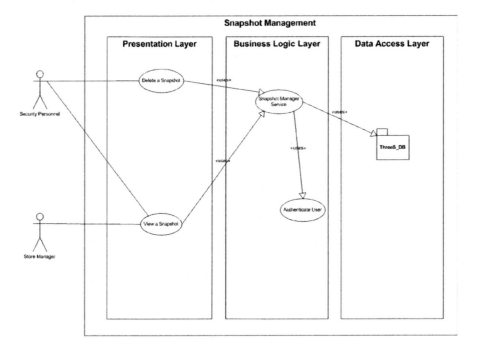

*(a) Use-case view with a use-case diagram for 3S*

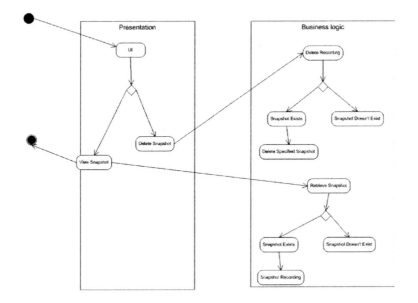

*(b) Use-case view with an activity diagram for 3S*

*Figure 7. Design view of the POS*

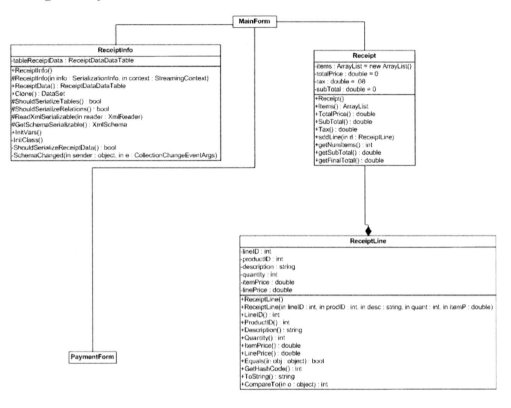

the snapshot or deletes the snapshot depending on the input received. If the snapshot does not exist, the service method returns.

## Design View

The class diagram in Figure 7 shows a design view by showing the class hierarchy of the main form, which contains all the interfaces for the POS. The main form uses two local classes, ReceiptInfo and Receipt, for managing receipts and receipt lines. Also, there is a payment form that allows cashiers to enter the designated payment and calculate the change required.

## Process View

A collaboration diagram can represent the process view by showing the interactions among objects and atomic Web services. An activity diagram can represent the process view of a composite service by showing the interactions among Web services. In Figure 8, the recognition subsystem of EAS is not responsible for anything other than recognizing fingerprints and contacting the appropriate Web service FingerprintProcess with the results. As such, it is much simpler than the other EAS components. The top class is a form; this class uses various other classes on the client side within the same application.

The collaboration diagram in Figure 9 depicts the relationship between a Web service object Ser-

*Figure 8. Process view of the recognition subsystem of EAS*

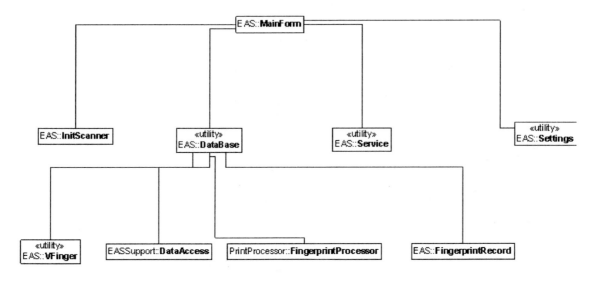

*Figure 9. Process views of 3S*

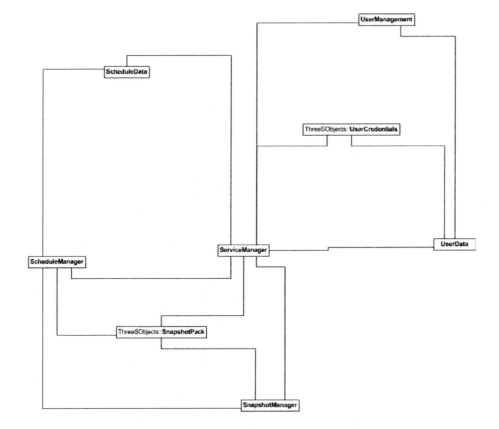

viceManager and other objects. ServiceManager is a central class that is a Web service, and there are various other classes and data sets that the Web service class uses. There are data transport classes such as ScheduleData and UserData as well as normal data type classes such as SnapshotPack, which are generally found in the class library called ThreeSObjects.

## Implementation View

A component diagram can represent an implementation view of a system by showing what physical components are used and how they are interrelated. Figure 10 contains a fairly simple overview of the components of the Employee-Hours subsystem required to generate a simple report containing information about how long an employee worked for a given day. There is one Web form EmployeeHours.aspx, its corresponding

code in C# behind page EmployeeHours.aspx.cs, two Web services used by the code behind pages UserValidation.asmx and EmployeeHours.asmx, and the employee database used by both Web services EAS_DataBase.mdf.

Figure 11 depicts the dependencies that exist in the 3S client application. The executable client depends on the Active X control called AxisMediaControl.dll used to interface with the camera. The MainForm that launches other forms depends upon the 3S class library Three-3Objects.dll as well as a few other forms such as UserManagementFrm, ScheduleViewerFrm, SnapshotSelecterFrm, and so forth.

## Deployment View

A deployment diagram can show which executable components are deployed at which tier. Figure 12 outlines which components are executable by

*Figure 10. Implementation view of the EmployeeHours subsystem of EAS*

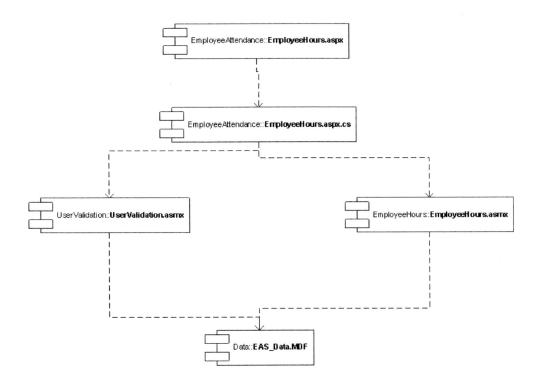

*Figure 11. Implementation view of the 3S client*

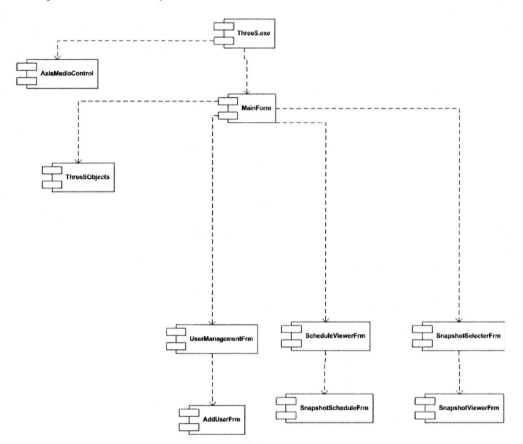

end users and where they are stored within the EAS architecture. There are two types of client: two client-side Windows applications called EASRecognition.exe and EASAdministration. exe, and two server-side Web applications called EASLogin.aspx and EmployeeHours.aspa, which require network connections on both the client and server sides, as well as the physical fingerprint scanning device on the client side. Figure 12b shows the three-tier architecture for the recognition component deployed, requiring a client workstation, an application server, and a database server as well as an internal network between each workstation. The service consumer EASRecognition.exe invokes the FingerprintProcess.asmx Web service.

Figure 13 illustrates the four-tier architecture used to deploy 3S. There is the actual camera called Axis 2100, the 3S Windows application client called 3S.exe, the 3S database ThreeS_Data. mdf, the 3S server ThreeSSever.exe, and a service ThreeS.asmx. Since the 3S sever can access the remote camera through the Web service, the legacy camera interface is hidden and any applicant can access the camera easily.

## ANALYSES AND DISCUSSIONS

One of the main goals of this project is to see the benefits of service-oriented architecture by applying it in particular to the area of retail

*Figure 12. Deployment view of EAS*

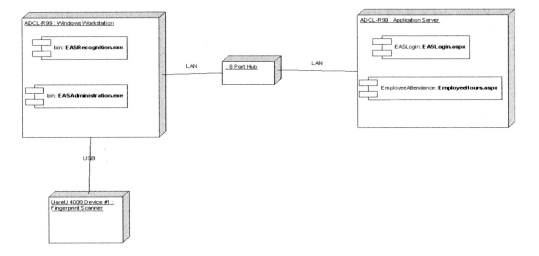

*(a) Deployment view of all subsystems of EAS*

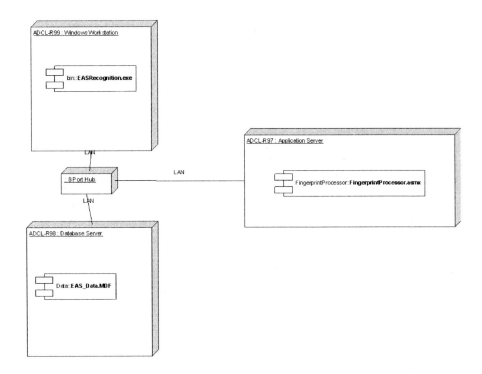

*(b) Deployment view of the recognition subsystem of EAS*

*Figure 13. Deployment of 3S*

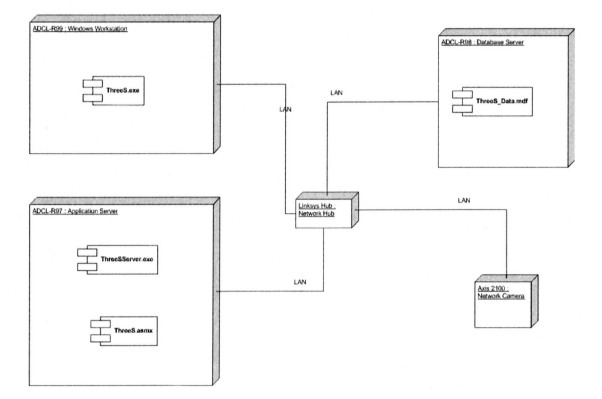

business applications. This section describes the main findings of this project that concern the effectiveness of SODI using SOA. The three systems being built for this project were each developed in similar fashion overall, using similar modeling strategies and development schedules. However, the main difference between these projects is the extent that Web services were used.

## The Role of Web Services

Using Web services for the EAS project was beneficial but less so than for the 3S project. With the EAS project, Web services were used in a fashion similar to traditional middleware, whereas the 3S project used Web services in new and interesting ways: Some functions that do not have to be changed very often are designed as a class library. Only functions that may be changed

very often and need to be shared are designed as Web services. The POS project also benefited from the use of Web services.

When building the EAS project, Web services were used for virtually all core functionality, on both the input and output side. In addition to the standard Windows forms classes and the Web services being used, a small class library was developed that contained a few helper classes. Since the main interface for EAS was written in Visual Basic (VB) .NET, and some client-side functionality was written in C#, this additional client-side functionality was stored in a class library as a proxy. A class library is just a collection of classes and interfaces bundled together into a single file. The use of this library was mainly for communicating with the Web service as a proxy and reusing existing code built in C# that would take too much time to convert to VB .NET. However, for the most part, EAS was a true wrapper for various Web service methods.

In relation to the EAS project, 3S used around 50% less Web service calls and 75% more functionality stored in class libraries. Since the 3S project was supposed to be used remotely, making long Web service calls frequently could have a significant performance impact. Instead, some of this prior Web service functionality was moved into a class library. Then, instead of using this class library on just the client side as in EAS, this new library was used by both the Web services and the Windows client. Using the class library in this fashion allowed the two main system components to share functionality without the worry of versioning conflicts or the need to manually copy classes from project to project. In a vein similar to Web services that are useful for sharing application logic between different systems, class libraries can also be used similarly. For example, in the EAS project, scheduling components were built directly into classes in the application. In the 3S project, the scheduling components were built to be more general and were stored in the class library. If developers wanted to build an extension to EAS that used scheduling, they would have to copy the scheduling classes and manually rewrite and rename them to fit into the new project. By including the scheduling functions in a class library, all you need to do is reference the .dll file and start developing. This is very efficient and a great way to increase productivity while Web services are being used as the main integration interfaces.

The POS project relied on Web services for even less than the 3S project. Since the POS system is used heavily for input purposes, it is unlikely to be used from outside your local area network; less benefit is gained from using Web services. However, Web services are still useful for logging in cashiers and authorizing them to use the POS system, and reporting on items sold and cashier activity. Without using a Web service, a rewrite of the EAS recognition system would

be required. Since this functionality already exists in a Web service, only minor modifications need to be made before the fingerprinting Web services are ready to be consumed in the POS application. Instead of depending on class libraries so much, the system interfaced directly with the SQL server being used as the data store by adopting a tightly coupled approach so as not to lose performance. This direct interface provided better transactional support then a series of Web services could provide and allowed the interface to react much more quickly to user input.

## Lessons Learned

First of all, software developers and integrators can easily transition from object-oriented analysis and design to service-oriented analysis and design since the valuable experiences of the developers in object-oriented architectures and design methodology are naturally streamlined with the service-oriented architecture and design methodology supporting loose couplings of software components. Both three-layered and multitier architectural patterns have been used to design and deploy object-oriented applications. The service-oriented architecture simply enhances the interoperability of some components that may be integrated with other remote components within organizations or across organizations by allowing standard interfaces and interaction protocols. Also, one of the design methods that have been adopted in object-oriented software developments, RUP, can be applied to service-oriented software developments with the revised 4+1 view.

Secondly, SODI goes much smoother without having to rewrite the same code you had written before on previous projects. This is possible because the Web service technology brings a BIS in which software components can easily be integrated with others (they are loosely coupled through Web services). You could just change your external interface for each situation, and then convert data into the formats supported by the system's Web services.

SODI allows developers to separate the presentation logic from the business logic in applications. A Windows desktop application or a Web application is located on the platforms of service consumers. The Web services are deployed onto the Web service platforms of service producers. The presentation tier, a Windows desktop application or a Web application, interacts with Web services that are located at remote Web service platforms. The Web services interact with various business objects. Also, the Web service can interact with other Web services that located at other platforms recursively.

Using Web services can explicitly represent the process view, which describes dynamic features of a software system. If only atomic services are used (like this project), the services are considered special classes, that is, interfaces. The objects of the interfaces are deployed onto a tier on which a Web service engine resides. Class and deployment diagrams are used to show the process view of the atomic services. If several atomic services can be composed to a composite service, the composite service can be described in an activity diagram to show the process view.

## FUTURE WORK

Web services allowed the vision of flexible interfaces to become reality rather than remain in the world of fantasy. The customization of business information systems to reflect the brand of a business will be of great benefit to anyone who runs a business. Web services hold strong promise in the future of business and the future of our daily lives as development on services progresses, taking us one step closer to a totally connected world.

Future work on improving the SOA architecture can progress as soon as new technologies are in place. As semantic description languages for services become commonplace, people can start describing the services with meaningful semantic descriptions. Once service orchestration languages start becoming feasible to implement, orchestration tool kits could be developed specifically for SODI systems. The composition of Web services will bring more explicit process views to the 4+1 view, and the dynamics of the integrated system can be understood in terms of business process workflow.

## REFERENCES

Alonso, G., Casati, F., Kuno, H., & Machiraju, V. (2004). *Web services concepts, architectures and applications.* Springer Verlag.

Association for Retail Technology Standards (ARTS). (2003). *Data model.* Retrieved from http://www.nrf-arts.org

Booch, G., Rumbaugh, J., & Jacobson, I. (1999). *The unified modeling language user guide.* Addison Wesley.

Connolly, T., & Begg, C. (2005). *Database systems: A practical approach to design, implementation, and management* (4th ed.). Addison-Wesley.

Coulouris, G., Dollimore, J., & Kindberg, T. (2002). *Distributed systems: Concepts and design.* Addison Wesley.

Erl, T. (2004). *Service-oriented architecture: A field guide to integrating XML and Web services.* Upper Saddle River, NJ: Prentice Hall Professional Technical Reference.

IBM. (2006a). *Rational unified process (RUP).* Retrieved from http://www-306.ibm.com/software/awdtools/rup/index.html

IBM. (2006b). *Web services policy framework (WS-Policy).* Retrieved from http://www-128.ibm.com/developerworks/library/specification/ws-polfram/

Jacobson, I., Booch, G., & Rumbaugh. J. (n.d.). *The unified software development process.* Addison Wesley.

Kerner, L. (2000). Biometrics and time & attendance. *Integrated Solutions.* Retrieved from http://www.integratedsolutionsmag.com/Articles/2000_11/001109.htm

Kruchten, P. B. (1995). The 4+1 view model of architecture. *IEEE Software, 12*(6), 42-50.

Linthicum, D. S. (2003a). *Next generation application integration.* Addison-Wesley.

Linthicum, D.S (2003b). Where enterprise application integration fits with Web servic*es. Software Magazine.* Retrieved from http://www.softwaremag.com/L.cfm?Doc=2003-April/Web-Services

Object Management Group (OMG). (2007). *Unified modeling language (UML).* Retrieved from http://www.uml.org/

Organization for the Advancement of Structured Information Standards (OASIS). (n.d.). *Universal description, discovery and integration (UDDI).* Retrieved from http://www.uddi.org/

Organization for the Advancement of Structured Information Standards (OASIS). (2006). *Web services security (WS-Security).* Retrieved from http://www.oasis-open.org/committees/tc_cat.php?cat=security

World Wide Web Consortium Extensible Markup Language (W3C XML) Protocol Working Group. (2006). *Simple object access protocol (SIAP).* Retrieved from http://www.w3.org/2000/xp/Group/

World Wide Web Consortium (W3C) Semantic Web Services Interest Group. (2006a). *Semantic Web services framework (SWSF).* Retrieved from http://www.w3.org/Submission/2004/07/

World Wide Web Consortium (W3C) Semantic Web Services Interest Group. (2006b). *Web ontology language for services (OWL-S).* Retrieved from http://www.w3.org/Submission/2004/07/

World Wide Web Consortium (W3C) Semantic Web Services Interest Group. (2006c). *Web services modeling ontology (WSMO).* Retrieved from http://www.w3.org/Submission/2005/06/

World Wide Web Consortium (W3C) Semantic Web Services Interest Group. (2006d). *Web services semantic (WSDL-S).* Retrieved from http://www.w3.org/Submission/2005/10/

World Wide Web Consortium (W3C) Web Services Addressing Working Group. (n.d.). *Web services addressing (WS-Addressing).* Retrieved from http://www.w3.org/Submission/ws-addressing/

World Wide Web Consortium (W3C) Web Services Description Working Group. (2006). *Web services description language (WSDL).* Retrieved from http://www.w3.org/2002/ws/desc/

# Chapter XVIII
# Realizing the Promise of RFID:
## Insights from Early Adopters and the Future Potential

**Velan Thillairajah**
*EAI Technologies, USA*

**Sanjay Gosain**
*The Capital Group Companies, USA*

**Dave Clarke**
*GXS, USA*

## ABSTRACT

*This chapter touches on some basics on RFID and covers business opportunities as well as some of the relevant challenges to be faced. Although it is a new technology, standard obstacles of organizational and technological barriers still have to be overcome and mediums crossed that were previously not traversed as on warehouse and distribution center floors yield for challenging environments from a business process as well as a technological adoption perspective. It provides a combined outlook sprinkled with key lessons from early adopters and service providers close to some of the emerging trends and implementations happening in the field. A range of benefits will evolve with the adoption of RFID within an organization and especially across the entire supply network. For now, the value is only slowing coming into focus. Various industry segments, like pharmaceutical and military applications, will provide a smoother supply chain trail for others to follow.*

## INTRODUCTION

Radio frequency identification (RFID) refers to a set of technologies that use radio waves to identify and transmit information from tagged objects. While there are several mechanisms to identify objects using RFID (http://archive.epcglobalinc.org/new_media/brochures/Technol-ogy_Guide.pdf), an important approach is to store a serial number that identifies a product, along with other product information, on a microchip that is attached to an antenna. The chip and the antenna together constitute an RFID transponder or an RFID tag. The antenna enables the chip to transmit identification information to a reader. Microchip-based RFID tags first started appearing in the late 1980s, with initial applications in areas such as access systems for office buildings and toll roads (http://archive.epcglobalinc.org/privacy_hearing.asp).

RFID technologies range from very-short-range passive RFID and short-range passive RFID to active-beacon, two-way active, and real-time locating systems (RTLS; http://www.savi.com/rfid.shtml). Low-frequency (30 KHz to 500 KHz) systems have a short reading range and are commonly used in asset tracking and security access implementations. High-frequency (850 MHz to 950 MHz, and 2.4 GHz to 2.5 GHz) systems offer long read ranges (greater than 90 feet) and high reading speeds. The range of frequencies and their general distance ranges are noted in Figure 1.

## BUSINESS OPPORTUNITIES

The opportunities from leveraging RFID technologies are varied and span a gamut of application areas. For example, in retail settings, radio tagging can help to reduce theft and loss, more easily locate items, provide suppliers with better information on real-time demand for products, and improve the speed of product distribution. While conventional bar codes need to be passed in front

(line of sight) of a scanner, RFID tags can be read remotely by a device up to 20 yards away, reducing the time and labor needed to recognize and process objects. RFID tags can also be encoded with data in addition to the basic identification. Some of the areas in which RFID tagging is expected to improve supply chain performance include the following.

*Forecasting Information:* RFID tagging has significant implications for the generation of better forecasts. Downstream data can be accurately assembled and processed for use in driving multitier forecasting processes (Lapide, 2004). Such data can include warehouse inventories and withdrawals and inventory replenishments, as well as product consumption. Retailers such as Wal-Mart, which are pushing RFID technologies, are promising their suppliers information on products as they arrive and leave their warehouses, stores, and store stockrooms, in addition to the point-of-sale data.

*On-Shelf Availability:* Past research indicates significant incidences of out-of-stock (OOS) issues in retail settings. While a considerable proportion of the problem occurs due to inadequate order processing, studies suggest a significant fraction of OOS occurs when products are actually in the store. According to a study conducted by Ander-

*Figure 1. Source: http://www.electrocom.com.au/rfid_frequencytable.htm*

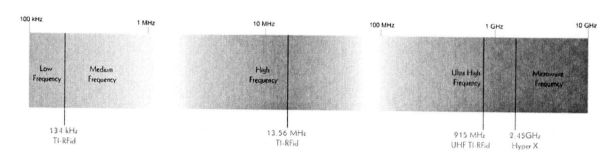

*Table 1. Source: http://members.surfbest.net/eaglesnest/rfidspct.html*

| FREQUENCY | LF<br>30-300KHz | HF<br>3-30MHz | VHF<br>30-300MHz | UHF<br>300-1000MHz | MICROWAVE<br>1GHz and Up |
|---|---|---|---|---|---|
| Distance | Less than 2 meters<br>1cm - 1.5m typical | Less than 1 meter<br>1cm - .7m typical | 1-100 meters<br>1-3m typical | 1-100 meters<br>1-3m typical | 1-300 meters<br>1-10m typical |

sen Consulting, 53% of OOS situations are based on inefficiencies in the store ordering process. Another 8% of OOS situations happen while the necessary supplies are in the backdoor inventory, but have not been shelved. Large retailers, such as U.K. supermarket chain Tesco, have begun using RFID technology to track shipments of high-value nonfood goods. To reduce OOS issues, the retailer will attach RFID tags to its own shipping trays and dollies at its national distribution center before they are loaded and sent through its supply chain to retail stores.

*Automating Proof of Delivery (PoD):* Savings for suppliers can also be found in the automation of many administrative tasks associated with product delivery (Kärkkäinen, n.d.). For example, even crate-level tagging would enable automating the proof of delivery and the invoicing process. Automatic proof of delivery would eliminate errors caused by manual processes, thus reducing costs associated with negotiating retailer claims for incomplete deliveries. Also, the arrival of tagged crates to the point of delivery could initiate an automatic invoicing process. The savings potential of automating invoice handling is much larger than what is usually imagined. For example, it is estimated to cost Procter & Gamble between $35 and $75 to process a customer invoice by traditional means due to the high number of manual interventions needed with order, billing, and shipment systems.

*Improving Product Security:* RFID tags can be used to ensure that a product is what it represents itself to be and also to identify product movement. For example, in the pharmaceutical industry, an EPC (electronic product code) and RFID pedigree system could enable authorized users to automatically identify and account for each unit of authentic medicine in real time as it enters and moves through the distribution system. It could identify the current location of all suspect products in the event of a recall, track the disposal of damaged and out-of-date products, and allow law enforcement agencies full and accurate supply chain visibility if terrorists were to launch an attack using tainted medicine (Kontnik & Dahod, 2004).

*Eliminating Stock Verification:* RFID tags can automate the process of inventory verification. As an example, Mitsubishi's high-tech library system utilizes Microchip's MCRF450 13.56 MHz tagging ICs that enable a library to monitor the distribution of books, and video and audio materials; track sign-outs and due dates, log sign-out histories and hold requests for library patrons, and prevent the theft of materials by identifying each item in the library's inventory with an irremovable RFID tag ("Mitsubishi Materials Corporation," 2003).

*Incorporating Shelf Life of Products and Product Self-Management:* Product recycling and reuse could be made easier and cheaper by shifting the responsibility for product management to the product itself. That is, a combination of information technology and product design could allow products to more or less automatically manage their end of life. RFID tags on products could be read by garbage or recycling trucks at the same time that the main can tag is being read. Using such a configuration, the collection service could, for example, provide rebates for the recycling of items (Thomas, 2003).

*Reducing Inventory Levels:* RFID is expected to lead to reduced inventory levels as firms have both the benefit of knowing what is on their shelves and better real-time forecasts based on accurate data. Thus, there is a reduced need for buffer stocks.

*Mass Customization:* RFID tags can be used to provide individual shoppers a customized product or service offering. As an example, Prada, an upscale clothing boutique, is planning an RFID pilot at one store. It announced its flagship store in Manhattan would experiment with RFID at that location (McGinity, 2004). Shoppers could choose a blouse that, once brought into the dressing room, would cause the associated RFID tag to trigger a fashion show displayed on a plasma screen that shows the potential buyer exactly what other items match that blouse, and what that blouse looked like on the runway on a model.

*Benefits from Component Tagging:* At a component level, tagging would allow better management of work-in-process inventory and also aid in the reverse logistics process.

*Integrated Forward Supply Chain Management:* In some industries, where the ability to supply a time- or application-sensitive product is a competitive advantage, tagging will allow suppliers to manage customer inventories proactively, thereby maintaining a competitive barrier to entry and a strong value proposition for the customer.

## THE CHALLENGES

### Cost

RFID reader costs can range from $100 to $3,000 or more depending on the frequency and range. Companies would need hundreds to thousands of readers to cover all their factories, warehouses, and stores. Readers typically operate at one radio frequency, and there is no consensus standard. If tags from three different manufacturers used three different frequencies, a store might have to have three readers in some locations, further increasing the cost.

RFID tags are also fairly expensive—$0.50 or more—which makes them impractical for identifying millions of items that may cost only a few dollars. RFID systems lack widely accepted and implemented standards for communication and functionality, thereby limiting their practical usefulness and keeping their system costs too high for many applications. In order to achieve significant item-level penetration within most supply chain applications, transponders will need to be priced well under $0.05.

These cost targets may be achieved only with a system-level approach that encompasses every aspect of the RFID technology, from IC design to RF protocols, from reader design to back-end data systems, and from IC manufacturing to antenna manufacturing. The challenge has been to develop a complete open-standards-based system that enables the design and manufacture of low-cost RFID systems (http://archive.epcglobalinc.org/publishedresearch/MIT-AUTOID-WH-014.pdf). Much of the focus surrounding RFID costs has been on chip or tag prices. However, implementing a fully functional system incurs multiple costs, including tags, readers, printers, middleware, infrastructure, consulting, research and development, system changes, implementation, training, change management, service-provider fees, and additional labor (Shutzberg, 2004). For early adopters, it is likely that implementations will prove more costly in the beginning stages given the likelihood of first-time mistakes and the lack of industry best practices.

## Disproportionate Investments and Benefits in Supply Chain

Studies that have simulated the financial impact of RFID implementations suggest that in most industries and scenarios, manufacturers will end up losing money, especially those that produce low-value grocery-like products. This occurs because they have to carry the cost of tagging the products. A probable exception is the electronics industry, where products have high value relative to the cost of tags. For manufacturers, losses increase as the scenarios shift from case-level tagging to blended tagging to item-level tagging, again with the exception of electronics, where the manufacturers actually increase their gains (http://workingknowledge.hbs.edu/item. jhtml?id=3651&t=dispatch ).

Retailers, on the other hand, face a very different scenario. In almost all situations, large retailers are winners. As the scenarios shift from cases to blends to items, retailers tend to realize a greater proportion of the gains. However, the retailers experience a lot of risk as the required capital expenditure of the blended and item scenarios approaches a half-billion dollars for the $10 billion retailer model. Big distributors experience sizable gains, which are greatest in the items scenario, but their suppliers experience even greater losses. Given this disparity in financial gains and required investments across supply chain players, mechanisms need to be developed for equitable redistributions that preserve incentives for all players.

## Security and Privacy Issues

RFID systems are different from other means of identification because RF communication requires no contact and does not require a clear line of sight, whereas other means of identification are either contact based or require line of sight. In other words, it is more difficult for the owner of the RF tag to physically impede communication with the tag. Consumers have expressed concerns that they do not have a choice as to when or where the technology is used or as to how it will impact them.

A second major concern is that consumers believe that the system will be abused and that this will have a negative effect on them, especially in regard to their privacy. Consumers are also concerned about the health effects of the network's radio waves, as well as shifts in employment caused by these technologies (http://archive. epcglobalinc.org/publishedresearch/cam-autoid-eb002.pdf ).

As an example, Italian clothier Benetton caused a ruckus last January when it said its apparel would be fitted with RFID tags. "People will know when I shop! What I bought...what size I wear!" were the kind of statements made by consumers. The proponents of RFID tried to point out that this information would only be available if someone was holding a scanner to that tag, which would only work inside a store. Still, given the public outcry, Benetton discontinued the trial (McGinity, 2004).

In terms of security, RFID tags are inherently less secure, mainly because of the lack of processing capacity on these tiny devices to handle much more than their core functions. A somewhat compensating aspect is that most of the data transmitted by an RFID tag are not all that meaningful unless they are considered in context. However, as location-based or personal information is linked in to that data, the possibility of a security threat emerges.

## Information Sharing Issues

A significant challenge to RFID implementations is the lack of clear standards or conventions for the sharing of data with other firms. As an example, Pfizer Inc., maker of the much-counterfeited Viagra, will place RFID tags on cases and retail packages of the impotence drug by the end of next year. The company says it is in discussions with wholesalers and retailers to work out hurdles, including data sharing. That issue has been one of the sticking points with the technology: Manufacturers, wholesalers, and retailers are looking for ways to create a detailed pedigree of the drugs' shipment history while also guarding proprietary information, such as sales volume and shipment frequency (Tesoriero, 2004). There are also industry-specific issues that may make some organizations slow to adopt RFID, for example, the need for regulatory approval of new technologies. In some cases, new labeling standards are emerging that may be in conflict with developing RFID standards.

## Compatibility, Integration, and Standardization

RFID implementations require new standards to be agreed upon for identifying objects and linking them to other related information. The Auto-ID Center, initially a project based at MIT, has proposed a universal standard for product license plates: the electronic product code. Like a bar code, the EPC is divided into numbers that identify the manufacturer, product, version, and serial number. Unlike the bar code, EPC uses an extra set of digits to identify unique items. The EPC is the only information stored on the RFID tag's microchip. This keeps the cost of the tag down and provides flexibility since an infinite amount of dynamic data can be associated with the serial number in a database. To help computer systems find and understand information about a product, the Auto-ID Center has developed associated standards and technologies. The first key element is called the object name service (ONS). ONS points a computer to an address on the Internet where information about a product is stored. The concept is based on the domain name service, which points computers to the address of particular Web sites on the World Wide Web.

## Physical and Environmental Challenges

Some implementations of RFID, in the food and drug industries, for instance, may encounter physical and environmental challenges corresponding to extremes in temperature, packaging constraints, labeling standards, and interactions between products and RFID signals.

## Data Storage and Management

The use of RFID technology will require the high-speed handling of very large data streams from readers. The processing, storage, and management of these streams and aggregated large data sets will pose a new challenge for hardware and software vendors. Much of the data stream will be repetitive information and can be ignored, but processing algorithms must still be developed to perform filtering operations, and these may be specific to applications or industries. Different industries may also have regulatory requirements that drive specific data processing and retention requirements that vary from other industries.

## MOVING FORWARD

## Global Standards Under Development

EPCGlobal, a member-driven organization, is leading the development of industry-driven standards for EPC to support the use of RFID.

## Costs are Falling

The costs of tags and readers are likely to fall in the future, particularly as standardization processes gain greater momentum, creating economies of scale. However, integration costs may still be prohibitive until system integrators develop industry-vertical solutions they can replicate.

## Consumer Education

It appears that consumers' concerns about privacy can be overcome by the following (http://archive. epcglobalinc.org/publishedresearch/cam-autoid-eb002.pdf).

- Offering consumers greater control by ensuring that they will be made aware of when and where the network is being used and offering them an option to kill the tag
- Creating governance mechanisms around the use of the network through appropriate guidelines, policies, regulations, or controls
- Offering information about the issues relating to health effects and the impact on employment

EPC RFID tag specifications specifically include the requirement that the tag will deactivate irrevocably if it receives a kill command (http://archive.epcglobalinc.org/privacy_hearing.asp).

## The Information Supply Chain

RFID technologies herald the emergence of disposable computing that will offer new challenges in scalability, flexibility, and security of the next generation as they create the possibility of tracking millions of new transactions, sending and receiving small amounts of data that track the flow of goods and services throughout the supply chain (http://cio-asia.com/pcio.nsf/unidlookup/218EE519E41C886248256D8300484409?OpenD

ocument). As more and more data become available in usable forms—particularly data generated after a product leaves the store—the information supply chain will begin to function as an independent source of revenue, generating invaluable data about product performance, consumer behavior, and logistics. This will enable businesses to be more responsive to market trends and their customers. Bundling information services with physical products, such as smart appliances, will be another key source of new value.

## Industry Adoption: Supplier Pains

Not all industries are expected to be equally likely to adopt RFID technologies. Some industries (like consumer electronics, pharmaceuticals, and toys) are likely to roll out their implementations earlier due to suitable product attributes and category economics.

In June 2003, Wal-Mart greatly advanced the process of adoption when it announced that starting in January 2005, 100 of its top suppliers would need to tag all pallets and cases going to some Wal-Mart warehouses. This mandate would be expanded throughout 2005 to include other parts of its distribution network. Wal-Mart began its RFID pilot on April 30, 2004. The eight suppliers that started shipping a handful of RFID-enabled pallets were Gillette, Hewlett-Packard, Johnson & Johnson, Kimberly-Clark, Kraft Foods, Nestle Purina PetCare, Procter & Gamble, and Unilever. The Department of Defense (DoD), Target, and Albertsons announced their future RFID-tagging requirements to their suppliers. For now, these mandates focus on pallets and cases, not on item-level tagging.

The U.S. Department of Homeland Security will soon begin using RFID technology at U.S. border checkpoints. Internationally, officials in Great Britain are discussing proposals to embed tags in vehicle license plates. The U.S. Food and Drug Administration (FDA) is pushing the pharmaceutical industry to tag medicines by

2007. Delta Airlines recently announced that it will invest $25 million to deploy disposable radio tags to track and locate lost luggage, which costs the airline $100 million annually. Las Vegas' Mc-Carran International Airport has also announced that it will begin attaching radio tags this fall to all checked luggage (Swartz, 2004).

Recently, several companies across the drug supply chain have been testing RFID deployment. Three drug makers announced plans for limited RFID rollouts. Closely held Purdue Pharma LP, maker of OxyContin, a potent opioid painkiller that is often diverted to the black market, will ship 100-tablet bottles with RFID tags to Wal-Mart and H. D. Smith Wholesale Drug Co.

Despite all of these announcements, the early adoption is not without pain for the participating companies. Wal-Mart's suppliers are split into two camps. About 30% of them are going the whole nine yards and integrating RFID into their infrastructures now. The rest are practicing a method known as "slap and ship." Essentially, to meet Wal-Mart's mandates, these companies are sticking an RFID tag on only a certain percentage of cases and pallets in warehouses that are closest to Wal-Mart's Texas distribution centers. Slap and ship involves minimal data integration and continues to leave the retail supply chain blind to product movement (Wailgum, 2004). It remains to be seen what kind of cost-benefit outcomes are realized by the two sets of suppliers.

## LESSONS FROM EARLY IMPLEMENTATIONS

Based on the experience of companies that have been rolling out early pilot implementations, we outline lessons for other companies considering RFID technologies.

## Need for Complementary Change in Business Processes

There is a clear need to integrate RFID technology within appropriate business processes. In some cases, this requires significant change in how companies manage processes such as ordering and inventory management. For example, moving from ordering to proactive replenishment would be needed to better leverage the power of instantaneous information availability. That would enable suppliers to respond quickly to changes in demand instead of accumulating information to form reasonably sized orders. The biggest value or impact will only result with a redesigned business process flow that takes advantage of the newly available information, especially at the edge locations. This traditionally is a notoriously slow and problematic process. It will be easier for new companies to start up in this mode then for existing companies to adopt it, and therein will lie some of the disruptors who emerge as a result of this technology.

## Think about Impact on Consumer Value

RFID applications in many industries are being driven by consumer needs. In the automotive industry, where RFID applications are gaining traction, the deployment of RFID solutions promises clear new functional capabilities. Michelin is starting to use tags in its tires, and there is great interest in next-generation consumer products, such as passive keyless entry systems, theft-deterrent immobilizers, and parts-marking initiatives. As in the infancy of any technology cycle, it is difficult to ascertain whether car makers are overspecifying demand and technical product content prior to the need or development of the real customer demand.

## Vertical Applications are the Key but a Current Weakness

An RFID solution consists of multiple components such as chips, tags, readers, antennas, software, and vertical market applications. Industry experts suggest that vertical market applications have been a weakness so far (Anonymous, 2003). Thus, greater effort needs to be expended in developing and deploying relevant business applications on top of the technology stack. One positive movement in this direction is the attention that has been given to the pharmaceutical industry's efforts with consumer packaged goods.

## Prepare for the Unexpected: Bug Zappers Anyone?

As with all new technologies, RFID is expected to have its share of teething issues. For example, IBM has tested RFID equipment in the backroom grocery sections of seven pilot Wal-Mart stores in support of the retailer's RFID project. During the deployment, IBM consultants encountered interference from handheld devices such as walkie-talkies, forklifts, and other devices typically found in distribution facilities. Nearby cell-phone towers, which transmit at the high end of the frequency band, sometimes leak unwanted radio waves into the RFID readers. Bug zappers in the backrooms of the test stores also caused interference (Sullivan, 2004). Organizations need to be cognizant of such teething issues that may affect their implementations. Table 2 illustrates an early sampling of the RFID spectra with the potential for interference.

## Galvanize Industry Collaboration

The lack of industry standards can be a significant roadblock to the use of RFID technologies. Some industries have taken a collaborative approach to this issue. The major pharmaceutical players, such as the world's largest pharmaceutical manufacturers, wholesale distributors, and chain drugstores, are working together as a group to explore key uses for EPC applications. This pharmaceutical industry group has also created three working teams to build a whole-industry perspective on issues relating to counterfeiting, shrinkage, and theft. EPCGlobal and its associated working groups (hardware, software, and industry verticals) are collaborating and working through real-life scenarios on what is needed and laying the groundwork for realistic standards. An example of such collaboration is the working group tasked with hammering out the specifics needed for EPC information services (ISs). VeriSign also offers a prototypical EPC IS to allow for research and development efforts and increasing familiarity with the EPC network.

Each new compliance driver, such as Wal-Mart, the DoD, and the FDA drug tracking, will both drive additional innovation and give practical industry insight that will be beneficial for all parties.

## Explore Applications through Small Pilots

Organizations should start by directing their implementations toward addressing specific goals (e.g., POS [point-of-sale] visibility for a manufacturer, or inventory shrinkage for a retailer) that create the most value for them and their customers. Certain companies with a targeted customer-service focus may find item-level RFID has operational cost savings. Marks and Spencer guarantees that it will have every size of pants and suit so that the customer will always find his or her required size. It does a nightly inventory of each item sold and replenishes accordingly. Automating inventory tracking would mean a reduction in the operational cost impact as well as an improvement in the percentage of having all stock on hand. The driver for Marks and Spencer

*Table 2. Source: http://members.surfbest.net/eaglesnest/rfidspct.htm*

| Spectrum Characteristics for RFID | | | | | |
|---|---|---|---|---|---|
| **FREQUENCY** | **LF** 30-300KHz | **HF** 3-30MHz | **VHF** 30-300MHz | **UHF** 300-1000MHz | **MICROWAVE** 1GHz and Up |
| **Reflection / Nulling** | None | Low | High | Higher | Highest |
| **"Skip" Interference Tropo Ducting** | None | High | Low | Lower | None |
| **Electrical Interference** | Very High | High | Medium | Med to Low | Low |
| **Distance** | Less than 2 meters 1cm - 1.5m typical | Less than 1 meter 1cm - .7m typical | 1-100 meters 1-3m typical | 1-100 meters 1-3m typical | 1-300 meters 1-10m typical |
| **Regulations** | Part 15 | Part 15 | Part 15 | Part 15 | Part 15 |
| **Data Rate** | 1-10KB/s | 1-3KB/s* | 1-20KB/s | 1KB-10MB/s | 1KB-10+MB/s |

**\* (Depends on regulations and distance. At very close spacing like smart cards, higher data rates are possible.)**

**Reflection is signal reinforcement or cancellation (nulling) due to direct and reflected signals being in or out of phase based on the distance between the source and target and the length of the reflected path. Enhancement is usually only 3 to 9 dB, while nulling can be total. Nulls can be -10 to -30 dB; however, signal fill from multiple reflections in most indoor environments will generally reduce nulling losses to -6 to -12 dB or so.**

**Blocking (not in the chart), on the other hand, can be total in some locations. Blocking occurs when an object larger than half a wavelength gets between the source and the target and shadows the target completely from the source. This will be more of a problem at higher frequencies (microwave) than at lower frequencies (LF, HF, or VHF).**

**Skip Interference/Tropo Ducting refers to out-of-area signals that may arrive at levels quite strong compared to the desired local signal due to refraction from the earth's ionosphere (HF) or tropospheric ducting (VHF/UHF). Tropospheric ducting occurs along the boundary between air masses of different temperatures. Consult the ARRL VHF Manual or RSGB VHF/UHF Manual for more complete descriptions of these propagation modes.**

*(http://members.surfbest.net/eaglesnest/rfidspct.htm )*

is its business rule requirement, designed to please the customers. With a business objective clearly defined, organizations should then focus their early pilots on tracking pallets and cases before even thinking of putting a tag on an item.

One of the most popular ways for piloting is limiting the magnitude and number of SKUs (stock-keeping units). Unfortunately, a key aspect, the business improvement process—identifying and mapping out changes that can have dramatic improvements—may not always surface or become visible in these pilots. Early and/or limited rollouts will definitely assist with learning how to use the information within and beyond organizational boundaries.

The pharmaceutical industry has some specific needs that allow for an easier road map for requirements implementation, but only on a state-by-state basis. There are certainly state differences, but there is an overall robust set of requirements for product tracking and traceability in CFR 21, as administered by the FDA. The FDA is often very cautious with respect to new technologies and will likely be quite slow in approving a new RFID standard for product tracking. That said, there is a clear mandate in the pharmaceutical industry for better tracking and supply chain integration that will make RFID attractive once some of the challenges are addressed.

The pedigree directive (2129) from the state of Florida is an example of a relatively simple record-keeping form for shipped drugs. RFID technologies could simplify and improve the process and reliability of tracking drug pedigrees by providing records of point of origin and destination, which are required to comply with state and federal laws. Manufacturers are hoping for clearer overall mandates from the FDA that will point the way for a consistent approach across all states, rather than requiring manual or semiautomated processes for individual states.

## OBSERVATIONS AND CONCLUSION

Analysts and end users agree that the required functions, such as EPC commissioning and central data routing, reader management, data routing, high-volume data management, advanced integration, and filtering, will get easier. Readers and related devices will get smarter and smaller, and will more easily incorporate RFID or tag data into edge and enterprise systems (ERP [enterprise resource planning], SCM, and WMS).

Enterprises will look for more flexible business processes and evolve beyond the initial static applications that allowed for RFID entry, integration, and implementation. Natural evolutionary steps will be tools like trading-partner management, process automation, and data management, which are critical components for developing and managing more dynamic composite business applications.

The military's use of active RFID in their supply chain in field operations yields some ideas for standard commercial use. In one such implementation, large payloads, such as shipping containers, have been outfitted with $70 to $100 of large-size active RFID units. With a battery-powered signal that can be read up to 100 feet, the system uses the bin or RFID number to look up the contents of the bin from the central repository database. In a theatre of operations, quickly assessing which container has the necessary supplies and accessing it are critical for supporting the mission. Similarly, as the most critical and vulnerable parts of supply chains are identified, both within an enterprise and across enterprises, RFID will evolve to provide cost savings, robustness, and operational improvements to current implementations of these vital processes.

There will be an obvious tendency to rely on the large providers and vendors because of their knowledge of the Fortune 1000 companies' business and back-end processes. An enterprise-wide

SAP infrastructure will tend to lean toward an SAP RFID extension or an SAP-centric RFID add-on. The ongoing investment in and acquisition of smaller companies in the space are indicative of the desire and needs of larger companies to expand their skill set in delivering expertise where there is not enough to go around. Different vendors are choosing various investment paths and associated road maps. Microsoft, as one example, is making some early attempts to create or forecast the need for an RFID services platform that will fit nicely within their technology stack. BizTalk and .NET provide the underlying fundamental components while Microsoft Business Solutions provides ancillary components that can be driven through Axapta. Overall, Microsoft hopes to capitalize on the medium-size business entities as they enter the RFID space after the market has matured overall and the price points are manageable for the second-tier companies.

The use of RFID information locally or at the edge should provide some real-time operational efficiency benefits. Limiting certain types of SKUs to certain dock doors or warehouse locations with immediate notification when a rule is violated will minimize errors in shipping and routing. Time intervals for expected reads can also help to qualify when expected shipments and movements should occur within an internal supply chain view.

Various tools have already been designed and other components are evolving that filter, alert, and transform RFID data reads to make intelligent use of the data for decision choices or controls that may have previously not been available. Initial deployments will help to pinpoint best practice and future uses of additional information being collected, as well as how to optimally view, filter, and act upon these data being made available on the edge (shop floor, warehouse, distribution center, etc.).

It will prove interesting as the supply and demand chain converge with the use of EPC tags and the EPC network and as we learn more about changes in customer behavior patterns. The increased visibility potentially offers new value-creation opportunities that will leverage targeted marketing and enhance the overall customer experience.

The challenge of integrating new technologies with existing business processes, or alternatively changing those business processes, will prove daunting if not prohibitive to many early adopters of RFID. As market channels begin to develop, including system integrators and enterprise software vendors who package industry solutions that they can scale and replicate, these challenges will diminish somewhat. In the interim, the big wins with RFID may belong to the disruptors who learn how to leverage this technology to challenge an industry and its dominant players. Who will emerge as the Southwest Airlines or Amazon.com of the RFID world? This may be the most interesting development to watch.

## REFERENCES

http://archive.epcglobalinc.org/new_media/brochures/Technology_Guide.pdf

http://archive.epcglobalinc.org/privacy_hearing.asp

http://www.electrocom.com.au/rfid_frequency-table.htm

http://members.surfbest.net/eaglesnest/rfidspct.htm

http://www.savi.com/rfid.shtml

http://archive.epcglobalinc.org/publishedresearch/MIT-AUTOID-WH-014.pdf

http://workingknowledge.hbs.edu/item.jhtml?id=3651&t=dispatch

http://archive.epcglobalinc.org/publishedresearch/cam-autoid-eb002.pdf

http://archive.epcglobalinc.org/publishedresearch/cam-autoid-eb002.pdf

http://archive.epcglobalinc.org/privacy_hearing.asp

http://cio-asia.com/pcio.nsf/unidlookup/

E519E41C886248256D8300484409?OpenDocument

http://members.surfbest.net/eaglesnest/rfidspct.htm

Anonymous. (2003). Swatch group makes Sokymat an automotive offer it can't refuse. *Frontline Solutions, 12*(2), 13.

Kärkkäinen, M. (n.d.). *RFID in the grocery supply chain: A remedy for logistics problems or mere hype?* Retrieved from http://www.ecr-academics.org/partnership/pdf/award/Kaerkkaeinen%20-%20ECR%20Student%20Award%202002%20-%20Bronze.pdf

Kontnik, L. T., & Dahod, S. (2004). Safe and secure. *Pharmaceutical Executive, 24*(9), 58-66.

Lapide, L. (2004). RFID: What's in it for the forecaster? *The Journal of Business Forecasting Methods & Systems, 23*(2), 16-20.

McGinity, M. (2004). RFID: Is this game of tag fair play? *Communications of the ACM, 47*(1), 15.

Mitsubishi Materials Corporation selects microchip technology RFID tagging IC for high-tech library system. (2003). *Assembly Automation, 23*(1), 91.

Shutzberg, L. (2004). Early adopters should be wary of RFID costs. *InformationWeek, 1012,* 98.

Sullivan, L. (2004). IBM shares RFID lessons. *InformationWeek, 1011,* 64.

Swartz, N. (2004). Tagging toothpaste and toddlers. *Information Management Journal, 38*(5), 22.

Tesoriero, H. W. (2004, November 16). Radio ID tags will help monitor drug supplies. *Wall Street Journal,* p. D9.

Thomas, V. M. (2003). Product self-management: Evolution in recycling and reuse. *Environmental Science & Technology, 37*(23), 5297.

Wailgum, T. (2004). Tag, you're late: Why Wal-Mart's suppliers won't make the Jan. 1 deadline for RFID tagging. *CIO, 18*(4), 1.

# Compilation of References

Acuna, S. T., & Juristo, N. (2005). *Software process modeling: International series in software engineering.* Springer.

Adhikari, R. (2002a). 10 rules for modeling business processes. *DMReview.* Retrieved from http://adtmag.com/article.asp?id=6300

Adhikari, R. (2002b). Putting the business in business process modeling. *DMReview.* Retrieved from http://adtmag.com/article.asp?id=6323

Akkiraju, R., Keskinocak, P., Murthy, S., & Wu, F. (1998). A new decision support system for paper manufacturing. *Sixth International Workshop on Project Management and Scheduling,* Istanbul, Turkey.

Alonso, G., Casati, F., Kuno, H., & Machiraju, V. (2004). *Web services concepts, architectures and applications.* Springer Verlag.

Ambrose, C., & Morello, D. (2004). *Designing the agile organization: Design principles and practices.* Gartner Group.

AMR Research. (2004). *ERP market.* Retrieved March 21, 2005, from http://www.amrresearch.com

Andrews, T., Curbera, F., Dholakia, H., Goland, Y., Klein, J., Leymann, F., et al. (2003). *Business process execution language for Web services, version 1.1.* Retrieved April 10, 2006, from http://www-128.ibm.com/developerworks/library/specification/ws-bpel/

Ankolekar, A., Burstein, M., Hobbs, J., Lassila, O., Martin, D., McIlraith, S., et al. (2001). DAML-S: Semantic markup for Web services. *Proceedings of the International Semantic Web Working Symposium (SWWS),* 411-430.

Anonymous. (2003). Swatch group makes Sokymat an automotive offer it can't refuse. *Frontline Solutions, 12*(2), 13.

Application servers. (2006). *MobileInfo.com.* Retrieved April 17, 2006, from http://www.mobileinfo.com/application_servers.htm

Aslan, E. (2002). *A COTS-software requirements elicitation method from business process models.* Unpublished master's thesis, Department of Information Systems, Informatics Institute of the Middle East Technical University, Ankara, Turkey.

Association for Retail Technology Standards (ARTS). (2003). *Data model.* Retrieved from http://www.nrf-arts.org

Aust, H. (2003). *Einführung von EAI bei der PostFinance.* Proceedings of the St. Galler Anwenderforum, St. Gallen, Switzerland.

Aversano, L., & Canfora, G. (2002). Process and workflow management: Introducing eservices in business process models. *Proceedings of the 14th International Conference on Software Engineering and Knowledge Engineering.*

Ball, M. O., Ma, M., Raschid, L., & Zhao, Z. (2002). Supply chain infrastructures: System integration and information sharing. *ACM SIGMOD Record, 31*(1).

Basili, V. R., Caldiera, C., & Rombach, H. D. (1994). Goal question metric paradigm. *Encyclopaedia of Software Engineering, 1*, 528-532.

Basu, A., & Kumar, A. (2002). Research commentary: Workflow management issues in e-business. *Information Systems Research, 13*(1).

Bath, U. (2003). *Web services als teil einer serviceorientierten architektur.* Proceedings of the EAI Forum Schweiz, Regensdorf, Switzerland.

Batory, D., & Geraci, B. J. (1997). Composition validation and subjectivity in GenVoca Generatos. *IEEE Transactions on Software Engineering, 23*(2).

Baumgarten, U. (2004). *Mobile distributed systems.* John Wiley & Sons.

Bayer, J. (2004). View-based software documentation. In *PhD theses in experimental software engineering* (Vol. 15). Stuttgart, Germany: Fraunhofer IRB Verlag.

Bayer, J., et al. (2004). *Definition of reference architectures based on existing systems, WP 5.2, lifecycle and process for family integration.* Proceedings of the Eureka Σ! 2023 Programme, ITEA Project ip02009.

Bayer, J., Flege, O., et al. (1999). PuLSE: A methodology to develop software product lines. *Proceedings of the Fifth ACM SIGSOFT Symposium on Software Reusability (SSR'99)*, 122-131.

Beall, J. (1997). Valuing difference and working with diversity. In J. Beall (Ed.), *A city for all: Valuing difference and working with diversity* (pp. 2-37). London: Zed Books Ltd.

Berners-Lee, T., Hendler, J., & Lassila, O. (2001). The semantic Web. *Scientific American.*

Besson, P. (1999). Les ERP a l'epreuve de l'organisaton. *Systemes D'Information et Management, 4*(4), 21-52.

Besson, P., & Rowe, F. (2001). ERP project dynamics and enacted dialogue: Perceived understanding, perceived leeway, and the nature of task-related conflicts. *The Data Base for Advances in Information Systems, 32*(4), 47-66.

Bhabha, H. K. (1986). The other question: Difference, discrimination and the discourse of colonialism. In F. Barker, P. Hulme, M. Iversen, & D. Loxley (Eds.), *Literature, politics and theory* (pp. 148-172). Methuen & Co. Ltd.

Bieber, M., Bartolacci, M., Fierrnestad, J., Kurfess, F., Liu, Q., Nakayama, M., et al. (1997). Electronic enterprise engineering: An outline of an architecture. *Proceedings of the Workshop on Engineering of Computer-Based Systems,* Princeton, NJ.

Bijker, W. E., & Law, J. (Eds.). (1997). *Shaping technology/building society: Studies in sociotechnical change* (2nd ed.). Cambridge, MA: The MIT Press.

Bloomfield, B. P., Coombs, R., Knights, D., & Littler, D. (Eds.). (1997). *Information technology and organizations: Strategies, networks, and integration.* Oxford University Press.

Boag, S., Chamberlin, D., Fernández, M. F., Florescu, D., Robie, J., & Siméon, J. (2005). XQuery 1.0: An XML query language. *W3C candidate recommendation.* Retrieved November 3, 2005, from http://www.w3.org/TR/2005/CR-xquery-20051103/

Boar, C. (2003). *XML Web services in the organization.* Microsoft Press.

Boehm, B., Abts, C., Brown, A. W., Chulani, S., Clark, B. K., Horowitz, E., et al. (2000). *Software cost estimation with Cocomo II.* NJ: Prentice Hall.

Booch, G., Rumbaugh, J., & Jacobson, I. (1999). *The unified modeling language user guide.* Addison Wesley.

Boyer, J., Landwehr, D., Merrick, R., Raman, T. V., Dubinko, M., & Klotz, L. (2006). XForms 1.0 (2nd ed.). *W3C recommendation.* Retrieved March 14, 2006, from http://www.w3.org/TR/xforms/

*BPMI.org releases business process modeling notation (BPMN) version 1.0.* (n.d.). Retrieved April 5, 2005, from http://xml.coverpages.org/ni2003-08-29-a.html

Brehm, L., Heinzl, A., & Markus, L. M. (2001). *Tailoring ERP systems: A spectrum of choices and their implications.* Paper presented at the 34th Hawaii Conference on Systems Sciences, HI.

Brodie, M., & Stonebraker, M. (1995). *Migrating legacy systems.* San Francisco: Morgan Kaufmann Publishers.

Brooke, C., & Maguire, S. (1998). Systems development: A restrictive practice? *International Journal of Information Management, 18*(3), 165-180.

Brynjolfsson, E., Yu, H., & Smith, M. D. (2003). Consumer surplus in the digital economy: Estimating the value of increased product variety at online booksellers. *Management Science, 49*(11), 1580-1596.

Business Process Management Initiative (BPMI). (2004). *Business process modeling notation (BPMN) specification* (Version 1.0).

Business Process Modeling Notation (BPMN). (2004, May 4). *Business process modeling notation (BPMN) information.* Retrieved April 5, 2006, from http://www.bpmn.org

Cadili, S., & Whitley, E. A. (2005). *On the interpretive flexibility of hosted ERP systems* (Working Paper Series No. 131). London: Department of Information Systems, The London School of Economics and Political Science.

Callon, M. (1986). Some elements of a sociology of translation: Domestication of the scallops and the fishermen of St Brieuc Bay. In J. Law (Ed.), *Power, action and belief: A new sociology of knowledge* (pp. 196-233). London: Routledge & Kegan Paul.

Capability Maturity Model Integration (CMMI) Product Team. (2001). *CMMI-SE/SW/IPPD, v.1.1. CMMI^SM for systems engineering, software engineering, and integrated product and process development: Staged representation* (Tech. Rep. No. CMU/SEI-2002-TR-004). Carnegie Mellon University, Software Engineering Institute.

Carlis, J., & Maguire, J. (2000). *Mastering data modeling: A user driven approach* (1st ed.). Addison- Wesley.

Carlson, D. (2001). *Modeling XML applications with UML.* Addison-Wesley.

Castano, S., et al. (2005). Ontology-based interoperability services for semantic collaboration in open networked systems. In D. Konstantas, J.-P. Bourrières, M. Léonard, & N. Boudjlida (Eds.), *Interoperability of enterprise software and applications.* Springer-Verlag.

Cavaye, A., & Christiansen, J. (1996). Understanding IS implementation by estimating power of subunits. *European Journal of Information Systems, 5,* 222-232.

Champion, M., Ferris, C., Newcomer, E., & Orchard, D. (2002). *Web services architecture* (Working draft). W3C. Retrieved from http://www.w3.org/TR/ws-arch/

Chappell, D. A. (2004). *Enterprise service bus.* Sebastopol: O'Reilly and Associates, Inc.

Charfi, A., & Mezini, M. (2004). *Service composition. Hybrid Web service composition: Business processes meet business rules.* Proceedings of the Second International Conference on Service Oriented Computing.

Chen, Q., Hsu, M., Dayal, U., & Griss, M. (2000). Multi-agent cooperation, dynamic workflow and XML for e-commerce automation. *Fourth International Conference on Autonomous Agents.*

Christensen, E., Curbera, F., Meredith, G., & Weerawarana, S. (2001). *Web services description language version 1.1* (W3C recommendation). Retrieved April 10, 2006, from http://www.w3.org/TR/wsdl

Clark, M., Fletcher, P., Hanson, J. J., Irani, R., & Thelin, J. (2002). *Web services business strategies and architectures.* Wrox Press.

Clements, P., Kazman, R., & Klein, M. (2002). *Evaluating software architectures: Methods and case studies.* Addison-Wesley.

Clemons, E. K., & Row, M. C. (1988). McKesson Drug Company. A case study of Economist: A strategic information system. *Journal of Management Information Systems, 5*(1), 36-50.

Cockburn, A. (2001). *Writing effective use cases.* Boston: Addison-Wesley.

Cockburn, A. (2004). *Writing effective use cases.* London: Addison-Wesley Professional.

Coffman, E. G., & Denning, P. J. (1973). *Operating systems theory.* Englewood Cliffs, NJ: Prentice-Hall.

Connolly, T., & Begg, C. (2005). *Database systems: A practical approach to design, implementation, and management* (4th ed.). Addison-Wesley.

Cooper, J., & Fisher, M. (2002). *Software acquisition capability maturity model (SA-CMM®) v.1.03* (Tech. Rep. No. CMU/SEI-2002-TR-010). Carnegie Mellon University, Software Engineering Institute.

Coulouris, G., Dollimore, J., & Kindberg, T. (2002). *Distributed systems: Concepts and design.* Addison Wesley.

Cummnis, F. A. (2002). *Enterprise integration: An architecture for enterprise application and systems integration.* John Wiley & Sons.

Curtis, B., Kellner, M. I., & Over, J. (1992). Process modeling. *Communications of the ACM, 35*(9), 75-90.

Date, C. J. (2000). *What not how: The business rules approach to application development.* Boston: Addison-Wesley.

Davenport, T. (1993). *Process innovation: Reengineering work through information technology.* Ernst & Young.

Davenport, T. H. (1998). Putting the enterprise into the enterprise system. *Harvard Business Review,* 121-131.

Davenport, T. H. (2000). *Mission critical: Realizing the promise of enterprise systems.* Boston: Harvard Business School Press.

Davenport, T. H., De Long, D. W., & Beers, M. C. (1999). Successful knowledge management projects. *Sloan Management Review, 39*(2), 43-57.

Dean, M., Hendler, J., Horrocks, I., McGuinness, D., Patel-Schneider, P. F., & Stein, L. A. (2004). *Semantic markup for Web services: OWL-S version 1.1.* Retrieved April 10, 2006, from http://www.daml.org/services/owl-s/1.1/

Delcambre, S. N., & Tanik, M. M. (1998). Using task system templates to support process description and evolution. *Journal of System Integration, 8,* 83-111.

Delphi. (2001). *In process: The changing role of business process management in today's economy.* Retrieved from http://www.ie.psu.edu/advisoryboards/sse/articles/a4bd42eb1.delphi-ip-oct2001.pdf

Demazeau, Y., & Müller, J.-P. (1989). Decentralized artificial intelligence. *Proceedings of the First European Workshop on Modelling Autonomous Agents in a Multi-Agent World.*

Demirörs, O., & Gencel, Ç. (2004). A comparison of size estimation techniques applied early in the life cycle. In *Lecture notes in computer science: Vol. Proceedings of the European Software Process Improvement Conference* (p. 184). Springer.

Demirörs, O., Demirörs, E., & Tarhan, A. (2001). Managing instructional software acquisition. *Software Process Improvement and Practice Journal, 6,* 189-203.

Demirörs, O., Demirörs, E., Tanik, M. M., Cooke, D., Gates, A., & Krämer, B. (1996). Languages for the specification of software. *Journal of Systems Software, 32,* 269-308.

Demirörs, O., Gencel, Ç., & Tarhan, A. (2003). Utilizing business process models for requirements elicitation. *Proceedings of the 29th Euromicro Conference,* 409-412.

Department of Defense (DoD) Architecture Working Group. (1997). *C4ISR architecture framework, version 2.0.*

Department of Defense (DoD) General Services Administration. (2001). *Federal acquisition regulation.*

Dong, L. (2000). *A model for enterprise systems implementation: Top management influences on implementation effectiveness.* Paper presented at the Americas Conference on Information Systems, Long Beach, CA.

Doolin, B. (1999). Sociotechnical networks and information management in health care. *Accounting, Management and Information Technology, 9*(2), 95-114.

Dumas, A. (2002). *Select the best approach for your EAI initiative* (Sunopsis White Paper). Sunopsis, Inc.

Dutton, J. E. (1993). Commonsense approach to process modeling. *IEEE Software, 10*(4), 56-64.

Endries, T. (2003). *Schenker AG: EAI.* Proceedings of Integration Management Day, St. Gallen, Switzerland.

Erl, T. (2004). *Service-oriented architecture: A field guide to integrating XML and Web services.* Upper Saddle River, NJ: Prentice Hall Professional Technical Reference.

Ettlinger, B. (2002, March 5). The future of data modeling. *DMReview.* Retrieved from http://www.dmreview.com/article_sub.cfm?articleid=4840

Ewalt, D. W. (2002, December 12). *BPML promises business revolution.* Retrieved from http://www.computing.co.uk/analysis/1137556

Finin, T., et al. (1993). *Specification of the KQML agent communication language: DARPA knowledge sharing initiative external interfaces working group.*

Fischer, L. (Ed.). (2005). *Workflow handbook 2005.* Lighthouse Point, FL: Future Strategies Inc.

Flege, O. (2000). *System family architecture description using the UML* (Fraunhofer IESE Report No. 092.00/E). Kaiserslautern, Germany.

Foremski, T. (1998, September 2). Enterprise resource planning: A way to open up new areas of business. *Financial Times,* p. 6.

Foundation for Intelligent Physical Agents (FIPA). (1997). Agent communication language. In *FIPA 97 specification, version 2.0.* Geneva, Switzerland: Author.

Fowler, M. (n.d.). *Patterns in enterprise software.* Retrieved November 2005 from http://www.martinfowler.com

Frankel, D. S. (2003). *Model driven architecture: Applying MDA to enterprise computing.* Wiley.

Fridgen, M., & Heinrich, B. (2004). *Investitionen in die unternehmensweite anwendungssystemintegration: Der einfluss der kundenzentrierung auf die gestaltung der anwendungslandschaft* (Working paper). Augsburg, Germany.

Friederich, M. (2003). *Zusammenspiel verschiedener integrationstechnologien und werkzeuge bei der Züricher Kantonalbank.* Proceedings of the EAI Forum Schweiz, Regensdorf, Switzerland.

Genesereth, M. R. (1997). An agent-based framework for interoperability. In Bradshaw (Ed.), *Software agents.* AAAI Press/MIT Press.

Georgeff, M., et al. (1999). The belief-desire-intention model of agency. In J.-p. Müller, M. Singh, & A. S. Rao (Eds.), *Lecture notes in artificial intelligence: Vol. 1555. Intelligent Agents V.* Berlin, Germany: Springer.

Goethals, F., Vandenbulcke, J., Lemahieu, W., Snoeck, M., De Backer, M., & Haesen, R. (2004). *Communication and enterprise architecture in extended enterprise integration.* Proceedings of the Sixth International Conference on Enterprise Information Systems, Porto, Portugal.

Gröger, S. (2003). *Enterprise application integration in the financial services industry.* Proceedings of Integration Management Day, St. Gallen, Switzerland.

Gudgin, M., Hadley, M., Mendelsohn, N., Moreau, J., & Nielsen, H. F. (2003). *Simple object access protocol version 1.2* (W3C recommendation). Retrieved April 10, 2006, from http://www.w3.org/TR/soap/

Guzmán Ruiz, F. (2001). *Mecanismos de comunicación externa para un sistema integrador de comercio electrónico de negocio a negocio.* Unpublished master's thesis, Instituto Tecnológico y de Estudios Superiores de Monterrey, Monterrey, Nuevo León, Mexico.

Haas, H., & Brown, A. (2004). *Web services glossary*. W3C. Retrieved from http://www.w3.org/TR/ws-gloss/

Hagel, J., & Brown, J. S. (2001). Your next IT strategy. *Harvard Business Review, 79*(9), 105-113.

Hagen, C., & Schwinn, A. (2006). Measured integration: Metriken für die integrationsarchitektur. In J. Schelp & R. Winter (Eds.), *Integrationsmanagement* (pp. 268-292). Berlin, Germany: Springer.

Hall, S. (Ed.). (1997). *Representation: Cultural representations and signifying practices*. Sage Publications.

Hammer, M., & Champy, J. (2001). *Reengineering the corporation: A manifesto for business revolution*. New York: HarperCollins Publishers.

Harrington, H. J. (1991). *Business process improvement: The breakthrough strategy for total quality, productivity, and competitiveness*. New York: McGraw-Hill.

Havenstein, H. (2005). Sabre replacing EDI with Web services. *Computerworld, 39*(34), 12.

Havey, M. (2005). *Essential business process modeling*. O'Reilly Media Inc.

Hofer, A. (2003). *Projekt SBB CUS: EAI ermöglicht eine erhöhung der kundeninformations-qualität im öffentlichen verkehr*. Proceedings of the St. Galler Anwenderforum, St. Gallen, Switzerland.

Hohpe, G., & Woolf, B. (2004). *Enterprise integration patterns*. Addison-Wesley.

Holland, C. P., Light, B., & Gibson, N. (1999, June). *A critical success factors model for enterprise resource planning implementation*. Paper presented at the the Seventh European Conference on Information Systems, Copenhagen, Denmark.

Humphrey, W. S. (1990). *Managing the software process*. Reading, MA: Addison-Wesley Publishing Company.

Iacovou, C. L., Benbasat, I., & Dexter, A. S. (1995). Electronic data interchange and small organizations: Adoption and impact of technology. *MIS Quarterly, 19*(4), 465-485.

IBM. (2006a). *Rational unified process (RUP)*. Retrieved from http://www-306.ibm.com/software/awdtools/rup/index.html

IBM. (2006b). *Web services policy framework (WS-Policy)*. Retrieved from http://www-128.ibm.com/developerworks/library/specification/ws-polfram/

IDC. (2004). *Worldwide ERP application market 2004-2008 forecast: First look at top 10 vendors*. Retrieved June 9, 2005, from http://www.IDC.com

IEC TC 65/290/DC. (2002). *Device profile guideline. TC65: Industrial process measurement and control*.

IEEE. (1998a). *IEEE software engineering standards*.

IEEE. (1998b). *IEEE Std. 1062: IEEE recommended practice for software acquisition*. New York.

IEEE. (1998c). *IEEE Std. 1220: IEEE standard for application and management of the system engineering process*.

Inmon, W. H. (2000). A brief history of integration. *EAI Journal*. Retrieved from http://www.eaijournal.com/applicationintegration/BriefHistory.asp

Intel Corporation. (2006). *Intel® mobile application architecture guide*. Retrieved April 17, 2006, from http://www.intel.com/cd/ids/developer/asmo-na/eng/61193.htm

ISO/IEC. (1991). *ISO/IEC 9126: Information technology. Software product evaluation: Quality characteristics and guidelines for their use*.

Jackson, M., & Twaddle, G. (1997). *Business process implementation: Building workflow systems*. Harlow, England: Addison Wesley Longman Limited.

Jackson, T., Majumdar, R., & Wheat, S. (2005). *OpenEAI methodology, version 1.0*. OpenEAI Software Foundation.

Jacobson, I., Booch, G., & Rumbaugh. J. (n.d.). *The unified software development process*. Addison Wesley.

Jena. (2003). Retrieved April 10, 2006, from http://jena.sourceforge.net

Compilation of References

Jennings, N. R. (2001). An agent-based approach for building complex software systems. *Communications of the ACM, 44*(4).

Jiménez, G. (2003). *Configuration wizards and software product lines.* Unpublished doctoral dissertation, Instituto Tecnológico y de Estudios Superiores de Monterrey, Monterrey, Nuevo León, Mexico.

Juárez Lara, N. (2001). *Integración visual de aplicaciones en sistemas de comercio electrónico de negocio a negocio.* Unpublished master's thesis, Instituto Tecnológico y de Estudios Superiores de Monterrey, Monterrey, Nuevo León, Mexico.

Juric, B. M., Mathew, & Sarang, P. (2004). *Business process execution language for Web services BPEL and BPEL4WS.* Birmingham, UK: Packt Publishing.

Kaib, M. (2002). *Enterprise application integration: Grundlagen, integrationsprodukte, anwendungsbeispiele.* Wiesbaden, Germany: DUV.

Kaplan, B., & Maxwell, J. A. (1994). Qualitative research methods for evaluating computer information systems. In J. G. Anderson, C. E. Aydin, & S. J. Jay (Eds.), *Evaluating health care information systems: Methods and applications* (pp. 45-68). Thousand Oaks, CA: Sage.

Kärkkäinen, M. (n.d.). *RFID in the grocery supply chain: A remedy for logistics problems or mere hype?* Retrieved from http://www.ecr-academics.org/partnership/pdf/award/Kaerkkaeinen%20-%20ECR%20Student%20Award%202002%20-%20Bronze.pdf

Kavantzas, N., Burdett, D., Ritzinger, G., & Lafon, Y. (2004). *Web services choreography description language (WS-CDL) version 1.0.* Retrieved April 10, 2006, from http://www.w3.org/TR/2004/WE-ws-cdl-10-20041217/

Kay, E. (1996, February 15). Desperately seeking SAP support. *Datamation*, pp. 42-45.

Keller, W. (2002). *Enterprise application integration.* Dpunkt Verlag.

Kerner, L. (2000). Biometrics and time & attendance. *Integrated Solutions.* Retrieved from http://www.integratedsolutionsmag.com/Articles/2000_11/001109.htm

Khan, R. N. (2004). *Business process management: A practical guide.* Tampa, FL: Meghan-Kiffer Press.

Kishore, R., Zhang, H., & Ramesh, R. (in press). Enterprise integration using the agent paradigm: Foundations of multi-agent-based integrative business information systems. In *Decision support systems.* Elsevier.

Klein, H. K., & Myers, M. D. (1998). *A set of principles for conducting and evaluating interpretive field studies in information systems.* Retrieved December 11, 1998, from http://www.auckland.ac.nz/msis/isworld/mmyers/klien-myers.html

Klesse, M., Wortmann, F., & Schelp, J. (2005). Erfolgsfaktoren der applikationsintegration. *Wirtschaftsinformatik, 47*(4), 259-267.

Klesse, M., Wortmann, F., & Winter, R. (2005). *Success factors of application integration: An exploratory analysis* (Working paper). St. Gallen, Switzerland: University of St. Gallen.

Knecht, R. (2003). *Application architecture framework UBS-WMBB.* Proceedings of Integration Management Day, St. Gallen, Switzerland.

Kontnik, L. T., & Dahod, S. (2004). Safe and secure. *Pharmaceutical Executive, 24*(9), 58-66.

Krafzig, D., Banke, K., & Slama, D. (2005). *Enterprise SOA: Service-oriented architecture best practices.* Upper Saddle River, NJ: Prentice Hall Professional Technical Reference.

Krallmann, H. (2003). Transformation einer industriell geprägten unternehmensstruktur zur einer service-orientierten organisation. *Proceedings des Symposiums des Instituts für Wirtschaftsinformatik "Herausforderungen der Wirtschaftsinformatik in der Informationsgesellschaft"* (pp. 1-12).

Krcmar, H. (1990). Bedeutung und ziele von informationssystemarchitekturen. *Wirtschaftsinformatik, 32*(5), 395-402.

Kreger, H. (2001). *Web services conceptual architecture.* IBM Software Group. Retrieved from http://www-106.ibm.com/developerworks

325

Kruchten, P. B. (1995). The 4+1 view model of architecture. *IEEE Software, 12*(6), 42-50.

Krumbholz, M., Galliers, J., Coulianos, N., & Maiden, N. A. M. (2000). Implementing enterprise resource planning packages in different corporate and national cultures. *Journal of Information Technology, 15*, 267-279.

Kuster, S., & Schneider, M. (2003). *Banking bus EAI-plattform der raiffeisengruppe schweiz.* Proceedings of Integration Management Day, St. Gallen, Switzerland.

Laguna, M., & Marklund, J. (2004). *Business process modeling, simulation, and design.* Prentice Hall.

Laitenberger, O. (2000). Cost-effective detection of software defects through perspective-based inspections. In *PhD theses in experimental software engineering* (Vol. 1). Stuttgart, Germany: Fraunhofer IRB Verlag

Lam, W. (2005). Exploring success factors in enterprise application integration: A case-driven analysis. *European Journal of Information Systems, 14*(2), 175-187.

Lam, W., & Shankararaman, V. (2004). An enterprise integration methodology. *IT Pro.*

Lapide, L. (2004). RFID: What's in it for the forecaster? *The Journal of Business Forecasting Methods & Systems, 23*(2), 16-20.

Latour, B. (1987). *Science in action: How to follow scientists and engineers through society.* Cambridge, MA: Harvard University Press.

Latour, B. (1988). *The pasteurization of France* (A. Sheridan & J. Law, Trans.). Harvard University Press.

Latour, B. (1991). Technology is society made durable. In J. Law (Ed.), *Sociology of monsters: Essays on power, technology and domination* (pp. 103-131). London: Routledge.

Latour, B. (1999). *Pandora's Hope: Essays on the reality of science studies.* Cambridge, MA: Harvard University Press.

Law, J. (1992). Notes on the theory of the actor-network: Ordering, strategy, and heterogeneity. *Systems Practice, 5*(4), 379-393.

Lawrence, C. P. (2005). *Make work make sense: An introduction to business process architecture.* Cape Town, South Africa: Future Managers (Pty) Ltd.

Lee, V., Schell, R., & Scheneider, H. (2004). *Mobile applications: Architecture, design, and development.* Hewlett-Packard Professional Books.

Leymann, F., Roller, D., & Schmidt, M.-T. (2002). Web services and business process management. *IBM Systems Journal, 41*(2), 198-211.

Linthicum, D. (2001). *B2B application integration.* Reading, MA: Addison Wesley.

Linthicum, D. S. (2000). *Enterprise application integration.* Reading, MA: Addison-Wesley.

Linthicum, D. S. (2003). *Next generation application integration: From simple information to Web services.* Addison-Wesley Information Technology Series.

Linthicum, D.S (2003b). Where enterprise application integration fits with Web services. *Software Magazine.* Retrieved from http://www.softwaremag.com/L.cfm?Doc=2003-April/WebServices

Liske, C. (2003). *Advanced supply chain collaboration enabled bei EAI.* Proceedings of the St. Galler Anwenderforum, St. Gallen, Switzerland.

Longueuil, D. (2003). *Wireless messaging demystified: SMS, EMS, MMS, IM, and others.* McGraw-Hill.

Madachy, R. J., & Boehm, B. W. (2006). *Software process modeling with system dynamics.* Wiley-IEEE Press.

Malhotra, Y. (1996). *Enterprise architecture: An overview.* @BRINT Research Institute. Retrieved from http://www.brint.com/papers/enterarch.htm

Mallick, M. (2003). *Mobile and wireless design essentials.* Wiley Publishing Inc.

Mandell, D., & McIlraith, S. (2003). Adapting BPEL4WS for the semantic Web: The bottom-up approach to Web service interoperation. *Proceedings of the 2nd International Semantic Web Conference (ISWC2003).*

Mangan, P., & Sadiq, S. (2002). On building workflow models for flexible processes. *Australian Computer Science Communications, Proceedings of the Thirteenth Australasian Conference on Database Technologies, 24*(2).

Manzer, A. (2002, June). Towards formal foundations for value-added chains through core-competency integration over Internet. *The Sixth World Conference on Integrated Design and Process Technology.*

Manzer, A., & Dogru, A. (2002, April). Formal modeling for the composition of virtual enterprises. *Proceedings of IEEE TC-ECBS & IFIP WG10.1, International Conference on Engineering of Computer-Based Systems,* Lund, Sweden.

Markus, M. L. (1983). Power, politics and MIS implementation. *Communication of the ACM, 26*(6), 430-444.

Markus, M. L., & Tanis, C. (2000). The enterprise system experience: From adoption to success. In R. W. Zmud (Ed.), *Framing the domains of IT research: Glimpsing the future through the past.* Cincinnati, OH: Pinnaflex Educational Resources, Inc.

Markus, M. L., Axline, S., Petrie, D., & Tanis, C. (2000). Learning from adopters' experiences with ERP: Problems encountered and success achieved. *Journal of Information Technology, 15,* 245-265.

Markus, M. L., Tanis, C., & Fenema, P. C. v. (2000). Multisite ERP implementations. *Communications of the ACM, 43*(4), 42-46.

Martin, R., & Robertson, E. (2000). *A formal enterprise architecture framework to support multi-model analysis.* Proceedings of the Fifth CAiSE/IFIP8.1 International Workshop on Evaluation of Modeling Methods in Systems Analysis and Design, Stockholm, Sweden.

Martin, W. (2005). *Business integration 2005: Status quo and trends.* Proceedings of EAI Competence Days Road Show.

McAfee, A. (2005). Will Web services really transform collaboration? *MIT Sloan Management Review, 46*(2), 78-84.

McDavid, D. W. (1999). A standard for business architecture description. *IBM Systems Journal, 38*(1), 12-31.

McGinity, M. (2004). RFID: Is this game of tag fair play? *Communications of the ACM, 47*(1), 15.

McIlraith, S., & Son, T. C. (2002). Adapting Golog for composition of semantic Web services. *Proceedings of the Eighth International Conference on Knowledge Representation and Reasoning (KR2002),* 482-493.

McIlraith, S., Son, T. C., & Zeng, H. (2001). Semantic Web services. *IEEE Intelligent Systems, 16*(2), 46-53.

McKeen, J. D., & Smith, H. A. (2002). New developments in practice II: Enterprise application integration. *Communications of the Association for Information Systems, 8,* 451-466.

Meadows, B., & Seaburg, L. (2004, September 15). *Universal business language 1.0.* OASIS. Retrieved February 14, 2006, from http://docs.oasis-open.org/ubl/cd-UBL-1.0/

Medjahed, B., Benatallah, B., Bouguettaya, A., Ngu, A. H., & Elmagarmid, A. K. (2003). Business-to-business interactions: Issues and enabling technologies. *The International Journal on Very Large Data Bases, 12*(1).

Mendel, B. (1999). Overcoming ERP projects hurdles. *InfoWorld, 21*(29).

Mitra, N. (2003, June 24). *SOAP version 1.2 part 0: Primer.* W3C. Retrieved June 10, 2004, from http://www.w3.org/TR/2003/REC-soap12-part0-20030624/

Mitsubishi Materials Corporation selects microchip technology RFID tagging IC for high-tech library system. (2003). *Assembly Automation, 23*(1), 91.

Molina, A., Mejía, R., Galeano, N., & Velandia, M. (2006). The broker as an enabling strategy to achieve smart organizations. In *Integration of ICT in smart organizations.* Idea Group Publishing.

Moll, T. (2003). *Firmenübergreifendes EAI-netzwerk: Integrierte umsetzung von geschäftsprozessen über einen marktplatz als EAI-hub.* Proceedings of Integration Management Day, St. Gallen, Switzerland.

Monteiro, E. (2000a). Actor-network theory and information infrastructure. In C. U. a. o. Ciborra (Ed.), *From control to drift* (pp. 71-83). New York: Oxford University Press.

Monteiro, E. (2000b). Monsters: From systems to actor-networks. In K. Braa, C. Sorensen, & B. Dahlbom (Eds.), *Planet Internet.* Lund, Sweden: Studentlitteratur.

Murphy, G., & Notkin, D. (1995). Software reflexion models: Bridging the gap between source and high-level models. *Proceedings of the Third Symposium on the Foundations of Software Engineering (FSE3).*

Nau, D. S., Cao, Y., Lotem, A., & Muñoz-Avila, H. (1999). SHOP: Simple hierarchical ordered planner. *Proceedings of the International Joint Conference on Artificial Intelligence (IJCAI-99),* 968-973.

Norris, G. (1998). *SAP: An executive's comprehensive guide.* New York: J. Wiley.

O'Riordan, D. (2002). Business process standards for Web services. *Web Services Architect.* http://www.webservicesarchitect.com.

Object Management Group (OMG). (2007). *Unified modeling language (UML).* Retrieved from http://www.uml.org/

*Oracle BPEL PM.* (2005). Retrieved April 10, 2006, from http://www.oracle.com/technology/products/ias/bpel/index.html

Organization for the Advancement of Structured Information Standards (OASIS). (n.d.). *Universal description, discovery and integration (UDDI).* Retrieved from http://www.uddi.org/

Organization for the Advancement of Structured Information Standards (OASIS). (2006). *Web services security (WS-Security).* Retrieved from http://www.oasis-open.org/committees/tc_cat.php?cat=security

Orlikowski, W. J., & Baroudi, J. J. (1991). Studying information technology in organizations: Research approaches and assumptions. *Information Systems Research, 2*(1), 1-28.

Osman, T., Wagealla, W., & Bargiela, A. (2004). An approach to rollback recovery of collaborating mobile agents. *IEEE Transactions on Systems, Man, and Cybernetics, 34,* 48-57.

Österle, H., Brenner, W., & Hilbers, K. (1992). *Unternehmensführung und informationssystem: Der ansatz des St. Galler informationssystem-managements.* Stuttgart, Germany: Teubner.

Pande, P. S., Neuman, R. P., & Cavanagh, R. R. (2000). *The six sigma way: How GE, Motorola, and other top companies are honing their performance.* New York: Graw-Hill.

Papazoglou, M. P. (2003). Service-oriented computing: Concepts, characteristics and directions. *Proceedings of the Fourth International Conference on Web Information Systems Engineering (WISE 2003),* 3-12.

Pechoucek, M., Vokrinek, J., & Becvar, P. (2005). ExPlanTech: Multiagent support for manufacturing decision making. *IEEE Intelligent Systems, 20*(1).

*Pellet.* (2004). Retrieved April 10, 2006, from http://www.minswap.org/2003/pellet/

*Process Family Engineering in Service-Oriented Applications (PESOA).* (n.d.). Retrieved November 2005 from http://www.pesoa.org

*PuLSE™.* (n.d.). Retrieved November 2005 from http://www.iese.fhg.de/pulse

Quattrone, P., & Hopper, T. (2005). A "time-space odyssey": Management control systems in two multinational organisations. *Accounting, Organizations & Society, 30*(7/8), 735-764.

Reingruber, M. C., & Gregory, W. W. (1994). *The data modeling handbook: A best practice approach to building quality data models.* Wiley & Sons.

Ross, J. W., & Vitale, M. R. (2000). The ERP revolution: Surviving vs. thriving. *Information Systems Frontiers, 2*(2), 233-241.

Ross, R. G. (2003). *Principles of the business rule approach.* Boston: Addison-Wesley.

Ruh, W. A., Maginnis, F. X., & Brown, W. J. (2001). *Enterprise application integration.* New York: John Wiley & Sons Inc.

Sadiq, S., Orlowska, M., Sadiq, W., & Foulger, C. (2004). Data flow and validation in workflow modeling. *Proceedings of the Fifteenth Conference on Australasian Database, 27.*

Sambamurthy, V., Bharadwaj, A., & Grover, V. (2003). Shaping agility through digital options: Reconceptualizing the role of information technology in contemporary firms. *MIS Quarterly, 27*(2), 237-263.

SAP AG. (2004). *SAP annual report 2004.* Author.

SAP's rising in New York. (1998, August 1). *The Economist.*

Sauter, J. A., & Van Dyke Parunak, H. (1999). *ANTS in the supply chain.* Workshop on Agent Based Decision Support for Managing the Internet-Enabled Supply Chain, Agents 99, Seattle, WA.

Sawhney, M. (2001). Don't homogenize, synchronize. *Harvard Business Review.*

Sawyer, S. (2001a). Effects of intra-group conflict on packaged software development team performance. *Information Systems Journal, 11,* 155-178.

Sawyer, S. (2001b). *Socio-technical structures in enterprise information systems implementation: Evidence from a five year study.* Paper presented at the IEEE EMS International Engineering Management Conference, Albany, NY.

Scheer, A. G. (2003). *ARIS toolset, version 6.2.*

Scheer, A. W. (Ed.). (1994). *Business process engineering: Reference models for industrial enterprises.* Berlin, Germany: Springer.

Schiller, J. (2000). *Mobile communications.* Addison Wesley.

Schmidt, J. G. (2002). Transforming EAI from art to science. *EAI Journal.*

Sharif, A. M., Elliman, T., Love, P. E. D., & Badii, A. (2004). Integrating the IS with the enterprise: Key EAI research challenges. *The Journal of Enterprise Information Management, 17*(2), 164-170.

Sharp, A., & Mcdermott, P. (2001). *Workflow modeling: Tools for process improvement and application development.* Norwood, MA: Artech House.

Shegalov, G., Gillmann, M., & Weikum, M. (2001). XML-enabled workflow management for e-services across heterogeneous platforms. *The VLDB Journal: The International Journal on Very Large Data Bases, 10*(1).

Shutzberg, L. (2004). Early adopters should be wary of RFID costs. *InformationWeek, 1012,* 98.

Siau, K., & Tian, Y. (2004). *Supply chains integration: Architecture and enabling technologies.*

Siegal, J. (2002). *Mobile: The art of portable architecture.* Princeton Architectural Press.

Simsion, G. (2000). *Data modeling essentials: A comprehensive guide to data analysis, design, and innovation* (2nd ed.). Coriolis Group Books.

Singh, M. P., & Huhns, M. N. (2005). *Service-oriented computing: Semantics, processes, agents.* John Wiley & Sons.

Sirin, E., Hendler, J., & Parsia, B. (2003). Semi-automatic composition of Web services using semantic descriptions. *Web Services: Modeling, Architecture and Infrastructure Workshop in ICEIS.*

Smith, H. (2003, September 22). *Business process management 101.* Retrieved from http://www.ebizq.net/topics/bpm/features/2830.html

Smith, H., & Fingar, P. (2003a). *Business process management (BPM): The third wave* (1st ed.). Meghan-Kiffer Press.

Smith, H., & Fingar, P. (2003b). *Workflow is just a pi process.* Retrieved from http://www.fairdene.com/picalculus/workflow-is-just-a-pi-process.pdf

Smith, H., & Fingar, P. (2004, February 1). *BPM is not about people, culture and change: It's about technology.*

Retrieved from http://www.avoka.com/bpm/bpm_articles_dynamic.shtml

Smith, R. G. (1980). The contract net protocol. *IEEE Transactions on Computers, C29*(12).

*Software inspektion.* (n.d.). Retrieved March 2006 from http://www.software-kompetenz.de

Su, O. (2004). *Business process modeling based computer-aided software functional requirements generation.* Unpublished master's thesis, Department of Information Systems, Informatics Institute of the Middle East Technical University, Ankara, Turkey.

Sullivan, L. (2004). IBM shares RFID lessons. *InformationWeek, 1011,* 64.

Swanson, E. B., & Ramiller, N. C. (2004). Innovating mindfully with information technology. *MIS Quarterly, 28*(4), 553-583.

Swartz, N. (2004). Tagging toothpaste and toddlers. *Information Management Journal, 38*(5), 22.

Tanik, U. (2001). *A framework for Internet enterprise engineering based on t-strategy under zero-time operations.* Unpublished master's thesis, Department of Electrical Engineering, University of Alabama at Birmingham.

Teo, H. H., Wei, K. K., & Benbasat, I. (2003). Predicting intention to adopt interorganizational linkages: An institutional perspective. *MIS Quarterly, 27*(1), 19-49.

Tesoriero, H. W. (2004, November 16). Radio ID tags will help monitor drug supplies. *Wall Street Journal,* p. D9.

Thakker, D., Osman, T., & Al-Dabass, D. (2005). Web services composition: A pragmatic view of the present and the future. *Nineteenth European Conference on Modelling and Simulation: Vol. 1. Simulation in wider Europe,* 826-832.

*The European enterprise application integration market.* (2001). Frost & Sullivan Studies.

The Standish Group. (1994). *CHAOS report.* Retrieved from http://www.standishgroup.com/sample_research/chaos_1994_1.php

Themistocleous, M., & Irani, Z. (2001). Benchmarking the benefits and barriers of application integration. *Journal of Benchmarking, 8*(4), 317-331.

Themistocleous, M., & Irani, Z. (2003). Towards a novel framework for the assessment of enterprise application integration packages. *Proceedings of the 36th Hawaii International Conference on System Sciences (HICSS'03).*

Themistocleous, M., Irani, Z., & O'Keefe, R. M. (2001). ERP and application integration: Exploratory survey. *Business Process Management Journal, 7*(3), 195-204.

Thomas, M., Redmond, R., Yoon, V., & Singh, R. (2005). A semantic approach to monitor business process performance. *Communications of the ACM, 48*(12).

Thomas, V. M. (2003). Product self-management: Evolution in recycling and reuse. *Environmental Science & Technology, 37*(23), 5297.

Torchiano, M., & Bruno, G. (2003). Article abstracts with full text online: Enterprise modeling by means of UML instance models. *ACM Sigsoft Software Engineering Notes, 28*(2).

Traverso, P., & Pistore, M. (2004). Automated composition of semantic Web services into executable processes. *Proceedings of Third International Semantic Web Conference (ISWC2004)* (pp. 380-394).

United Kingdom Software Metrics Association (UKSMA). (1998). *MkII function point analysis counting practices manual, v.1.3.1.*

Van der Aalst, W. M. P., Dumas, M., ter Hofstede, A. H. M., & Wohed, P. (2002). *Pattern-based analysis of BPML (and WSCI)* (Tech. Rep. No. FIT-TR-2002-05). Brisbane, Australia: Queensland University of Technology.

Van Dyke Parunak, H. (1999). Industrial and practical applications of DAI. In G. Weiss (Ed.), *Multi-agent systems.* Cambridge, MA: MIT Press.

Vinoski, S. (2002). Middleware "dark matter." *IEEE Distributed Systems Online.* Retrieved from http://dsonline.computer.org/0210/d/wp5midd.html

Wailgum, T. (2004). Tag, you're late: Why Wal-Mart's suppliers won't make the Jan. 1 deadline for RFID tagging. *CIO, 18*(4), 1.

Walker, S. S., Brennan, R. W., & Norrie, D. H. (2005). Holonic job shop scheduling using a multiagent system. *IEEE Intelligent Systems, 20*(1).

Walsham, G. (1995). The emergence of interpretivism in IS research. *Information Systems Research, 6*(4), 376-394.

Walsham, G. (2001). *Making a world of difference: IT in a global context.* John Wiley & Sons Ltd.

Ward, J., & Peppard, J. (1996). Reconciling the IT/ business relationship: A troubled marriage in need of guidance. *Journal of Strategic Information Systems, 5*, 37-65.

Weske, M., Goesmann, T., Holten, R., & Striemer, R. (1999). A reference model for workflow application development processes. *ACM Sigsoft Software Engineering Notes, Proceedings of the International Joint Conference on Work Activities Coordination and Collaboration, 24*(2).

Whalen, M. W., & Heimdahl, M. P. E. (1999). On the requirements of high-integrity code generation. *High-Assurance Systems Engineering: Proceedings of the Fourth IEEE International Symposium*, 217-224.

White, P., & Grundy, J. (2001). Experiences developing a collaborative travel planning application with .NET Web services. *Proceedings of the 2003 International Conference on Web Services (ICWS)*.

Wiegers, K. E. (1999). *Software requirements.* Microsoft Press.

Wigand, R. T., Steinfield, C. W., & Markus, M. L. (2005). Information technology standards choices and industry structure outcomes: The case of the U.S. home mortgage industry. *Journal of Management Information Systems, 22*(2), 165-191.

Wing, L., & Shankararaman, V. (2004). An enterprise integration methodology. *IT Professional, 6*(2), 40-48.

Winter, R. (2003a). *An architecture model for supporting application integration decisions.* Proceedings of 11th European Conference on Information Systems (ECIS), Naples, Italy.

Winter, R. (2003b). Modelle, techniken und werkzeuge im business engineering. In H. Österle & R. Winter (Eds.), *Business engineering: Auf dem weg zum unternehmen des informationszeitalters* (pp. 87-118). Berlin, Germany: Springer.

Winter, R. (2006). Ein modell zur visualisierung der anwendungslandschaft als grundlage der informations-system-architekturplanung. In J. Schelp & R. Winter (Eds.), *Integrationsmanagement* (pp. 1-29). Berlin, Germany: Springer.

Wohed, P., et al. (2003, April). *Pattern based analysis of EAI languages: The case of the business modeling language.* Proceedings of the International Conference on Enterprise Information Systems (ICEIS), Angers, France.

Wooldridge, M. (2002). *An introduction to multiagent systems.* John Wiley & Sons.

World Wide Web Consortium (W3C) Semantic Web Services Interest Group. (2006a). *Semantic Web services framework (SWSF).* Retrieved from http://www.w3.org/Submission/2004/07/

World Wide Web Consortium (W3C) Semantic Web Services Interest Group. (2006b). *Web ontology language for services (OWL-S).* Retrieved from http://www.w3.org/Submission/2004/07/

World Wide Web Consortium (W3C) Semantic Web Services Interest Group. (2006c). *Web services modeling ontology (WSMO).* Retrieved from http://www.w3.org/Submission/2005/06/

World Wide Web Consortium (W3C) Semantic Web Services Interest Group. (2006d). *Web services semantic (WSDL-S).* Retrieved from http://www.w3.org/Submission/2005/10/

World Wide Web Consortium (W3C) Web Services Addressing Working Group. (n.d.). *Web services ad-*

dressing (WS-Addressing). Retrieved from http://www.w3.org/Submission/ws-addressing/

World Wide Web Consortium (W3C) Web Services Description Working Group. (2006). *Web services description language (WSDL)*. Retrieved from http://www.w3.org/2002/ws/desc/

World Wide Web Consortium (W3C). (2005). *Extensible markup language (XML) activity statement.* Retrieved March 10, 2006, from http://www.w3.org/XML/Activity.html

World Wide Web Consortium Extensible Markup Language (W3C XML) Protocol Working Group. (2006). *Simple object access protocol (SIAP)*. Retrieved from http://www.w3.org/2000/xp/Group/

Wu, D., Parsia, B., Sirin, E., Hendler, J., & Nau, D. (2003). Automating DAML-S Web services composition using SHOP2. *Proceedings of 2nd International Semantic Web Conference (ISWC2003).*

Xia, W., & Lee, G. (2004). Grasping the complexity of IS development projects. *Communications of the ACM, 47*(5), 95-74.

Yoo, M.-J. (2004, October). *Enterprise application integration and agent-oriented software integration.* IEEE International Conference on Systems, Man, and Cybernetics, The Hague, Netherlands.

Yoo, M.-J., Sangwan, R., & Qiu, R. (2005). Enterprise integration: Methods and technologies. In P. Laplante & T. Costello (Eds.), *CIO wisdom II: More best practices.* Prentice-Hall.

Youngs, R., Redmond-Pyle, D., Spass, P., & Kahan, E. (1999). A standard for architecture description. *IBM Systems Journal, 38*(1), 32-50.

Yourdon, E. (2000). *Managing software requirements.* Addison Wesley Publishing Company.

Zachman, J. A. (1987). A framework for information systems architecture. *IBM Systems Journal, 26*(3), 276-292.

Zachman, J. A. (2001). You can't "cost-justify" architecture. *DataToKnowledge Newsletter (Business Rule Solutions LLC), 29*, 3.

Zahavi, R. (2000). *Enterprise application integration with CORBA.* New York: John Wiley & Sons.

Zambonelli, F., Jennings, N. R., & Wooldridge, M. (2000). Organizational abstractions for the analysis and design of multi-agent systems. In *Lecture notes in computer science: Vol. 1957. First International Workshop on Agent-Oriented Software Engineering.* Springer-Verlag.

# About the Contributors

**Wing Lam** is an associate professor at Universitas 21 Global, a leading online graduate school, where he is also program director for the master's in management in information technology. In addition to academic positions in universities in Singapore and the United Kingdom, Dr. Lam has held consultancy positions with Logica-CMG, ICL (now Fujitsu), and Accenture. His current research interests include enterprise integration, knowledge management, and software engineering management. Dr. Lam has over 80 publications in peer-reviewed journals and conference proceedings, and currently serves on the editorial boards of several journals.

**Venky Shankararaman** is a practice associate professor at the School of Information Systems, Singapore Management University, Singapore. His current areas of specialization include enterprise architecture, enterprise integration, service-oriented architecture, and business-process management. He has over 14 years of experience in the IT industry in various capacities as a researcher, academic faculty member, and industry consultant. Shankararaman has designed and delivered professional courses for governments and industries in areas such as enterprise architecture, technical architecture, enterprise integration, and business-process management. He also worked as a faculty member at several universities in the United Kingdom and Singapore, where he was actively involved in teaching and research in the areas of intelligent systems and distributed systems. He has published over 50 papers in academic journals and conferences.

\* \* \*

**David Al-Dabass** holds the chair of intelligent systems in the School of Computing and Informatics, Nottingham Trent University, UK. His research work explores algorithms and architecture for machine intelligence.

**Antonio de Amescua-Seco** holds a BS in computer science and a PhD in computer science from the Polytechnic University of Madrid. He has been a full professor in the Department of Computer Science at the Carlos III Technical University of Madrid since 1991. Previously, he worked as a researcher at the Polytechnic University of Madrid from 1983 to 1991. He has also been working as a software engineer for the public company Iberia Airlines, and as a software engineering consultant for the private company Novotec Consultores. Amescua-Seco's research in new software engineering methods has appeared in published papers and conferences. He was the research project leader for the development of the Information System Development Methodology for the Spanish Administration and participated in other projects sponsored by the European Union.

**Michalis Anastasopoulos** is a member of the Department of Product Line Architectures at the Fraunhofer Institute for Experimental Software Engineering (IESE). Since 2000, he has been involved in technology transfer projects in the fields of product line architectures and implementation, as well as configuration and variability management. He received a diploma in mathematics from the University of Patras, Greece, and in software engineering from the Technical University of Dresden.

**Zachary Bylin** received his AA degree the same day he graduated from Auburn High School. He then received his BS degree at the University of Washington, Tacoma (UWT), just 2 years later in 2003. He received his master's degree in computer software systems on his 21st birthday from the University of Washington, Tacoma, in June of 2004. He then went on to work for the Boeing Company as a software engineer until he passed away at the age of 22 on July 3, 2005, in Brewster, Washington. Mr. Bylin was an excellent student and software developer while he was working with Dr. Chung's Intelligent Service Oriented Computing Research Group.

**Brian H. Cameron** is an assistant professor of information sciences and technology at The Pennsylvania State University. Prior to joining Penn State, he was director of information technology for WorldStor, Inc., a storage service provider (SSP) in Fairfax, Virginia. He has also held a variety of technical and managerial positions within IBM and Penn State. His primary research and consulting interests include enterprise systems integration, storage networking, emerging wireless technologies, and the use of simulations and gaming in education. He has designed and taught a variety of courses on topics that include networking, systems integration, storage networking, project management, and IT consulting.

**Sam Chung** currently teaches at the Institute of Technology. Prior to joining the UWT faculty, he taught at Pacific Lutheran University and the University of Texas of the Permian Basin. His main research interests are in intelligent service-oriented computing.

**Andrew P. Ciganek** is approaching the final stages of his PhD program at the University of Wisconsin, Milwaukee. His research focus is on the adoption, diffusion, and implementation of service-oriented architectures (SOAs) and other Net-enabling initiatives, mobile computing devices, knowledge management systems, and the role that the social context of an organization has in information systems research. He has published works in a number of refereed conference proceedings and has manuscripts under various stages of review with scholarly IS journals.

**Dave Clarke** is a senior executive with more than 20 years of experience in information technology, manufacturing, product development, emergency management, and strategic planning. He has expertise in several industries, including pharmaceutics, automotives, chemicals, biotechnology, electronics, information technology, consumer products, and nonprofit. Prior to joining Tatum, Mr. Clarke served as CTO of the American Red Cross and CIO (chief information officer) for multiple business units at General Motors, where he propelled GM to a world leadership position in the use of high-performance computing for product engineering and created the governance framework for GM North America's $1 billion annual outsourcing deal with EDS. He was the first-ever CIO and chief knowledge officer for W. L. Gore & Associates and was instrumental in developing a knowledge- and technology-based product development strategy that doubled corporate revenues in just 3 years. Mr. Clarke was also a consultant and project manager with IBM. He has published articles in *CIO* and *Computerworld*, and his work has been featured in *CIO, Computerworld, Forbes,* and *Fast Company*. He is a recognized thought leader on collaboration, knowledge management, vendor management, innovation, and project management. Mr. Clarke is an adjunct professor of organizational development and knowledge management at the George Mason University School of Public Policy and has also lectured at the University of Virginia. He is president of the Washington Area CTO Roundtable and serves on the advisory boards of several organizations. Mr. Clarke earned his BS in electrical engineering from the University of Kentucky and his MS in computer science from The Johns Hopkins University.

**Sergio Davalos** currently teaches at the Milgard School of Business. Prior to joining the UWT faculty, he taught at Portland State University, the University of the Virgin Islands, and the University of Portland. His main research interests are in the organization and management of information systems and the application of machine learning and evolutionary computing.

**Onur Demirörs** is an associate professor and the chair of the Department of Information Systems at the Middle East Technical University (METU). He is also the strategy director of Bilgi Grubu Ltd. He holds a PhD in computer science from Southern Methodist University (SMU). He has been working in the domain of software engineering as a consultant, academician, and researcher for the last 15 years. His work focuses on software process improvement, software project management, software engineering education, software engineering standards, and organizational change management.

**Ali H. Dogru** is an associate professor of computer engineering at METU. His current research is mostly in the component orientation area. He took part in large-scale software development projects and technological training for engineering organizations in different countries. Dr. Dogru obtained his BS and MS in electrical engineering from the Istanbul Technical University and from the University of Texas at Arlington, and his PhD in computer science from SMU in 1992.

**Randall E. Duran** is the CEO (chief executive officer) of Catena Technologies Pte Ltd., a consulting firm geared toward business-process automation and enterprise integration architectures. Over the past 15 years, Duran has worked with financial institutions in the United States, Europe, Asia, Australia, and Africa, designing and implementing enterprise solutions. Prior to founding Catena Technologies, he led solution consulting at TIBCO Finance Technology Inc. and Reuters PLC. Duran was been awarded BS and MS degrees by the Massachusetts Institute of Technology (MIT) in computer science.

**Amany R. Elbanna** is an associate research fellow in the Department of Information Systems, The London School of Economics. She holds a PhD in information systems from The London School of Economics in addition to an MBA and MS in the analysis, design, and management of information systems. She has conducted research on the implementation of enterprise resource planning (ERP) systems and presented her work at several conferences including ECIS, IFIP 8.6, and Bled. Her research interests include the interaction between technology and organizations, the management of large IS projects, and the management of change associated with the adoption and appropriation of ICT.

**Javier M. Espadas-Pech** is a postgraduate student and research assistant in information technology at the Instituto Tecnológico y de Estudios Superiores de Monterrey (ITESM) in Monterrey, Nuevo León, México. He is also the project technical leader of the electronic services platform. His current interests are software architectures, service-oriented architectures, and application integration technologies.

**Javier García-Guzmán** received an engineering degree and PhD in computer science at Carlos III University of Madrid. He has 7 years of experience as a software engineer and consultant in public and private companies. He has participated in numerous research projects, financed with public (European and national) and private funds, in relation to software process improvement and its integration with organizational business processes. García-Guzmán has published books and international scientific papers related to software engineering and collaborative working environments. His research interests are collaborative working environments, the formal measurement of processes improvement, software process maturity assessments, software capacity rapid audits, and knowledge management related to process improvement, software engineering, and systems integration. He belongs to the Collaborative Working Environments Expert Group (Collaboration@Work), a task force founded by the European Commission (EC) Directorate General of Information Society Technologies and Media, in charge of defining the main research lines related to collaborative working environments for the definition of goals related to European research and development work programs.

**Çiğdem Gencel** works as a part-time instructor at the Department of Information Systems at Middle East Technical University. She holds a PhD degree in the same department with a focus on software size measurement. She has been working in the domain of software engineering as an academician for 5 years and as an instructor for 1 year. Her research interests involve software metrics, software size estimation, and software requirements elicitation.

**Sanjay Gosain** is a Wes strategy analyst for the Capital Group in their California offices. Previously, he was an assistant professor of information systems at the Robert H. Smith School of Business, University of Maryland, College Park. His research interests are in the area of the integration of enterprise information systems. He has consulted with companies on integration issues and taught MBA-level courses on e-business-process integration and information technology.

**Marc N. Haines** is an assistant professor in the School of Business Administration at the University of Wisconsin, Milwaukee. His research has been published in the *International Journal of Human Computer Interaction*, *Information Resources Management Journal*, *Journal of Data Warehousing*, *Computers and Operations Research*, and other publications. He is chairing the HICSS minitrack on service-oriented architectures and Web services. He is also a member of the Organization for the Advancement of Structured Information Standards (OASIS). His research interests include the adoption and impact of integration standards and technologies such as Web services, organizational issues concerning the implementation of enterprise systems (ERP), and the development and application of XML-based (extensible markup language) vocabularies.

**William D. Haseman** is the Wisconsin distinguished professor and director of the Center for Technology Innovation at the University of Wisconsin, Milwaukee. He received his PhD from the Krannert Graduate School of Management at Purdue University and served previously on the faculty at Carnegie-Mellon University. His research interests include groupware, Web services, decision support systems, services-oriented architecture, and emerging Internet technologies. Dr. Haseman has published a book and a number of research articles in journals such as *Accounting Review, Operations Research, MIS Quarterly, Decision Support Systems, Information Management, Information Systems,* and *Database Management.* He was conference chair for the Americas Conference on Information Systems (AMCIS) in 1999 and is the conference chair for the International Conference on Information Systems (ICIS) for 2006.

**Guillermo Jiménez-Pérez** has a PhD in computer science and currently is a professor of software engineering at ITESM in Monterrey, Nuevo León, México. His current interest is in applying component-based software development and software product-lines techniques in systems integration.

**Chris Lawrence**, a business architecture consultant, has designed and implemented solutions in the United Kingdom, United States, and Southern Africa over a 25-year career in financial-services IT. He has addressed conferences in the United Kingdom and United States, specializing in the area where process architecture meets holistic delivery and transition methodologies. In 1996 he left England for Cape Town to cofound Global Edge, a strategic business-enablement competency employing a version of Sungard's Amarta architecture to support financial-services group Old Mutual's international expansion. He based his book *Make Work Make Sense* (Future Managers, 2005) on the process-architectural delivery methodology he developed for Global Edge. Chris studied philosophy at Cambridge and London University. He and his wife Wendy have one son, Joe.

**So-Jung Lee** is principal of JLee Consulting. Mr. Lee has been providing IT consultancy services to financial services providers such as banks, brokerage houses, and insurance firms within the ASEAN (Association of Southeast Asian Nations) region for over 20 years. After spending many years working for various large consultancies, he recently established his own consultancy firm with an emphasis on strategic IT consulting. His expertise spans a number of different areas that includes systems integration, business-process analysis, and IT security. His is currently writing a book on the role of technology in globalization.

**Ayesha Manzer** earned her PhD in July 2002 from the Computer Engineering Department of Middle East Technical University, Ankara, Turkey. Her research and publications are related to software engineering, personalization for e-commerce, virtual enterprises, and process integration. Manzer holds MS and BS degrees in computer science from Gomal University, Pakistan, and from Jinnah College for Women, Pakistan, respectively.

**Dirk Muthig** heads the Department of Product Line Architectures at the Fraunhofer IESE in Kaiserslautern, Germany. He has been involved in the definition and development of Fraunhofer's PuLSE™

(Product Line Software Engineering) methodology since 1997, as well as in the transfer of product-line and architecture technology into diverse industrial organizations. Muthig received a diploma and a PhD, both in computer science, from the University of Kaiserslautern.

**Mariano Navarro** graduated as a telecommunications engineer at the Universidad Politécnica de Madrid 15 years ago. He is currently working as the technological innovation department manager. For the last 5 years, he has been working in forestry-related projects. He has been working also in traceability projects related to different product chains for 6 years. He belongs to the EC expert group Collaboration@Work and collaborates frequently with the Directorate General of Information Society Technologies and Media in defining the main research lines of European research and development work programs. At the present moment, he is working on the Strategic Innovation Plan for the next 5 years in the Tragsa Group, MAFF, and ME.

**Lin T. Ngo-Ye** is a PhD candidate at the University of Wisconsin, Milwaukee. He has a Master of Science in information management from Arizona State University. His research interests include the issue of software licensing, pricing and maintenance, service-oriented architectures, Web services, and electronic commerce. He has published a number of refereed conference proceedings and has several manuscripts under various stages of review.

**Taha Osman** is a senior lecturer at the Nottingham Trent University, United Kingdom. He received a BS honors degree in computing from Donetsk Polytechnical Institute, Ukraine, in 1992. He joined the Nottingham Trent University in 1993 where he received an MS in real-time systems in 1994 and a PhD in 1998. His current research investigates fault tolerance in open distributed systems.

**María-Isabel Sánchez-Segura** has been a faculty member of the Computer Science Department in the Carlos III Technical University of Madrid since 1998. Her research interests include software engineering, interactive systems, and usability in interactive systems. María-Isabel holds a BS in computer science (1997), an MS in software engineering (1999), and a PhD in computer science (2001) from the Universidad Politecnica of Madrid. She is the author of several papers related to the improvement of virtual environments development from the software engineering point of view, published recently in journals such as *Software Practice and Experience, Interacting with Computers,* and *Journal of Systems and Software*. She is also the author of more than 20 papers presented at several virtual-environments and software engineering conferences. Maria-Isabel is one of the instructors, joint Carnegie-Mellon and Technical University of Madrid researchers, at tutorials held at the ACM CHI conference, Vienna, April 2004, and the IEEE ICSE conference, Edinburgh, May 2004.

**Alexander Schwinn** was a research assistant at the Institute of Information Management at the University of St. Gallen (HSG), Switzerland, for 4 years after studying computer science and economics at Ludwig-Maximilians University in Munich, Germany. His research topics included EAI, enterprise architecture, and IT management. He wrote several journal publications and spoke at a number of conferences. He completed his doctoral studies successfully with a thesis titled *Designing Integration Architectures for Information Systems* in April 2006.

**Gokul Seshadri** is an architect with TIBCO Software Inc. His primary role is to design EAI (enterprise application integration) solutions and SOA middleware for large-scale enterprise customers based on his experience and best practices in the industry. Gokul has designed regional SOA architectures of significant magnitude spanning multiple geographic locations. He also plays a strategic advisory role to various customers, advising them on the right enterprise-wide architectural and technical strategies to adopt and in identifying areas of IT investment for the long-term growth and evolution of the organization. Gokul carries nearly 13 years of industry experience, the better part of which was spent in the financial services industry in retail, corporate, and investment banks, and in brokerage houses and stock exchanges. He is an avid author and has written many articles and a textbook. He has presented papers at various industry conferences as well.

**Ayça Tarhan** works for Bilgi Group Ltd. She is a part-time instructor at the Department of Software Management and a PhD candidate at the Department of Information Systems at Middle East Technical University. She has been working in the domain of software engineering as an academician for 7 years and as consultant and instructor for 3 years. She pursues her studies on software quality, software measurement, and software engineering standards. She has an MS degree in computer science.

**Dhavalkumar Thakker** obtained his master's degree in data communication systems at Brunel University, London. He also received his BE from Gujarat University, India. He is a PhD candidate at the School of Computing and Informatics, Nottingham Trent University, United Kingdom. His current research interests are distributed computing technologies, Web services composition, Semantic Web, and ontologies.

**Velan Thillairajah** is the founder and CEO of EAI Technologies (http://www.eaiti.com). Mr. Thillairajah has 20 years of experience, ranging from management consulting to information technology, telecommunications, and Web-based solutions. He has worked with or for PriceWaterhouseCoopers, Hewlett-Packard, Network Solutions, AOL, Verizon, BMC, VeriSign, KPMG, and Agilent. He studied operations research and industrial engineering at Cornell University, and then received an MBA at the University of Rochester's Simon School of Business, where he was featured in *Forbes*' "Best and Brightest." EAI Technologies leverages its clients' existing investments in enterprise-level (ERP, CRM [customer relationship management], e-commerce, supply chain, and e-marketplace) systems by integrating data and applications with Web services. Also, as an active board member of the Integration Consortium (http://www.integrationconsortium.org), Mr. Thillairajah speaks and represents the consortium at numerous conferences and events such as Gartner and contributes to content related to EAI best practices and strategies.

**Robert Winter** is a full professor of information systems at HSG, director of HSG's Institute of Information Management, and academic director of HSG's executive MBA program in business engineering. He received master's degrees in business administration and business education as well as a doctorate in social sciences from Goethe University, Frankfurt, Germany. After 11 years as a researcher and deputy chair in information systems, he was appointed chair of information management at HSG in 1996. His research interests include business engineering methods and models, information systems architectures and architecture management, and integration technologies and integration management.

**Min-Jung Yoo** is an assistant professor at HEC (Ecole des Hautes Etudes Commerciales), The University of Lausanne. Yoo's research interests include multiagent systems, extended enterprises, dynamic scheduling in collaborative supply chains, the semantic Web, and Web services. Her research is focused on the application of multiagent systems and other related technologies to designing practical enterprise information systems and e-learning environments as well. She received MS and PhD degrees in computer science from the University of Paris 6 (Pierre et Marie Curie, France) in 1999.

# Index

## A

abstraction 79–80
acquisition
  cycle 51
  planning 58
actor network theory 41
agent paradigm 212
agility 15, 25
American National Standards Institute 109
application
  integration 9, 167, 273
  programming interface (API) 10, 94, 188
architecture 145
artificial-intelligence (AI) 227
ATAM 153
automation 244

## B

batch integration 12
best practice (BP) 247–248
bottleneck 123
broker-based integration 12
browser 205
business
  -process integration 12
  -to-business (B2B)

integration (B2Bi) 5, 165, 286
  relationship 226
  transaction 92–106
domain 60
information system (BIS) 285
process
  integration 119–138, 217
  management (BPM) 166, 226, 242
  modeling 54, 107–118
    language (BPML) 109
  reengineering (BPR) 78
rule 126

## C

C4ISR 52
call center 241
campaign 191, 193
  designer 200
chief information officer (CIO) 3
COBOL 256
code generator 169
commercial off the shelf (COTS) 27, 274
communication infrastructure 215
computer-aided software engineering (CASE) 290
CORBA 95, 167, 226
correctness 153
coupling 28